Demography and Natural History of the Common Fruit Bat, *Artibeus jamaicensis,* on Barro Colorado Island, Panamá

Charles O. Handley, Jr.,
Don E. Wilson,
and
Alfred L. Gardner
EDITORS

SMITHSONIAN INSTITUTION PRESS
Washington, D.C.

Library of Congress Cataloging-in-Publication Data
Handley, Charles O. (Charles Overton), 1924–
Demography and natural history of the common fruit bat, Artibeus jamaicensis, on Barro Colorado Island,
Panama /Charles O. Handley, Jr., Don E. Wilson, and Alfred L. Gardner.
p.   cm. — (Smithsonian contributions to zoology, no. 511)
Includes bibliographical references (p.   ).
1. Artibeus jamaicensis—Panama—Barro Colorado Island.
I. Wilson, Don E.   II. Gardner, Alfred L.   III. Title.   IV. Series.
QL1.S54   no. 511   [QL737.C57]   591 s–dc20   [599.4]   91-13449

ISBN 1-56098-146-6

This publication is the trade edition of Smithsonian Contributions to Zoology, 511.

COVER IMAGE: *Artibeus jamaicensis*. Pencil sketch by Nancy Moran, 1976.

Manufactured in the United States of America.

Contents

FRONTISPIECE.—*Artibeus jamaicensis*. Pencil sketch by Nancy Moran, Barro Colorado Island, Panamá, October, 1976.

Demography and Natural History
of the Common Fruit Bat,
Artibeus jamaicensis,
on Barro Colorado Island, Panamá

1.   Introduction

Charles O. Handley, Jr., Don E. Wilson,
and Alfred L. Gardner

BCI Bat Project

In 1974, Charles Handley was invited to develop a project to monitor bats as part of the Smithsonian Tropical Research Institute's (STRI) long-term environmental monitoring program on Barro Colorado Island (BCI). The STRI monitoring program, launched in 1970 and supported by the Smithsonian Environmental Sciences Program (ESP), sought to monitor a wide array of biotic and physical environmental components of the island continuously over a long period of time.

The BCI Bat Project was born under the administrative title: "Biomass and energetics of selected populations in Panamá: Bats." We wanted to monitor demographic parameters and natural history of all the bats regularly found on BCI. Based on our earlier experiences, we thought the fauna might total 40 species of bats. The length of the project was designed to continue through a generation of bats, however long that might be. The only clue to possible duration was the report (Wilson and Tyson, 1970) of a seven-year-old *Artibeus jamaicensis* on BCI.

At the outset it was evident that ESP funds were spread over too many projects to be able to support a really meaningful monitoring project for bats. Clyde Jones, then Director of the National Fish and Wildlife Laboratory, U.S. Fish and Wildlife Service, offered to provide both financial and personnel support, and the Bat Project became a joint venture of the Smithsonian Institution and the U.S. Fish and Wildlife Service, with Michael A. Bogan, Alfred L. Gardner, and Don E. Wilson joining Handley as field crew leaders.

Handley made several trips to BCI in 1975 and 1976 to become familiar with the island and its bats, as well as to determine what was feasible and how to organize the project. On 2 July 1977 a year-round capture and marking program began. With the help of collaborators and dozens of volunteers, we took turns manning the field survey on BCI until November 1980 when this phase of the Project was completed. Thereafter, Handley continued work on BCI on a periodic basis—the fall of 1981, the fall of 1982, and the 12 months from September 1984 through August 1985—with the support of the Smithsonian's ESP, STRI, and National Museum of Natural History (NMNH). The purpose of the continuing study was to maintain the pool of marked bats, refine the demographic data, and gain further information on the biology of the bats, particularly their responses to food sources.

During the reconnoitering phase at the beginning of the project we focused much of our attention on developing a reliable, long-lasting, harmless marking system. We established colonies of bats at the National Zoological Park (NZP) that we used in marking experiments and in establishing

Charles O. Handley, Jr., and Don E. Wilson, National Museum of Natural History, Smithsonian Institution, Washington, D.C. 20560. Alfred L. Gardner, NERC, U.S. Fish and Wildlife Service, National Museum of Natural History, Washington, D.C. 20560.
Review Chairman: W. Ronald Heyer, Smithsonian Institution. Three anonymous reviewers are gratefully acknowledged.

standards for describing age and reproductive state. As a result, we discarded forearm banding, tattooing, and heat branding and settled on necklacing with stainless steel ball chain. Band-bearing necklaces were first attached to free-living bats on BCI on 18 October 1976. By the fall of 1980 we had placed 18,953 necklaces on bats.

The project evolved rapidly. We soon realized that our resources were not sufficient to monitor simultaneously all species of bats on BCI in an effective manner. We settled on studying the bat that had proved to be our principle catch—*Artibeus jamaicensis* Leach, the common fruit-eating bat.

At first we established netting stations at places that, based on experience elsewhere, "looked good for bats," such as stream valleys, trails through open forest, and gaps along ridges

where underbrush and other vegetation did not interfere with the nets or obstruct flyways. However, it became apparent that abundance of bats coincided with nearby fruit sources and that what at first appeared to be complicated distributional patterns for the island's bats proved to be nothing more that a direct correlation between foraging activity and the uneven distribution and availability of fruit. We improved our netting success by making systematic surveys along the trails on BCI to locate pellets of fruit pulp dropped by feeding bats and then setting our nets nearby where they were most likely to catch bats.

We realized early in the project that squeaking bats often attracted others, and in 1978 we began to use the wooden Audubon Bird Call as a substitute to attract bats into the nets. At a good site, an Audubon Bird Call*, an occasional squeaking bat, and one or two nets could produce enough bats to keep everyone busy for hours.

In the beginning we sometimes caught only two or three bats in a night, and we were satisfied with catches of 30 bats. Later,

---

The use of product brand names in this publication is not intended as an endorsement of the products by the Smithsonian Institution.

TABLE 1-1.—Bats recorded on Barro Colorado Island, Gatún Lake, Panamá.

Family EMBALLONURIDAE

Rhynchonycteris naso (Wied)
Saccopteryx bilineata (Temminck)
Saccopteryx leptura (Schreber)
Cormura brevirostris (Wagner)
Centronycteris maximiliani (Fischer)

Family NOCTILIONIDAE

Noctilio albiventris Desmarest
Noctilio leporinus Linnaeus

Family MORMOOPIDAE

Pteronotus gymnonotus Wagner
Pteronotus parnellii (Gray)

Family PHYLLOSTOMIDAE
Subfamily PHYLLOSTOMINAE

Micronycteris brachyotis (Dobson)
Micronycteris hirsuta (Peters)
Micronycteris megalotis (Gray)
Micronycteris nicefori Sanborn
Micronycteris schmidtorum Sanborn
Macrophyllum macrophyllum (Schinz)
Tonatia bidens (Spix)
Tonatia silvicola D'Orbigny
Mimon crenulatum (E. Geoffroy)
Phyllostomus discolor Wagner
Phyllostomus hastatus(Pallas)
Phylloderma stenops Peters
Trachops cirrhosus (Spix)
Chrotopterus auritus (Peters)
Vampyrum spectrum (Linnaeus)

Subfamily GLOSSOPHAGINAE

Glossophaga commissarisi Gardner
Glossophaga soricina (Pallas)
Lonchophylla robusta Miller

Subfamily CAROLLINAE

Carollia brevicauda (Schinz)

Carollia castanea H. Allen
Carollia perspicillata (Linnaeus)

Subfamily STURNIRINAE

Sturnira luisi Davis

Subfamily STENODERMATINAE

Uroderma bilobatum Peters
Uroderma magnirostrum Davis
Platyrrhinus helleri (Peters)
Vampyrodes caraccioli (Thomas)
Vampyressa nymphaea Thomas
Vampyressa pusilla (Wagner)
Chiroderma villosum Peters
Mesophylla macconnelli Thomas
Artibeus hartii Thomas
Artibeus jamaicensis Leach
Artibeus lituratus (Olfers)
Artibeus phaeotis (Miller)
Artibeus watsoni Thomas
Ametrida centurio Gray
Centurio senex Gray

Subfamily DESMODONTINAE

Desmodus rotundus (E. Geoffroy)

Family THYROPTERIDAE

Thyroptera discifera (Lichtenstein and Peters)
Thyroptera tricolor Spix

Family VESPERTILIONIDAE

Myotis albescens (E. Geoffroy)
Myotis nigricans (Schinz)
Rhogeessa tumida H. Allen

Family MOLOSSIDAE

Tadarida laticaudata (E. Geoffroy)
Molossus bondae J.A. Allen
Molossus coibensis J.A. Allen
Molossus molossus (Pallas)

we were disappointed with less than 100 bats per night, and we logged many nights with catches exceeding 200. Our best catch came on 25 October 1979 at a giant *Ficus dugandii* with ripe fruit where we netted 282 bats in about four hours.

As of 1985, we had found 56 species of bats on BCI (Table 1-1). Bonaccorso (1979) categorized the bats of the island into nine "feeding guilds." The distribution of the 56 species among Bonaccorso's guilds is: canopy frugivores (14), groundstory frugivores (4), scavenging frugivores (2), omnivores (4), sanguinivores (1), gleaning carnivores (12), slow-flying hawking insectivores (14), fast-flying hawking insectivores (4), and piscivores (1). *A. jamaicensis*, the major subject of this report, is a canopy frugivore.

By every measure *A. jamaicensis* is the most widespread and abundant bat on BCI (Table 1-2). On a yearly basis, it averaged 60% of the total catch of bats, and we caught it almost every night that nets were set. Altogether, in the period 1975–1980 (including bats captured before marking with necklaces began), we recorded 17,820 captures of *A. jamaicensis*.

By the end of 1980, we had learned enough about *A. jamaicensis* including its populations, reproduction, movements, foraging, and physiology to justify a pause to

TABLE 1-2.—Measures of abundance of bats captured on BCI during 1979. Bats were netted on 157 nights, and captures (including both marks and recaptures) totalled 9118 bats. Tabulations are by frequency of capture (number and percentage of nights caught), number caught (total), catch per species night (total of a species caught divided by number of nights it was caught), and catch per netting night (total of a species caught divided by total nights of netting).

Species	Mean catch per species night	Rank	Total	Mean catch per netting night	Nights caught	
					N	%
Artibeus jamaicensis	36.3	1	5484	34.9	151	96
Uroderma bilobatum	9.2	2	551	3.5	60	38
Artibeus lituratus	6.0	3	717	4.6	120	76
Chiroderma villosum	4.4	4	172	1.1	39	25
Carollia perspicillata	4.0	5	428	2.7	107	68
Vampyrodes caraccioli	3.9	6	325	2.1	83	53
Phyllostomus discolor	3.3	7	130	0.8	40	25
Artibeus phaeotis	2.2	8	235	1.5	108	69
Carollia castanea	2.0	9	154	1.0	78	50
Micronycteris hirsuta	1.8	10	91	0.6	52	33
Vampyressa pusilla	1.8	10	87	0.6	49	31
Vampyressa nymphaea	1.8	10	51	0.3	28	18
Pteronotus parnellii	1.7	13	86	0.6	51	32
Tonatia silvicola	1.6	14	113	0.7	71	45
Tonatia bidens	1.6	14	77	0.5	49	31
Artibeus watsoni	1.6	14	67	0.4	43	27
Micronycteris megalotis	1.6	14	35	0.2	22	14
Glossophaga soricina	1.6	14	28	0.2	17	11
Phyllostomus hastatus	1.6	14	26	0.2	16	10
Mimon crenulatum	1.5	20	50	0.3	33	21
Rhogeessa tumida	1.5	20	3	0.02	2	1
Trachops cirrhosus	1.4	22	70	0.4	49	31
Cormura brevirostris	1.4	22	7	0.05	5	3
Platyrrhinus helleri	1.3	24	31	0.2	23	15
Micronycteris brachyotis	1.2	25	13	0.08	11	7
Macrophyllum macrophyllum	1.2	25	11	0.07	9	6
Micronycteris nicefori	1.1	27	21	0.1	19	12
Saccopteryx bilineata	1.1	27	8	0.05	7	4
Desmodus rotundus	1.0	29	12	0.08	12	8
Glossophaga commissarisi	1.0	29	8	0.05	8	5
Myotis nigricans	1.0	29	7	0.05	7	4
Micronycteris schmidtorum	1.0	29	5	0.03	5	3
Phylloderma stenops	1.0	29	5	0.03	5	3
Centurio senex	1.0	29	3	0.02	3	2
Vampyrum spectrum	1.0	29	2	0.01	2	1
Uroderma magnirostrum	1.0	29	2	0.01	2	1
Lonchophylla robusta	1.0	29	1	0.01	1	1
Carollia brevicauda	1.0	29	1	0.01	1	1
Artibeus hartii	1.0	29	1	0.01	1	1

summarize the information for publication. Handley, Wilson, and Gardner shared the manuscript preparation. Fortunately, Cindy Taft, Gene Studier, Doug Morrison, and Bert Leigh, who were heavily involved with various aspects of the Bat Project, joined us in the authorship of several of the sections.

The Environment

Protected by its isolation in Gatún Lake, and by the vigilance of its caretakers, BCI long has attracted biologists seeking to understand life in the tropical forest. The island provides an ideal outdoor laboratory for the curious naturalist. The long-term protection provided by the Smithsonian Institution and the Government of Panamá, combined with easy access and excellent working and living conditions, make available a small, but manageable bit of tropical habitat where intensive and long-term studies of the ecosystem are practical (Figure 1-1).

Once known as West Hill, which sloped to the Río Chagres on the north, and the Río Gigantito to the east, and bordered by the Gigante and Peña Blanca swamps on the south and west, BCI came into being following completion of Gatún Dam in 1914 and the subsequent flooding that formed Gatún Lake. However, because Gatún Lake took about four years to fill (Chapman, 1938) we don't know exactly when West Hill became separated from the mainland.

The resulting island consists of about 1500 ha (3×5 km) and at its highest point (the Plateau) rises to 137 m above the lake. The remaining hilltop, which comprises the island, is a broad and flat basaltic cap covered with red clay soils (for which the island was named) up to 1 m thick.

About 90% of the yearly rainfall occurs between May and December, followed by a pronounced dry season extending from January to April. Rainfall is not spread evenly, and when it occurs it often pours. The monthly average during the rainy season is 31 cm, and the yearly average is about 260 cm (Rand and Rand, 1982). High rates of transpiration from the forest and evaporation of the surrounding water of Gatún Lake contribute to a high relative humidity throughout the year.

The vegetation of BCI has been characterized by Foster and Brokaw (1982) as semideciduous lowland forest. Some species of trees lose all of their leaves for a period of months while others remain evergreen. Although most species lose their leaves in the dry season, some do so in the rainy season. Just as BCI is intermediate in the amount of rainfall it receives, compared with both coasts of Panamá, so is the forest of an intermediate type, containing elements of both the wetter Caribbean and dryer Pacific slope forests (Foster and Brokaw, 1982).

Although the structure of forests is shaped by substrates and climate, the forest on BCI, like forests all over the world, has been reshaped by the activities of human beings. Foster and Brokaw (1982) suggested that even the older forest on BCI might have been cleared 300–400 years ago.

The younger forest on BCI has considerably more evidence of recent human settlements, including some commonly cultivated trees and an abundance of crockery and bottles. Much of the younger forest may date from the 1880's when the French were attempting to build a trans-isthmian canal (Foster and Brokaw, 1982).

Bananas and other fruits and vegetables were raised in small patches near the lab clearing up to the 1950's when all agriculture on the island was stopped. Trees now cover all of the former cleared land, except around the laboratory buildings and in a narrow strip between the canal navigational lights on Miller Ridge.

Methods

We conducted studies at three sites: (1) Mostly we worked on BCI, where we mist-netted bats on a large scale and gathered data on their basic demographic and biological variables. (2) To gain additional information on movements and fidelity, we netted on nearby islands and on the adjacent mainland, where forests are younger and prime age fig trees are more abundant than on BCI. From navigational lights on Buena Vista Island and on Palenguilla Point ("Peña Blanca" light) we regularly trapped the bats of a pair of maternity roosts (referred to in the text as the "canal marker roosts"). (3) In addition, to gather baseline data on reproduction and ontogeny and to establish standards for describing age and reproductive state in the field, we maintained colonies of bats at the National Zoological Park in Washington, D.C., from March 1975 through August 1981.

There was no precedent for the Bat Project, so we learned techniques and developed protocols and standards by trial and error throughout the life of the project. What we learned might be useful for reducing start-up time in future projects. Adoption of similar standards and terminology could make data comparable among projects, and greatly enhance the value of the data. For these reasons we have included only a brief summary of methods in the body of the text, but have provided a detailed and fully illustrated treatment of methods as an easily reprintable appendix.

CAPTURING BATS.—Bats were captured in 12 m nylon mist nets hung between 3 m tubular aluminum poles. We found it more productive to net in the vicinity of trees bearing ripe fruit than in sites selected at random. When it was practical we placed the nets along a trail near a fruiting tree, around the base of the tree, or on a ridge overlooking the tree. The nets were set where underbrush would not impede the flight of bats. As far as was possible, the nets were arranged in a zig-zag pattern with ends overlapping about one meter, thereby blocking as much of an underbrush-free area as possible. The standard netting station contained ten nets.

We stretched the shelf strings of the nets tight (but not as tight as possible) to minimize tangling of large *Artibeus,* and high enough to keep heavy bats entering the lower shelf from touching the ground. These tactics facilitated the capture of the

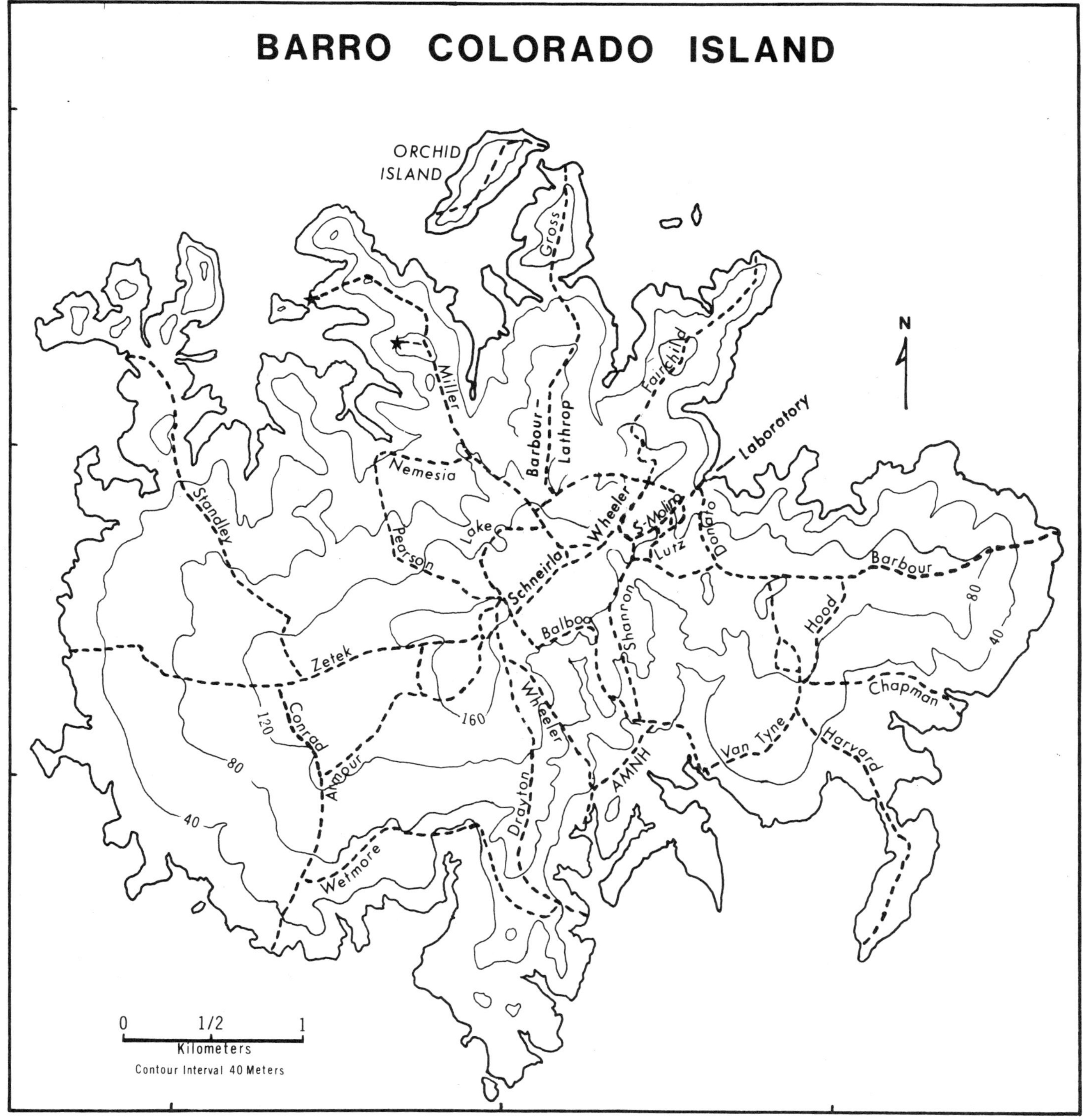

FIGURE 1-1.—Barro Colorado Island, Gatún Lake, Panamá, showing topography and approximate alignment of trails.

larger bats, but somewhat diminished our chances of catching smaller species.

We usually netted five nights each week. Normally the netting sessions began at dusk and ended when the bats quit flying, regardless of light conditions. Sometimes we closed up after an hour or two if netting was slow, but occasionally we netted until dawn. We furled the nets during heavy rains.

The distress calls of particularly vocal bats were useful for attracting other bats into the nets. In the absence of such bats, the squeaks of an Audubon Bird Call often were effective in bringing in bats.

For safety of the bats we preferred bare hands throughout the capture and marking process. Using folded capture bags as baffles that the bats might bite, we seldom were bitten ourselves. Everyone handling bats was protected with rabies vaccine.

We caught bats at 106 netting localities on BCI (plus 15 others on the mainland and other islands), which was far too many sites for convenient analysis. Consequently, we arbitrarily allocated the 106 localities on BCI to 20 regional groups of approximately equal area (Figure 13-5), which we refer to as "locality groups." We netted often where access was easy and fig trees were common, and locality groups that encompassed such areas often included many netting localities. In parts of the island where access was difficult or fig trees were few, a locality "group" sometimes included only a single netting locality. Complete details on marking bats and recording data can be found in the Appendix.

CANAL MARKER ROOSTS.—In the fall of 1977 we examined all of the land-based navigational lights (used in pairs to orient ship's pilots on straight "reaches" of the canal) between Gamboa and Gatún. Many contained maternity colonies of *Phyllostomus hastatus, Saccopteryx bilineata,* and *Micronycteris hirsuta.* Two (on Buena Vista Island and Palenguilla Point) contained harems of *A. jamaicensis.* Fortuitously, both of the *Artibeus* roosts were near BCI, and were used by bats that we mist-netted on BCI and on the mainland. We sampled these roosts at intervals throughout the project, using a plastic funnel-like trap, with which we always caught most or all of the bats in a roost. These roosts gave us information on harem composition, stability, and recruitment and were our best link between netting sites and roost locations.

THE ZOO BATS.—Studies on the captive bats at the NZP were indispensable to the field work on BCI. In 1975, we took *Artibeus jamaicensis, A. lituratus, Uroderma bilobatum, Glossophaga soricina, Carollia castanea,* and *C. perspicillata* to the NZP. The *Artibeus* died in an accident a few months after we established the colony, but the others survived and were the subjects of marking experiments that led to our necklacing technique. In 1978, we established a new colony, consisting entirely of *A. jamaicensis.* This colony is described in detail in Section 3, Reproduction in a Captive Colony.

Acknowledgments

In a project of such long duration it is inevitable that a great many people have contributed. We are indebted to every one of them.

From the inception of the project we enjoyed the interest and support of David Challinor, then Assistant Secretary of the Smithsonian Institution. Also, we sought help from offices of the Smithsonian and its bureaus: James Crockett, Rita Jordan, and Dante Piachesi, (Smithsonian); John Eisenberg and Devra Kleiman (National Zoological Park); Mercedes Arroyo, Tom Borges, Gloria Maggiori, and Donald M. Windsor (Smithsonian Tropical Research Institute). On BCI, many people helped in various ways year after year and were among our greatest benefactors: Pedro Acosta, Mario Bernal, Barbara Butler Cusatti, Ricardo Cortez, Miguel Estribi, Alberto Gonzales, Bonifacio de Leon, Egbert and Elizabeth Leigh, A. Stanley Rand, and Nicholas Smythe.

Michael A. Bogan and Clyde Jones of the U.S. Fish and Wildlife Service were closely involved with the project in the early years. Clyde was instrumental in securing funding and Mike contributed significantly to the field work. Carl Johnson, Gorgas Memorial Laboratory, helped us through medical emergencies.

The radio-telemetry and related roosting and foraging studies were conducted principally by Douglas H. Morrison. Special thanks are due to Susan Hagen-Morrison for her assistance. These studies were supported by STRI, Rutgers University, and the National Geographic Society. We are grateful to Jackie Belwood and James Fullard for testing bat calls and providing information on their acoustics. In addition, Jackie often joined us at capture stations to help with netting.

Eugene H. Studier led our studies of the physiology of *A. jamaicensis* and other bats on BCI. Support for these investigations came from Sigma Xi, the Penrose Fund of the American Philosophical Society, the Faculty Development Fund of the University of Michigan-Flint, the Vice-President for Research of the University of Michigan-Ann Arbor, the Rackham School of Graduate Studies at the University of Michigan-Ann Arbor, and the U.S. Fish and Wildlife Service, as well as the Smithsonian's Environmental Sciences Program. John Bassett and Douglas Morrison read the section on physiology and provided many useful and constructive criticisms. We thank Renee Seeke for typing the original draft of that section.

We are grateful to Katherine Ralls and Tracey Werner for capturing bats in Panamá and bringing them to Washington, D.C., to establish our experimental colonies at the National Zoological Park (NZP). Nathan Gale, Canal Veterinary Office, advised us on maintenance of the bats and helped us obtain the necessary permits for their export. At the NZP the bats were cared for at first by Todd Davis and Devra Kleiman. Later, in pursuit of a Master of Science degree at George Mason University, Cindy Taft devoted much of her time, day and night, for three years to the bats. The Department of Zoological Research, NZP, through Devra Kleiman, generously provided facilities, food for the bats, and expert keepers—Peter Beaman, John Hough, Eugene Maliniak, Larry Newman, and Tex Rowe. We were assisted in data gathering by Peter Beaman, Linda Gordon, Laura Grady, Betty Howser, Devra Kleiman, Eugene Maliniak, Katherine Ralls, and Jane Small. Laura Grady was especially helpful with compiling the data on wing measurements. The Department of Pathology, NZP, provided necropsies of bats that died. We appreciate the advice of Carl Ernst, Devra Kleiman, and Larry Rockwood, who read early drafts of the section on Reproduction in a Captive Colony. Luther Brown, Betty Howser, Karl Kranz, and John Wilson provided encouragement and indispensable help. The Friends of the National Zoo provided a summer traineeship and a predoctoral fellowship to Cindy Taft, who also had partial support from a National Science Foundation grant (DEB 79-05841) to Luther Brown.

The seemingly endless job of organizing and compiling the data was done by Jenny Banner, Greg Blair, Sally Hart Carter, Kristin Day, Carolyn Gamble, Darelyn Handley, Jennifer Hoffman, Molly Morton Mayfield, John Miles, Penny Nelson, Anita Przemieniecki, Cynthia Ramotnik, Becca Shad, Pippa Vanderstar, and Ellen Winchell. Without Peter Kauslick's programing skills we would have been hard put to analyze the data.

Claudia Angle prepared the graphics. For critical reading of the manuscript we are grateful to Thomas H. Fritts, Ronald W. Heyer, Paul A. Opler, R.W. Thorington, Jr., and anonymous reviewers. Final stages of manuscript assembly fell to Darelyn Handley. She did the word-processing, checked literature citations, collated figures and tables with text, and patiently coped with updates to produce several "final" drafts of the text.

Finally, there were the field crews who endured rabies inoculations, suffered innumerable bat and all manner of other bites and stings, survived the ticks and chiggers of BCI, lugged heavy net poles up and down slippery slopes and over interminable miles of trails, were seldom comfortable, were often rained on, and between looking for bats in the daytime and catching them at night, had little time for sleep or recreation. Except for the handful of us who are Smithsonian Institution or Fish and Wildlife Service staff, all of the field people were volunteers. They and the data compilers were the real heros of the project. By year, the field crews were as follows:

1975 (February and March)
Charles and Darelyn Handley, Katherine Ralls, and Merlin Tuttle.

1976 (February and March; October and November)
Edythe Anthony, Michael Bogan, Todd Davis, Charles and Darelyn Handley, Nancy Moran, Kim Mortensen, and Don Wilson.

1977–1978 (July 1977 through August 1978)
Barbara and Michael Bogan, Carolyn Gamble, Alfred Gardner, Charles Handley, Peter Kauzmann, Tad Lawrence, Molly Morton Mayfield, Patricia Mehlhop, David Sahagian, Eugene Studier, Tracey Werner, and Kate and Don Wilson.

1978–1979 (September 1978 through August 1979)
Greg Adler, Barbara and Michael Bogan, Carolyn Gamble, Alfred Gardner, Charles Handley, Marshall Hasbrouk, Douglas and Susan Morrison, Marion and Vern Read, Lucinda Taft, Merlin Tuttle, and Don Wilson.

1979–1980 (September 1979 to June 1980)
Michael Bogan, Dale Clayton, Adele Conover, Robert Fisher, Alfred Gardner, Linda Gordon, Charles and Darelyn Handley, Molly Morton Mayfield, Deborah Page, Charles Rupprecht, Norman Scott, Eugene Studier, and Don Wilson.

1980 (August to October)
Charles Handley, Hui Purdy, and Becca Schad.

2. Physiology

Eugene H. Studier and Don E. Wilson

Although most of our work on Barro Colorado Island (BCI) was dedicated to marking and recapturing large numbers of *Artibeus jamaicensis,* we continually faced questions regarding the animal's physiology. As our data on population dynamics, movements, reproductive cycle, and other natural history variables increased, we needed answers to our physiological questions. Clearly, the issues of energy, water, and mineral budgets are critical to understanding factors limiting the bat's populations. Thus, we conducted physiological studies on newly captured animals in the laboratories available on BCI. Most involved nutritional economy as it relates to feeding behavior, thermoregulation, metabolism, and other maintenance variables.

Energy Balance

FEEDING BEHAVIOR.—Although *A. jamaicensis* consumes a variety of foods (Gardner, 1977), the fruits of figs are at the fore. On BCI, where *Ficus insipida* and *F. yoponensis* are unusually abundant, *F. insipida* is the fig of choice (Morrison, 1978a); in the Llanos of Venezuela it is *F. trigonata* (August, 1981).

A. jamaicensis extracts juices of the fruit of these figs (Morrison, 1980b; Studier, Boyd et al., 1983). The bat bites off a chunk of pulp and skin and rolls it back and forth with its tongue against its heavily corrugated hard palate. Juice is extracted by chewing and by squeezing the shredded pulp against the hard palate with the tongue. Juices are swallowed and the shredded pulp is dropped as a "dry" pellet. These bats are highly efficient at extracting juice from pulp. The technique of sucking dry one small bite at a time extracted almost as much juice (91%) as Morrison (1980b) was able to obtain with a C-clamp press.

Fruits of *F. insipida* carried into mist nets by *A. jamaicensis* on BCI had an average wet mass of 7.0 g (see Section 11, Diet and Food Supply). However, note that fruit weighed by

Bonaccorso (1975) averaged 9.5 g. Variation in fruit mass from tree to tree, from season to season, or from year to year may account for this discrepancy. When humans sense that these figs are ripe, their average wet mass is 0.58 times less than when bats select them (Morrison, 1978d). In other words, bats consider figs ripe at a later stage than humans do. This fact is important in interpreting results of studies on energetics, feeding physiology, and fruit composition because there are marked rapid changes in the composition and concentrations of nutrients during terminal ripening in fruit (Hulme, 1970). We know the composition and concentrations of many of the nutritionally important components of the fruit of *F. insipida* (Hladik et al., 1971; Morrison, 1980b; Nagy and Milton, 1979; Studier, Boyd et al., 1983). Analysis of the juice of this fig is pertinent to understanding nutritional balance in *A. jamaicensis.*

The digestive ability of *A. jamaicensis* is comparable to that of other nonruminant mammals, at least when it is feeding on *F. insipida* (Morrison, 1980b). In this case the bat exhibits an overall digestive efficiency of 25%–30% of whole fruit calories. However, such efficiency is misleading because 57% of the metabolizable energy is dropped as pellets and never swallowed. Sixty-four per cent of potential energy in food actually swallowed is assimilated. This value is slightly lower than assimilation efficiencies of other mammals on low-fiber diets (Maynard and Loosli, 1969) and may correlate with the extremely rapid transit time through the bat's digestive tract. The passage of 25%–33% of swallowed energy as undigested seeds also contributes significantly to the lowered assimilation efficiency. On the other hand, 93% of swallowed soluble carbohydrates are absorbed (Morrison, 1980b), and these carbohydrates account for almost all retained calories in *A. jamaicensis.*

THERMOREGULATION.—We studied thermoregulation in many species of bats on BCI, but particularly in *A. jamaicensis.* We found this bat to be a heterothermic endotherm when we measured body temperature (T_b) as a function of ambient temperature (T_a) within 12 hours of capture (Studier and Wilson, 1970). We determined a relation between T_b and T_a expressed by

$$T_b = 8.8 + 0.933\ T_a$$

Eugene H. Studier, Department of Biology, University of Michigan—Flint, Flint, Michigan 48503.
Don E. Wilson, National Museum of Natural History, Smithsonian Institution, Washington, D.C. 20560.

in a T_a range of 8.0°–33.2°C. The regression coefficient for this relationship (0.933; $SE = 0.202$) does not differ significantly from 1.0 and we conclude that in *A. jamaicensis,* T_b parallels T_a. Individual bats maintain a constant range of T_b 6.6°–8.3°C higher than T_a throughout the tested T_a range.

Previously, Morrison and McNab (1967) reported little daily fluctuation in T_b cycles in captive *A. jamaicensis* at an estimated T_a of 27°C. During daylight, some of their data points (13/128 = 10.1%) were significantly lower than average. They demonstrated that *A. jamaicensis* exhibited the thermoregulatory pattern of a homeothermic endotherm throughout a T_a exposure range of 5°–39°C. McNab (1969) obtained similar results in captive *A. jamaicensis* held for up to two weeks before testing; no bats had reduced T_b relative to their mean temperatures.

We reinvestigated these contradictory results and concluded that although differences in thermoregulatory patterns might have been related to genetic differences in the populations studied, more likely the variation reflected different methodological approaches (Studier and Wilson, 1979). The primary significance of our 1979 study was the demonstration of a transition from a heterothermic pattern of T_b regulation on the day of capture to a homeothermic pattern after three days of captivity. Such a "captivity effect" for *A. jamaicensis* may help to resolve apparently conflicting thermoregulatory patterns in other mammals, for example, *Myotis lucifugus* (Stones and Wiebers, 1967; Studier and O'Farrell, 1972) and *Peromyscus leucopus* (Gaertner et al., 1973; Hill, 1977).

Although the captivity effect explains divergent data on thermoregulation, it does not show which data represent the natural thermoregulatory pattern. Studier and O'Farrell (1972) found that the T_b of *Myotis lucifugus* and *M. thysanodes* in their natural roosting sites was highly variable and similar to the data for newly caught, laboratory-tested individuals. Apparently, data from bats tested soon after capture better reflect natural thermoregulatory performance, whereas data from captive bats held longer reflect their greatest homeothermic capabilities.

We did not attempt to determine which components of the captivity effect are responsible, either singly or in combination, for the changeover in thermoregulatory performance, but several possibilities exist. The increased homeothermic response probably does not result from thermal acclimation. We did not try to hold captive bats at constant T_a. Instead, T_a for captive bats fluctuated in slightly muted fashion with that of BCI's natural environment (Studier and Wilson, 1970). Furthermore, the captivity effect does not seem to be a general stress response because stress would be greatest during the initial hours of captivity; thus bats would be expected to exhibit the most rigid homeothermy in day zero testing. The captivity effect may result from individual or combined actions of reduced activity while caged or the continual presence of excess food, which the bats can eat at will.

In its natural environment, *A. jamaicensis* is probably a homeothermic endotherm during periods of feeding and flight activity, but loosens T_b control (becoming a nonhomeothermic endotherm) during roosting (nonfeeding and nonflying episodes). The slight reduction in T_b at such times would conserve large amounts of energy (Studier, 1981). Heterothermy in *A. jamaicensis* conserves 38.7% and 67.4% of the energy that would be required of homeothermic individuals at ambient temperatures of 30°C and 25°C, respectively (Studier and Wilson, 1970). This would amount to a major energy cost reduction during a roosting period. We assume that the slight T_b reduction in *A. jamaicensis* (T_b of 35.2°C at T_a of 30°C and T_b of 32.5°C at T_a of 25°C) would not reduce responsiveness to environmental stimuli during roosting nor would it preclude initiation of flight.

Thermoregulation probably is dependent on nutritional state in captive bats. In the wild, *A. jamaicensis* feeds on fruit that varies seasonally from scarce to plentiful. Individual bats may undergo a natural period of diel torpor, whereas captive animals with unlimited food may never become torpid as long as their food supply is constant and plentiful. Captive bats routinely weigh more than wild-caught individuals (see Section 3, Reproduction in a Captive Colony).

The significant questions regarding possible effect of nutrition on the variability of T_b in *A. jamaicensis* are: Does the heterothermic pattern of bats tested immediately after capture reflect undernourished individuals? Or, is the homeothermic pattern of bats kept in captivity a response to overnourishment and inactivity? Undernourished or not, poor T_b regulation in newly caught *A. jamaicensis* is not related to physiological competence, but reflects reduced metabolic heat production and rate of depletion of energy stores. Manakins, which are small frugivorous birds, also exhibit heterothermy on BCI (Bartholomew et al., 1983).

McNab (1969) reported a resting metabolic rate of 1.70 cc/g/hr for *A. jamaicensis* within its thermal neutral zone (TNZ) and a thermal conductance of 0.17 cc/g/hr/°C (below its TNZ). As with other frugivorous bats, this mass-specific standard metabolic rate is slightly higher than would be predicted by Kleiber's (1932) classic relation between metabolism and mass in mammals. The energetic cost of high T_b homeothermy in *A. jamaicensis* is, therefore, higher than expected for a mammal of its size.

McNab (1983) presented extensive arguments concerning energetics, body sizes, and the limits to endothermy that can be expressed by a minimum boundary curve for high T_b homeothermy. This curve estimates the smallest mass at which continuous endothermy can occur for a given metabolic rate. When this minimum boundary curve is drawn by relating the T_b to T_a differential as a function of body mass, data for *A. jamaicensis* falls almost exactly on the curve (McNab, 1982). *A. jamaicensis,* therefore, would be predicted to be marginally able to maintain its reported high homeothermic T_b to T_a differential. It is, therefore, not surprising that undernourished (or normally nourished) bats tested immediately after capture maintain a markedly lower T_b and, consequently, lower T_b to T_a

differentials. As stated previously, the slight T_b reduction found in *A. jamaicensis* saves 38.7%–67.4% of the energy required for maintenance of higher T_b. Such energy conservation would seem critically important in a species such as *A. jamaicensis* that has severely limited reserves of stored energy.

BODY FAT.—Total body fat is a direct indication of overall nutritional status in vertebrates. Annual variability in body fat content in numerous Neotropical bats, including *A. jamaicensis,* has been reported (McNab, 1976; 1982). However, Neotropical frugivores and nectarivores demonstrate less of the seasonal variation and none of the gender-related differences that characterize temperate zone insectivores. All Neotropical bats exhibit low peak fat reserve levels when compared with temperate zone bats (Baker et al., 1968; Ewing et al., 1970; Pagels, 1975; Weber and Findley, 1970). The extremely low fat reserves reported by McNab (1976) for *A. jamaicensis* emphasize its need to reduce daily energy expenditures. Lack of fat reserves and high intake of dietary carbohydrate suggest that glycogen should be examined as the normal energy reserve of *A. jamaicensis.*

Glycogen levels have not been reported for *A. jamaicensis,* but comparable data are available from megachiropterans that, although not related to *A. jamaicensis,* are nutritional equivalents. Daily variation in glycogen and fat levels in liver and flight muscle tissue in *Eidolon helvum,* a Paleotropical frugivorous bat, have been reported by Okon et al. (1978). Their findings show that liver glycogen levels at sunrise (90.0 mg/g) are extremely high in comparison with levels seen in large domesticated mammals (Watt and Merrill, 1963). Liver glycogen then drops precipitously in *E. helvum* until sunset when levels (35.0 mg/g) reach a range normal for large mammals. Breast muscle glycogen in this bat remains low (6.0–8.1 mg/g) and constant throughout the roosting period. The extreme drop in liver glycogen suggests that glycogen (as glucose) is the primary energy source of *E. helvum* throughout its roosting period. Fat concentrations in liver and breast muscle in *E. helvum* show slight increases at sunset, but the range of all values (5.0–10.3 mg/100 g) is nearly two orders of magnitude less than fat levels in liver and muscle in other mammals, large and small (Kirkham and Allfrey, 1972; Watt and Merrill, 1963). These glycogen levels reflect the high carbohydrate, low lipid composition of the diet of these bats.

Van der Westhuyzen's (1978) report on another Paleotropical fruit bat, *Rousettus aegyptiacus,* provides additional support for the extreme importance of glucose or glycogen and the relative unimportance of fat as an energy source in tropical frugivorous bats. He reported the diurnal cycle of several metabolites including glucose, free fatty acids, lactic acid, and pyruvic acid in captive bats during normal feeding cycles as well as after a "prolonged" fast. The most salient features of this study are the cycles of blood glucose and free fatty acid levels. He found that blood glucose levels follow the expected pattern and fall within normal concentrations for mammals in general. Most of the bats studied died during 31–32 hour fasts;

however, bats that survived showed no further change in blood glucose level at 35.3 mg/100 ml. The diurnal pattern of free fatty acid (FFA) plasma levels follows the expected general inverse relation to blood glucose levels. Nighttime FFA's are essentially constant at about 0.5 milliequivalents/liter (mEq/L), which is quite normal for mammals. If food is withheld for three hours after sunset, plasma FFA concentrations rise to 4.0 mEq/L concomitant with the fall in blood glucose level.

Although histochemical studies of the muscles of bats (Armstrong et al., 1977; Talesara and Kumar, 1974) demonstrate the relative importance of fats and glucose as energy substrates, muscle enzyme profile studies such as those of Muller and Baldwin (1978) and, especially, those of Yacoe et al. (1982), are particularly germane to the present discussion. Yacoe et al. (1982) determined enzyme activity levels for citrate synthetase, hexokinase, 3-hydroxyacyl-CoA dehydrogenase (HOAD), and phosphorylase in two frugivorous species (one of which was *Artibeus lituratus*) and eight insectivorous species of bats. The four enzymes measured are indirect indicators (in sequence) of citric acid cycle capacity, blood glucose oxidation capacity, beta-oxidative capacity, and glycogenolytic capacity.

As expected, citrate synthetase activity was extremely high, among the highest reported for mammalian skeletal muscle, and there were no interspecific differences. Elevated HOAD activity levels in all species indicated the expected high capacity for fatty acid oxidation. Enzymes participating in glucose storage, mobilization, and cell entry, however, provided the most intriguing picture. Hexokinase activity in the frugivorous species was from two to three times higher than in the insectivorous species. Phosphorylase activity in all species was on the high end of the normal mammalian range, and, although not statistically significant, phosphorylase activity in *A. lituratus* was higher than in any other species tested.

Frugivorous bats, therefore, retain the capability of rapidly metabolizing fats as a fuel source but also have unusually high glycogenolytic ability. The combination of high aerobic and high glycogenolytic activities previously has been thought to be mutually exclusive in mammalian muscle fiber (Burleigh and Schimke, 1969). Such a combination defies easy classification in the slow (I) and fast (IIA) and fast (IIB) categories for mammalian muscle fiber types (Lamb, 1984). Although similar results have been reported for Australian bats (Muller and Baldwin, 1978), extremely high capacities for aerobic glucose oxidation have been reported primarily in flight muscle of insects, which have a normal diet with high glucose density (Beenakkers, 1969; Beenakkers et al., 1975; Heinrich, 1979).

Because of its presumed high dietary glucose density and extremely high glucose assimilation efficiency (Morrison, 1980b), *A. jamaicensis,* along with other frugivores and nectarivores, should have much more glucose available for oxidation than do bats of other dietary preferences. However, there is a significant energy penalty for converting dietary carbohydrate to fat (Martin and Lieb, 1979).

From the foregoing, it may be presumed that *A. jamaicensis* produces little fat from glucose and does not take advantage of the reduced weight and high caloric density gained by storing energy as fat. Furthermore, *A. jamaicensis* must exhibit an extreme facility in the storage, mobilization, and turnover of glycogen. Finally, the overall energy reservoir in *A. jamaicensis* is severely limited. In their natural environment, these bats maintain no significant positive daily energy balance that would allow for storage of surplus caloric energy.

SALIVATION.—Like many frugivores, *A. jamaicensis* has exceptionally large salivary glands (Phillips et al., 1977). Results of our histological examination of these glands in *A. jamaicensis* (Studier, Boyd et al., 1983) are in agreement with those of Wimsatt (1956). In spite of the large size of the glands there is no relative increase in number of ducts, nor is there evidence of relative excess production of serous secretion. The functional significance of these structural observations include possible increases in salivary amylase production, increased involvement in regulation of mineral and water balance, and overall increase in production of saliva, all related to the unusually large size of the glands.

Salivary components may act chemically to neutralize alkaloids in figs. Mucus may hold together the pellets that are dropped during feeding (Dalquest et al., 1952). We have suggested that the abundant saliva may strongly buffer gastric secretions, thus preventing gastric contents from becoming acidic, while simultaneously coating the gastric epithelium with an extensive buffering barrier (Studier, Boyd et al., 1983). If gastric fluid remains above pH 4, salivary amylase should continue to function. This may be of considerable importance in view of the rapid passage time and consequent brief digestion exhibited by *A. jamaicensis* (Morrison, 1980b; Studier, Boyd et al., 1983).

The gastric glands in the stomach and duodenum of many frugivorous bats contain typical parietal and zymogen cells (Bhide, 1980; Forman, 1972; Rouk and Glass, 1970). Brunner's glands are reduced or absent in frugivorous bats, including species of *Artibeus*. The function of Brunner's glands among mammals is still debated, but for a long time a connection with protection of the duodenum from damage by highly acidic chyme leaving the stomach has been suspected. If saliva of *A. jamaicensis* has sufficient buffering capacity to prevent chyme from becoming highly acidic, reduction or absence of Brunner's glands in these frugivores would be compensated. Salivary bicarbonate concentration has been shown to be directly proportional to the rate of saliva formation (Burgen and Emmelin, 1961). Because the bicarbonate buffer system accounts for the bulk of the buffering power of the saliva (Izutsu, 1981), increased or high relative rates of saliva production may indicate elevated salivary buffering capacity in *A. jamaicensis*. For further discussion of this possibility and for details of gastric ultrastructure see Phillips et al. (1984).

Caloric Balance

Total daily energy requirements can be estimated by various methods (Kunz and Nagy, 1988). Maintenance energy (ME) requirements are most simply calculated as a function of body mass (W). For endothermic mammals, which maintain constant high T_b, ME (Kcal/day) = 106 $W^{0.75}$, where W is in kilograms (National Research Council, 1978). For an *A. jamaicensis* weighing 45 g, estimated ME equals 10.3 Kcal (43.3 Kj)/day (Table 2-1). Daily energy requirements for an *A. jamaicensis* weighing 45 g also can be estimated based on time partitioning (Table 2-1). The first estimate is based on Morrison's (1978d) data, in which he partitioned flight time into various activities and flight distances. Metabolic cost of flight was based on the wind tunnel studies of Thomas (1975) on other frugivorous bats where metabolic cost of horizontal flight for a 45 g bat is 0.30 Kj. At an average flight speed of 5 m/sec, the cost of flying 100 m is 0.092 Kj. Appropriate increases in the energy cost of flight for transport of figs to feeding roosts also are included (Morrison, 1978d). These flight costs are added to energy needs for maintaining the basal metabolic rate.

Using data from McNab (1969) for well- or overfed *A. jamaicensis* (1.7 cc/g/h), we estimated a basal energy cost of

TABLE 2-1.—Daily energy expenditures for a 45 g *Artibeus jamaicensis*. All values are Kcal/days; equivalent Kj/day are given in parentheses. See text for further details.

Parameters	Kcal/day	Kj/day
Based on mass [a]	10.30	(43.30)
Based on time partitioning		
Flight costs [b]		
searching	0.20	(0.84)
commuting	0.28	(1.17)
between feeding passes	0.19	(0.79)
feeding passes	1.04	(4.33)
transmitter [c]	0.11	(0.50)
subtotal	1.82	(7.63)
Basal metabolism (well- or overfed) [d]	8.77	(36.70)
Total[b+d]	10.59	(44.33)
Basal metabolism (normally- or underfed) [e]	3.87	(16.20)
Total[b+e]	5.69	(23.83)
Based on time partitioning		
Flight costs [f]	3.20	(13.40)
Basal metabolism (well- or overfed) [d]	8.77	(36.70)
Total[f+d]	11.97	(50.10)
Basal metabolism (normally- or underfed) [e]	3.87	(16.20)
Total[f+e]	7.07	(29.60)

[a] National Research Council, 1978.
[b] Morrison, 1978d.
[c] Additional 7% increment as cost of carrying transmitter in flight.
[d] McNab, 1969.
[e] Studier and Wilson, 1979.
[f] Morrison, 1980b.

8.77 Kcal/day. Using metabolic rates of underfed (= normally fed) *A. jamaicensis* (0.75 cc/g/h; Studier and Wilson, 1979), reduces basal energy cost to 3.87 Kcal/day. In either circumstance, daily energy cost of basal metabolism represents the major fraction (67.9%–82.7%) of total daily energy requirements. An additional time-partitioning energy budget (Morrison, 1980b) is based simply on a minimum estimated flight time of 45 min/day (Table 2-1). Although this second time-partitioning budget markedly increases metabolic energy expenditure for flight (from 1.8 to 3.2 Kcal/day), basal metabolic costs still represent the major fraction (54.8%–73.1%) of total daily energy budgets. Kunz (1980) proposed a daily energy budget for bats in general in which caloric needs (in Kcal/day) equal $0.92m^{0.767}$ (where m is mass in grams). Kunz's (1980) estimate of daily energy needs, however, includes data from lactating and pregnant bats as well as from reproductively inactive individuals. Therefore, he overestimated caloric needs.

The estimates for total daily energy costs to a reproductively inactive, endothermic *A. jamaicensis*, maintaining a high T_b are remarkably consistent with a total range of 10.3–12.0 Kcal/day (43.3–50.1 Kj/day) and probably represent realistic estimates for a bat that remains constantly homeothermic. The minimal daily energy budget of 5.7 Kcal/day (23.8 Kj/day) for heterothermic individuals is surely an underestimate because the bats are not continuously heterothermic. The marked reduction in daily energy demand associated with heterothermy, however, may be invaluable to free-flying individuals that have a marginal energy intake. Fleming (1988) estimated daily energy budgets of 41.9–47.3 Kj/day for *Carollia perspicillata*, a smaller phyllostomid with a more varied diet and different foraging strategy. The similarity between these figures is striking.

A. jamaicensis extracts 55.5 g of juice per 100 g of fresh fruit of *F. insipida* (Morrison, 1980b). Morrison reported that juice from these ripe figs contained 0.315 Kcal/g; we found 0.415 Kcal/ml (Studier, Boyd et al., 1983). The specific gravity of an artificial fig juice solution (111 mg glucose per ml) is 1.042 g/cc. Most of the energy in fig juice is in dissolved glucose and the energy assimilation efficiency of fig juice by *A. jamaicensis* is 98.3% (Morrison, 1980b). Using the maximum estimate for a daily energy budget of 12.0 Kcal/day, an *A. jamaicensis* weighing 45 g would need to assimilate all the energy from 28.9 to 36.5 ml of fig juice. Given the 98.3% assimilation efficiency, this would require 29.4–37.2 ml of ingested fruit pulp juice or 55.1–69.8 g of fresh ripe fruit. At 7 g per fruit, an *A. jamaicensis* would require 8–10 whole fruits of *F. insipida* per day to meet its caloric requirement entirely from the ingestion of the fruits of figs. If average-weight ripe fruits weighing 5.6 g (Morrison, 1978a) were ingested, required intake would be 9.8–12.5 figs. These numbers of whole figs correspond nicely to the 7 ± 2 nightly feeding passes observed by Morrison (1978a) when *A. jamaicensis* was feeding

exclusively on fruits of *F. insipida*.

In summary, *A. jamaicensis* has meager, if any, fat reserves; probably exhibits extraordinary glucose assimilation, storage, mobilization, and glycolytic capacities; probably has extreme daily fluctuation in glycogen levels with little reserve capacity; and is marginally able to maintain caloric balance on a normal daily intake of 7 ± 2 *Ficus insipida* fruits. Ingestion of nine fruits probably would allow *A. jamaicensis* to maintain a high T_b for a 24-hour period but ingestion of fewer fruits would not allow caloric balance as a high T_b homeotherm. Such bats would conserve energy and thus remain in caloric balance by a drop in regulated T_b, and a corresponding drop in energy needs.

Nitrogen Balance

Nitrogen excretion is related to metabolic rate. Consequently, nitrogen requirements are appropriately related to metabolic body mass ($W^{0.75}$) (Brody, 1945; Kleiber, 1975). The daily nitrogen requirement (mg/day) for high T_b homeotherms is 200 $W^{0.75}$, where W is in kilograms (National Research Council, 1978). The nitrogen requirement can be converted to a minimum protein requirement by multiplying by 6.25 (Herbst, 1988). Alternatively, dietary protein requirement can be calculated as a function of ingested caloric intake. Minimum protein for maintenance is 10.7 mg protein per Kcal ingested. For a 45 g *A. jamaicensis*, the daily minimum protein requirement is 122 mg, based on body mass alone. Using maximal estimated daily energy expenditure (12.0 Kcal/day from Table 2-1), calculated daily minimum maintenance protein intake is 128 mg. The protein density of *F. insipida* fruit juice is 4.7 mg/ml (calculated from Morrison, 1980b). Assuming fig juice to be the only source of dietary nitrogen of *A. jamaicensis*, constantly homeothermic individuals maintaining high T_b would require 26.0–27.2 ml of fig juice/day (at 100% assimilation) to maintain nitrogen balance. This estimated required intake is less than the calculated daily volume of fig juice needed to maintain caloric balance in high T_b homeothermic individuals (29.4–37.2 ml/day).

Digestibility of protein in low-fiber diets in a variety of mammals ranges from 77% to 90% (Maynard and Loosli, 1969). Assuming an assimilation of 85% for *A. jamaicensis*, daily minimum intake of fig juice would rise to 30.6–32.0 ml/day. This would mean that the daily volume of fig juice necessary to meet protein needs is equal to or less than that needed for caloric balance and that maintenance of nitrogen balance is less of a problem for the bats than maintaining caloric economy. The calculated protein minimum, however, is for the "ideal" protein whose amino acid composition exactly reflects the needs of the subject. Rasweiler (1977) pointed out that animal proteins generally have amino acid compositions that correspond more closely to mammalian requirements and may be more readily digestible than proteins of plant origin, which often are incomplete in terms of essential amino acids.

The calculated requirement for fig juice volume given above is surely a minimum or low estimate of actual juice needs. As is the case for caloric economy, it seems likely that a high T_b homeotherm such as *A. jamaicensis* could maintain nitrogen balance only marginally on a diet of *F. insipida* fruit. Again, the ability of a free-living *A. jamaicensis* to reduce its regulated T_b acts as a safety valve, not only for caloric balance, but also by reducing dietary nitrogen requirements to a level sufficient to allow nitrogen balance.

Although studies of the structure and function of kidneys of Neotropical bats will be discussed in more detail with respect to water and mineral balance, some of the data are pertinent to nitrogen balance. Studier, Wisniewski et al. (1983) examined kidneys in 25 species of Neotropical bats and suggested that renal morphology is primarily a function of dietary protein density. Among the phyllostomids, members of the primarily frugivorous/nectarivorous subfamilies Glossophaginae, Carolliinae, and Stenodermatinae (including *A. jamaicensis*), most of which have low-protein diets, possess renal medullae that cannot be subdivided readily into inner and outer zones. Members of the phyllostomid subfamilies Phyllostominae and Desmodontinae, as well as all members thus far studied in all other families of New World bats, have renal medullae that are easily subdivided into inner and outer zones. The dietary preferences of this second group of bats are varied, but all species regularly consume some food of animal origin, which is high in protein.

Our data on urine composition further support the proposed relation between dietary protein density and renal function in Neotropical bats (Studier and Wilson, 1983). We found urinary ammonia and urea nitrogen levels for *A. jamaicensis* (495 mg% N), although highly variable, to be markedly lower than levels for the insectivorous species, *Myotis nigricans* (1887 mg% N). If the minimal nitrogen requirement for a 45 g *A. jamaicensis* is 19.54 mg/day (= $200 \times 0.045^{0.75}$), then total urinary N is 434 mg/kg/day, which approximates urinary nitrogen excretion rates in many mammalian herbivores (Altman and Dittmer, 1961). If the two abnormally high values for urinary ammonia and urea nitrogen we found in *A. jamaicensis* are disregarded, the recalculated average level is 370 mg% N ($n = 17$, Studier and Wilson, 1983).

Water Balance

We now have sufficient data to present a rough water balance account for *A. jamaicensis* (Table 2-2). Values in Table 2-2 represent estimates for a 45 g, high T_b, homeothermic individual. As previously discussed, daily intake of fig juice is 29.4–37.2 ml. Swallowed juice is 90% water (Morrison, 1980b). Daily intake of water in food, therefore, is 27.6–34.9 g. Estimated metabolic water assumes complete aerobic oxidation of all glucose in the consumed juice (111 mg/ml, from Studier, Boyd et al., 1983). The calculated value for metabolic water is probably slightly low because oxidation of the small amounts

TABLE 2-2.—Water economy budget for a 45 g, high T_b, homeothermic *Artibeus jamaicensis*. All values are g/day. EWL = evaporative water loss. See text for further details.

Parameters	g/day
WATER GAINS	
In food	27.6–34.9
From metabolism	1.9–2.4
Drunk	0
Total gain	29.5–37.3
WATER LOSSES	
EWL at rest	5.2
EWL in flight	1.3
In urine	9.4–11.8
In feces	11.8–14.9
Total loss	27.7–33.2
UNACCOUNTED LOSSES	1.8–4.1

of dietary fats and proteins will produce a little additional water. There is no indication that *A. jamaicensis* needs to consume free water when feeding on figs. The studies of Phillips et al. (1984) on gastric ultrastructure in *Artibeus* also indicated a lack of importance of free water ingestion.

Assuming constant temperature and humidity, evaporative water loss (EWL) at rest is calculated as

$$\log \text{EWL} = \log 0.398 + 0.672 \log W$$

where EWL is grams of water/animal/day and W is in grams (Studier, 1970). We estimated EWL in flight based on 45 minutes total flying time/day (Morrison, 1980b) and scaling up the EWL (957 mg/h) found by Carpenter (1969) for flying *Leptonycteris sanborni* (24.4 g). Estimates of urinary water loss and water lost in fecal matrix are derived from Morrison's (1980b) data showing that urinary water represents 31.8% and fecal water represents 40.1% of swallowed fig juice, respectively. If all ingested nitrogen appeared in the urine, *A. jamaicensis* would produce urine volumes of 4.0–5.3 ml/day, assuming a minimum daily required nitrogen intake of 19.54 mg/day and urinary ammonia and urea nitrogen levels of 370–495 mg%. These values are considerably lower than estimates based on Morrison's (1980b) data, but still represent remarkably high urine output.

Considering the multiple sources and assumptions used to construct the water economy budget for *A. jamaicensis* (Table 2-2), the values come remarkably close to balancing and are probably accurate. Some useful comparisons can be made with values for other mammals. Water turnover rates for a 45 g *A. jamaicensis* (655–829 ml/kg/day) are extreme in contrast to values of 40–273 ml/kg/day reported for a wide variety of mammals (Altman and Dittmer, 1961). Similarly, preformed water consumption in *A. jamaicensis* of 613–775 ml/kg/day is elevated in comparison with values for other mammals (35–211 ml/kg/day; Altman and Dittmer, 1961). These exceptional values relate to the essentially liquid diet of *A. jamaicensis,* coupled with its poor renal water conservation, and relatively small size. Comparative values for total body turnover and preformed water turnover are 330 ml/kg/day and 150 ml/kg/day, respectively, for an 8.4 g *Myotis thysanodes* (O'Farrell et al., 1971), and 250 ml/kg/day and 180 ml/kg/day, respectively, for *M. lucifugus* (O'Farrell et al., 1971). Total body water turnover in *Pizonyx* (= *Myotis*) *vivesi* is estimated at 480 ml/kg/day (Carpenter, 1968). Preformed water turnover for a common vampire, *Desmodus rotundus,* weighing 34.2 g is 410 ml/kg/day (Wimsatt, 1969), assuming bovine blood to be 78.5% water (McNab, 1973). Preformed water turnover also can be estimated from published data for the nectarivorous bat, *Leptonycteris sanborni,* at 465–712 ml/kg/day (Howell, 1974), and the nectarivorous bird, *Selasphorus flammula,* at 692 ml/kg/day (Hainsworth and Wolf, 1972).

Urinary output in *A. jamaicensis* varies from 209 to 262 ml/kg/day in contrast to many other mammals with values that vary from 2.5 to 74 ml/kg/day (Altman and Dittmer, 1961). These rates of urine production in *A. jamaicensis* (6.5–8.2 microliters/min) are roughly 30 times the maximal rate of urine production (0.23 microliters/min) found by Bassett and Wiebers (1979) for the insectivorous *M. lucifugus.* Having collected urine samples from many species of bats, we have no difficulty believing that rates of urine production in *A. jamaicensis* and other frugivorous/nectarivorous bats far exceed urine volumes considered normal in other species.

As mentioned earlier, *A. jamaicensis* and other frugivorous and nectarivorous bats possess kidneys with undivided renal medullae. Such species routinely produce natural urine of low osmotic pressure and low urinary nitrogen levels (Studier and Wilson, 1983) compared with those species with renal medulla divisible into inner and outer zones (Studier, Wisniewski et al., 1983). Mean maximal urine concentration (MMUC) in *A. jamaicensis* is 972 mOsm/kg (Studier, Boyd et al., 1983). The total medullary (M) thickness to cortical (C) thickness ratio (M/C) for *A. jamaicensis* (2.4), while typical for frugivores, is markedly lower than M/C for bats of other feeding preferences and does not vary with habitat aridity (Studier, Wisniewski et al., 1983).

Geluso (1980) presented a highly predictive equation relating MMUC of insectivorous bats to M/C in which MMUC (mOsm/kg) = 702 + 387 (M/C). Based on this equation, *A. jamaicensis* would produce MMUC of 1,620 mOsm/kg, a value nearly double the observed MMUC. Similar observations hold true for other frugivorous/nectarivorous species with undivided renal medullae (Carpenter, 1969; Studier and Wilson, 1983). None of these species produce natural urine concentrations approaching those predicted by Geluso's equation, suggesting that his formula does not apply to bats with undivided renal medullae (Studier and Wilson, 1983).

We made several attempts to induce MMUC in *A. jamaicensis,* that included dehydration/starvation, loading with strongly hyperosmotic salt solutions, and feeding dehydrated fruits of figs (Studier, Boyd et al., 1983). Although none of these methods worked well, ingestion of dehydrated figs, the functional equivalent of Geluso's (1975, 1978) "water denied" experiments with insectivorous bats, resulted in the most uniform and highest urine concentrations for this species. It is of particular interest that although dehydration and salt loading caused an expected and predictable increase in osmotic pressure of the blood, these treatments were not associated with the expected increase in osmotic pressure of the urine (Studier, Boyd et al., 1983). Dehydration in *Myotis lucifugus,* induced by Bassett and Wiebers (1979), however, resulted in expected increases of osmotic pressures in both blood and urine. There seems to be a slow or minimal release of antidiuretic hormone (ADH) or a slow or reduced renal tubular response to ADH in response to rising osmotic pressure in the blood of *A. jamaicensis.*

Urine samples taken from *A. jamaicensis* at sunset in May were significantly more concentrated and less variable than samples taken in November (Studier, Boyd et al., 1983). The May samples were taken at the end of the dry season and the November sampling occurred toward the end of the wet season (Smythe, 1974). Presumably heat/dehydration stress is greater in the dry season than in the wet. In May, there was a marked rapid decrease in total urine concentration 0.5–1.5 hours after sunset in free-flying *A. jamaicensis.* This decrease is associated with rapid food passage time in this species, ingestion of adequate hypotonic fluid for rehydration, and rapid assimilation and equilibration with the ingested fig pulp juice. Urine then became progressively more concentrated throughout the remainder of the night. Osmotic pressures of urine in captive, rehydrated (taken two hours after sunset) individuals were identical throughout the night with urine concentrations of free-flying *A. jamaicensis* (Studier, Boyd et al., 1983). This suggests that rehydration occurs early in the nightly feeding period and is unaffected by subsequent feeding bouts. We know from other studies that *A. jamaicensis* feeds sporadically throughout the night (Morrison 1978a; 1978b; 1978c).

Mineral Balance

How herbivorous mammals ingest adequate amounts of dietary sodium (Na+) has generated considerable interest (Blair-West et al., 1968; Cowan and Brink, 1949; Dalke et al., 1965; Herbert and Cowan, 1970; Jordan et al., 1973; Stockstad et al., 1953; Weeks and Kirkpatrick, 1976; 1978). Sodium levels in most plants and plant parts are typically low (Likens and Bormann, 1970; Sauchelli, 1969; Weeks, 1978) and related to soil Na+ levels, which in turn are highly affected by the Na+ levels in rainfall (Blair-West et al., 1968) and the frequency of rainfall. Tropical rain forest may be particularly susceptible to loss of nutrients by leaching due to rapid decomposition of litter and heavy, frequent rains (Jordan and Herrera, 1981). Although some data on characteristics and composition of the soils of BCI are available (Knight, 1975), there is no information on soil sodium levels.

Fig fruits contain little sodium (Diem, 1962; Heinz International Research Center, 1964; Oates, 1978). Sodium levels in ripe fruits of *F. insipida* and *F. yoponensis* have been measured at 0.49 and 0.48 mg/g dry weight, respectively (Nagy and Milton, 1979). We determined sodium levels in figs carried by *A. jamaicensis* to be similar to those found by Nagy and Milton (1979), but potassium concentrations were lower (Studier, Boyd et al., 1983). Sodium density in dried pulp is about 2.4 times the sodium level of dried seeds and is identical to the sodium level of pulp juices. Thus, the consumption of fig juice rather than the entire fruit greatly increases dietary Na+ density in *A. jamaicensis* and significantly reduces the weight of figs estimated to be required for maintenance of Na+ balance. On the other hand, the K+ level of fig juice is less than half the concentration of that ion in dried pulp. This implies that K+ is concentrated in specific organelles within the fig pulp and is not likely to be extracted by *A. jamaicensis*. It is released in the process of homogenization of the dried pulp for laboratory analysis. The lowered dietary K+ level obtained from pulp juices as opposed to whole pulp probably also lowers the Na+ requirements of *A. jamaicensis* as it does in other mammals (Meyer et al., 1950; Staaland et al., 1980; Weeks and Kirkpatrick, 1978).

Minimal sodium requirements have been estimated for few small mammals. Sodium required for growth in laboratory rats and mice is estimated at 10 and 18 mg/animal/day, respectively (National Research Council, 1978). Assuming the average requirement for growth in *A. jamaicensis* is 14 mg/animal/day, bats would require 69 ml (72 gms) of *F. insipida* fruit juice at 8.8 mEq/L sodium level to maintain sodium balance. Such amounts far exceeds the requirements of fig juice previously calculated for maintenance of caloric and nitrogen balance (29.4–37.2 and 30.6–32.0 ml/day, respectively). Estimated daily Na+ requirements for growth are probably higher than requirements for maintenance. However, it would appear that acquiring adequate sodium for daily requirements from fig juice is more likely to be a nutritional limiting factor than ingestion of sufficient juice to meet caloric and nitrogen needs.

Mammals suffering Na+ deficiency characteristically exhibit hypertrophy and hyperplasia of the zona glomerulosa of adrenal glands and greater development of striated and excretory ducts within salivary glands (Blair-West et al., 1968). These features, acting through the renin-angiotensin-aldosterone system, result in renal and salivary Na+ retention. The adrenal zona glomerulosa of *A. jamaicensis* shows no obvious hypertrophy or hyperplasia when compared with the condition in Neotropical insectivorous bats, although possible subtle differences may yet be detected (Studier, Boyd et al., 1983). We have not found an increase in the number of ducts within salivary glands, another fact that argues against a probable Na+ deficiency in *A. jamaicensis*. The inordinate size of these glands, however, indicates a proportionate increase in absolute number of ducts per gram of body mass of the bat.

Energy and Water Balance during Lactation

Because the foregoing discussion has concerned nutritional requirements and balances needed for maintenance, it is useful to estimate nutritional increments needed under stress such as during lactation (see Section 3, Reproduction in a Captive Colony). A 45 g *A. jamaicensis* should produce milk at a rate of 12.3 gm/day based on Linzel's (1972) measure of milk production as a function of body mass (daily milk production in Kg/day = $0.126^{0.75}$ Kg/Kg). Milk of *A. jamaicensis* contains 2.3 Kcal/g (Jenness and Studier, 1976). This energy level is comparable to milk energy content for other bats, but is higher than milk energy levels for many large mammals. The high energy content of *A. jamaicensis* milk is in accord with Ben Shaul's (1962) suggestion that mammals that nurse their young on a scheduled basis produce milk of higher energy content than mammals that nurse continuously or on demand.

If one assumes that food energy can be converted to milk energy with no cost, the production of 12.3 g of milk per day at 2.3 Kcal/g imposes a minimal additional caloric requirement for *A. jamaicensis* of 28.3 Kcal/day. This would raise the total daily caloric requirement to about 40 Kcal/day for a high T_b, homeothermic, lactating female compared to 12.0 Kcal/day for a nonlactating individual. The total milk protein level for *A. jamaicensis* is 4.7 g% (1.1 g% casein and 3.6 g% whey protein) (Jenness and Studier, 1976). Again assuming no energy costs for milk protein synthesis, a high T_b, homeothermic, lactating female would require an additional incremental daily protein intake of 578 mg for a total daily protein requirement of 700–706 mg compared to 122–128 mg/day for a nonlactating individual. Milk of *A. jamaicensis* is about 70% water; therefore, production of 12.3 g of milk would require an additional intake of 8.6 ml of water for a total daily water need of 38.1–45.9 ml in lactating females compared to the 29.5–37.3 ml water requirements for nonlactating bats.

During lactation, therefore, the additional incremental needs for extra caloric and nitrogen intake are massive in comparison to additional water needs. Whereas estimated milk production may seem somewhat high, energy efficiency for milk production is certainly not 100%. Brody (1945) calculated gross energetic efficiency of milk production to be 28%–34% for humans and 44%–48% for rats. We suggest that during periods of high nutritional demands, such as growth and especially pregnancy and lactation, *A. jamaicensis* would be expected to lower their regulated T_b to reduce caloric and nitrogen needs and that during such times, these bats may supplement their staple fig diet with food items of higher caloric and protein density. Studies directed at shifts in food habits during pregnancy and lactation could test this hypothesis.

Summary

Studies on energy balance in *A. jamaicensis* were focused on this bat's use of the fruits of *Ficus insipida*, the food that it used most on BCI. Probably these bats always are in slight food stress in the wild and are more likely to behave as facultative heterotherms. Poor body temperature regulation in newly captured bats is not related to physiological competence, but reflects reduced metabolic heat production and rate of depletion of energy stores. *A. jamaicensis* falls very near the minimum boundary curve for high body temperature homeothermy, suggesting that energy savings from heterothermy may be critical to this bat in the wild.

Artibeus jamaicensis has extremely low fat reserves and a high intake of dietary carbohydrate, suggesting that glycogen may be its normal energy reserve. This bat, in its natural environment, maintains no significant positive daily energy balance that would allow for storage of surplus caloric energy.

The energy cost of basal metabolism represents the major fraction (55%–83%) of total daily energy requirements. Daily energy budgets of 10–12 Kcal/day combined with the caloric content of daily food intake requires an intake of 9.8–12.5 figs per day, a figure not inconsistent with the 7 ± 2 nightly feeding passes observed in radiotelemetry studies.

Nitrogen and protein requirements can be met by the amount of daily food intake necessary to maintain caloric balance, although questions remain about the digestibility of plant proteins. Urinary ammonia and urea nitrogen levels are markedly lower in *A. jamaicensis* than in bats that use animal protein.

Based on our studies of caloric, nitrogen, and water balances, there is no indication that *A. jamaicensis* needs to consume free water when feeding on figs. Water turnover rates for *A. jamaicensis* (655–829 ml/kg/day) are extreme in contrast to values reported for a wide variety of other mammals. Mean maximal urine concentration in *A. jamaicensis* is 972 mOsm/kg, a strikingly low value indicating poor urine concentrating ability. Published predictive equations for maximal urine concentration do not apply to frugivorous bats. Dehydration in *A. jamaicensis* results in increased blood osmotic pressure, but without a concomitant increase in urine osmotic pressure.

Sodium is potentially limiting in tropical animals restricted to a frugivorous diet, but structural features of the kidneys and salivary glands of *A. jamaicensis* argue against any probable chronic sodium deficiency in these animals. During periods of high nutritional demands such as growth, pregnancy, and lactation, *A. jamaicensis* probably lowers its regulated body temperature to reduce caloric and nitrogen needs and may use other, more energy- and protein-rich food resources.

3. Reproduction in a Captive Colony

Lucinda Keast Taft and Charles O. Handley, Jr.

We report on the reproductive biology of *Artibeus jamaicensis* based upon studies of a colony held captive at the National Zoological Park (NZP). Although information is available on many aspects of the reproduction of bats, most is based on temperate zone Vespertilionidae. The most recent reviews of pre- and postnatal development among the vespertilionids are those of Orr (1970) and Tuttle and Stevenson (1982). Racey (1988) summarized methodology for reproductive assessment and Wilson (1988) provided information on maintaining bats for captive studies. There is little information, however, on reproduction and ontogeny of other bats. Kleiman and Davis (1979) reviewed the available literature on development and maternal care in phyllostomids. The most detailed information on ontogeny is for *Carollia perspicillata* (Kleiman and Davis, 1979) and *Desmodus rotundus* (Schmidt and Manske, 1973).

Phyllostomid bats show far greater diversity of social systems, dietary habits, reproductive strategies, and selection of roost sites than do the vespertilionids (Baker et al., 1976, 1977, 1979). The selective pressures with which phyllostomids must cope are quite different from those encountered by the better-known vespertilionids of the temperate zone. Comparisons between the two families should be productive.

Each year most female *A. jamaicensis* give birth to a single young during two reproductive episodes. This pattern of bimodal polyestry is common to many Neotropical frugivorous and nectarivorous bats (Wilson, 1979). *A. jamaicensis* is unusual, however, because development of the implanted blastocyst is delayed for 2.5–3 months during one of the birth episodes (Fleming, 1971; Fleming et al., 1972). Delayed development has been documented only in Panamanian populations of *A. jamaicensis,* but it is likely to prove widespread. Delayed development also occurs in a phylogenetically and ecologically diverse array of bats, *Macrotus* (Bradshaw, 1962), *Miniopterus* (Medway, 1971), *Hipposideros* (Bernard and Meester, 1982), and *Haplonycteris* (Heideman, 1988).

Parturition and neonatal appearance have been described (Bhatnagar, 1978; Jones, 1945, 1946), but little is known about early development of the young. In Panamá, where the young of *A. jamaicensis* usually are born in tree holes (Morrison, 1975), the study of early development is difficult because many tree hole roosts are inaccessible and the bats show a strong tendency to desert a roost if disturbed. Therefore, observations on captives is still the best means of gaining information on reproduction and development.

Novick (1960), who was successful in breeding and long-term maintenance of *A. jamaicensis* in captivity, believed that freedom from handling was necessary for successful reproduction. Improved artificial diets (Rasweiler and de Bonilla, 1972) have enhanced the successful maintenance of a number of Neotropical frugivorous and nectarivorous phyllostomids (Greenhall, 1976; Rasweiler, 1975, 1977; Rasweiler and de Bonilla, 1972; Rasweiler and Ishiyama, 1973). Kleiman and Davis (1979) found that captive *Carollia perspicillata* successfully bred in spite of the handling necessary for weekly examinations of the adults and young.

History of the Research Colony at NZP

We established a research colony of *A. jamaicensis* at the NZP in Washington, D.C., with 24 bats captured in June 1978 near Corozal, Panamá, 35 km SE of Barro Colorado Island (BCI). We studied reproduction and development in this colony from July 1978 through August 1981 and, consequently, we can describe several aspects of the biology of *A. jamaicensis* in considerable detail.

Among the original 24 bats, three males and 12 females were subadults, born early in 1978. Four males and five females were adults. The females were neither lactating nor obviously pregnant at capture. Prior to the first births in captivity (January 1979), two adult males and four subadult females died. One of these was a female that apparently did not adjust to the captive feeding regime. On the 47th day of captivity all bats were wing-banded with plastic bands and three days later another female died, possibly of complications from banding-associated trauma. The other four bats died because of

Lucinda Keast Taft, National Zoological Park, Smithsonian Institution, Washington, D.C. 20560.
Charles O. Handley, Jr., National Museum of Natural History, Smithsonian Institution, Washington, D.C. 20560.

unspecified pathological conditions. Subsequently, no additional deaths occurred among the original animals during the next 2.5 years.

Three captive-born males (one adult and two subadult) were removed to another cage on 18 February 1980 to reduce crowding and aggression in the flight cage. On 10 October 1980 and 24 June 1981, ten and 12 bats, respectively (including two original females that had repeated reproductive failures), were removed for the same reasons.

Routine handling did not appear to harm the bats in our colony as conception rates were extremely high, particularly for parous females. Exceptions were cases where young were handled frequently during the first three days postpartum. The accompanying disturbance to the mothers sometimes caused maternal abandonment and death of the young. Bats that died before the end of the study were necropsied by the NZP Department of Pathology to determine cause of death.

At the NZP, we kept our bats in two adjoining, climate-controlled rooms, each measuring $3 \times 3 \times 2.5$ m with free access between them through a 1×2 m door. The temperature was maintained at approximately 29°C (25°–31°C) and the relative humidity kept above 50% (range, 50%–80%). After one month of habituation, the bats were clock-shifted a half hour every other day until they were on a reversed light cycle (light, 2000–0800 hours).

Each room contained a burlap-lined wire mesh box, $0.5 \times 0.5 \times 1$ m, open at the bottom, about 1.5 m above the floor. These were intended to be roosts, but the bats preferred the open wire mesh-covered ceilings. Food and water cups were hung on wire brackets on one of the roost boxes, and tree branches were laid across the top of the box. At the beginning of each dark cycle, the bats were provided with a supplemented, peach nectar-based diet (Rasweiler and de Bonilla, 1972) and water *ad libitum*. Partially opened, ripe bananas were suspended from the branches by rubber-coated wire.

Adult bats were marked with one or two plastic bird bands (A.C. Hughes, 1 High Street, Hampton Hill, Middlesex, TW12 1NA, England) applied to the left forearm of males, and the right of females. Each bat received a unique color combination so that individuals could be recognized without capture. Bands were not applied to young bats until skeletal growth was complete (about 12 weeks of age). Soon after birth the young were punch-marked with a number on the wing membrane using a tattooing device (Bonaccorso and Smythe, 1972). It was usually necessary to repeat the tattooing process two or three times before the bats were old enough to be banded because the punch-mark holes healed and faded rapidly (Kleiman and Davis, 1974).

For periodic physical examinations, one of the two adjoining flight rooms was closed off with a burlap curtain and kept darkened. Bats were captured in butterfly nets in the lighted room and released into the dark room after examination. We conducted bimonthly examinations of the adults. The young were examined every two or three days until they were nine weeks old, weekly until 12 weeks old, and then bimonthly thereafter. After pregnant females were found, the colony was carefully inspected each day for births.

In each capture session, the bats were placed in snug-fitting paper tubes and weighed to the nearest 0.1 g. Several external characteristics that vary with reproductive condition were examined and described. These included vulval coloration, mammary size and appearance, and testis size. Lactation was confirmed by gently massaging the areola until milk was expressed from the nipple. Palpation of the abdomen revealed pregnancy.

From the records of these regular examinations we developed criteria for categorizing nipple size and condition, and testis size that we used at the NZP and on BCI to describe age and reproductive state. At the NZP we were able to assign real time to the age categories that we used in the field on BCI (see Appendix). In our study at the NZP we added two categories to distinguish younger juveniles: (A) neonate (day one), babies during the first 24 hours after birth; and (B) infant (two to about 30 days of age), a category to distinguish nonvolant young from juveniles capable of independent flight.

The forearm was measured with dial calipers to the nearest 0.1 mm and wing span was measured with a metric ruler to the nearest 5.0 mm. Wing area was determined by outlining the completely outstretched right wing on paper. A straight line was drawn from the point where the leading edge of the wing joined the body to the point where the trailing edge joined the foot (Davis, 1969a). Area to the nearest 0.1 cm^2 was then determined with a Hewlett-Packard 9874A digitizer and doubled to account for the area of both wings.

General physical condition, genital appearance, degree of epiphyseal fusion, pelage growth, and dental development were monitored at each examination of the young bats. The ability of neonates to emit ultrasonic vocalizations was determined with a QMC Mini bat detector (QMC Instruments, Ltd., 229 Mile End Road, London E14A A, England).

Flight development was determined by tests in which the young were dropped from a height of six feet into a net and scored according to the following criteria:

Drop: fell into the net without spreading the wings.

Spread: wings were extended, but not flapped, while the bat fell straight down.

Flap: wings were flapped while the bat fell straight down.

Glide: some forward motion was achieved with flapping, but no altitude was gained.

Gain: altitude was gained, but landing and maneuvering skills were not exhibited.

Hang up: showed ability to maneuver around obstacles and land by flipping the feet over the head to hang inverted in the roost.

We observed individual behavior through windows in the walls of the flight rooms. Rheostat-controlled, low-level incandescent lighting facilitated observations after short periods of habituation. Prior to the birth of young, the colony was watched for a minimum of six hours per week. Offspring of the first birth group (BG-I), early spring 1979, were each observed for at least three hours per week, with the result that the general behavior of all the colony individuals was incidentally monitored for six to 30 hours per week from 29 January to 25 July 1979. We used the focal animal technique for observations on infants and juveniles and recorded on a checksheet. Here we only report the behavioral patterns, such as flight, that are dependent on the physical development of the young.

Statistical analyses followed techniques recommended by Sokol and Rohlf (1969). We used appropriate parametric tests when required assumptions were met; otherwise nonparametric tests were used. Mean and standard deviation or variance were used as standard measures of central tendency and variation. Some statistical analyses and the plots of growth of several measured characters were made with the Statistical Analysis System (Helwig and Council, 1979) at the George Mason University Computer Center, Fairfax, Virginia.

Characteristics of Adults

SIZE.—Measurements of the original wild-caught individuals were taken after 125 and 305 days of captivity when all definitely had reached the adult age-class. Females were larger (but not significantly) than males in mass and forearm length. Other measures of wing size (span, area, width, and length of tip) were significantly larger in females (Table 3-1).

We found less difference in forearm lengths of adult males ($\bar{X}$ = 62.1 mm, range 57–68, n = 30) and females ($\bar{X}$ = 62.3, range 59–68, n = 30) on BCI than in the NZP colony ($\bar{X}$ = 61.3, n = 7 in males; $\bar{X}$ = 62.4, n = 12 in females). The average mass of adult males from BCI was consistently less than 50 g (48.5 g, n = 100), while averages of nonpregnant females were greater than 50 g (51.1 g, n = 100). Mean mass of the NZP bats was 51.1 g in males and 54.3 g in females. Female-to-male size ratios among the captives show that females were from two to nine percent larger than males in various wing measurements

and about six percent heavier in mass (Table 3-1).

Forearm and mass measurements indicating that females are larger than males in this species have been published (Goodwin and Greenhall, 1961; McManus and Nellis, 1972; Ralls, 1976; Silva Taboada, 1979), but the magnitude and significance of the differences were not mentioned in those reports. Wing-loading values (body mass in g/wing area in cm^2) were significantly greater in males over similar-sized females in *A. jamaicensis* from the Virgin Islands (McManus and Nellis, 1972).

Female vespertilionids generally are larger than males in mass and forearm length, but significant differences in skull measurements are not evident (Myers, 1978; Williams and Findley, 1979). Species where females must carry relatively greater litter weights were predicted to show greater degrees of dimorphism with respect to wing size (Myers, 1978).

Although their data did not support the Myers' (1978) hypothesis, Williams and Findley (1979) agreed that the need for greater weight bearing capacity should select for larger size in female bats. Comparable studies of sexual dimorphism have not been reported for the phyllostomids.

APPEARANCE OF NIPPLES.—1. Preparturient Condition: The eight wild-caught females recognized as subadults in the summer of 1978 produced offspring in Birth Group I (BG-I) in the spring of 1979. Their tiny (~ 0.5 mm diameter), unpigmented, hairy nipples (hairy and naked in this discussion refer to the condition of the areola) began to show enlargement (to 1.0–2.0 mm diameter) almost four weeks before we were able to detect fetuses ($\bar{X}$ = 3.75 weeks prior to fetal detection; SD = 3.105; range, 0.0–8.0 weeks before fetal detection; all statistics ± one week). Over the following two weeks the nipples grew in size and shed hair until attaining the large (~6 mm diameter) and naked condition at an average of 1.5 weeks prepartum (range, 0.5–3.0 weeks prepartum; SD = 0.926; all statistics ± one week). Once the large size was reached, the nipple skin appeared pigmented (darkened) in all but one of the females. The correlation between the first detectable change in size of nipples and first pregnancy in subadult females has not been

TABLE 3-1.—Measurements of wild-caught adult *Artibeus jamaicensis* from central Panamá.

Variable	Female, $N = 12$		Male, $N = 7$		Student's t-test		
	mean	SD	mean	SD	t_s	$P*$	Female/Male
Mass (g)	54.3†	3.888	51.1	4.771	1.594	ns	1.063
Forearm (mm)	62.4	1.859	61.3	1.746	1.271	ns	1.018
Wing span (mm)	472.5	12.154	448.6	8.997	4.512	<0.0005	1.053
Wing area (cm^2)	281.6	25.518	258.8	19.381	2.038	<0.05	1.088
Wing width (mm)	93.1	4.641	85.6	6.024	3.051	<0.005	1.088
Wing tip (mm)	117.5	4.101	107.7	8.077	3.540‡	<0.005	1.091

* One-tailed.

† An average non-pregnant fasting mass was determined for each female and then these were averaged.

‡ The variances of wing tip measures were not homoscedastic; therefore a Wilcoxan two-sample test was also performed (U_s = 75.5, P <0.005).

noted for other species of bats, although Kleiman and Davis (1979) reported darkened nipples in parturient *Carollia perspicillata*.

2. Postlactating Condition: We examined regression of nipples after lactation to a smaller size following each of the 51 births involving the original 13 females. In 14 out of 38 cases where weaning was completed, nipples regressed to a small size (~1.0–2.0 mm diameter) after an average of 124.3 days (80–188 days; SD = 32.511). However, they did not regress below medium size (~4.0 mm diameter) in the other 24 cases (63%). Out of 13 instances in which infants died before weaning, there was one example of nipples not shrinking below medium size and 12 where the nipples reached small size by an average of 20.2 days (5–36 days; SD = 8.167) after death of the infant. Nipples of postlactating females did not regress to the tiny size (0.5 mm diameter) characteristic of subadults.

In all cases where nipples regressed to the small size, there were other characteristics indicative of adult age (e.g., pigmentation of the nipple skin, and/or wholly or partially denuded condition of the areolar region). Apparently, the pigmentation of the nipples does not completely disappear after lactation. Regrowth of fur around the nipple was slow, and evidently it did not surround the nipple as closely as in nulliparous females. The fur surrounding the nipple was noticeably sparser than the rest of the ventral fur, even after regrowth was completed, but the new hair was conspicuous because it was darker than the surrounding old hair. It was a sure indication of a postlactating condition.

3. Comment: The appearance of the nipples has been used in bats to assign females to various age classes (e.g., Dwyer, 1963; Pearson et al., 1952). Enlarged, darkened, and denuded nipples typically indicate previous birth experience and thus adulthood. In addition, size and color of nipples and condition of areolar hair, together with vulvar coloration and other reproductive indicators such as an embryo or milk can pinpoint any stage in the reproductive cycle. Our observations indicated that these criteria were valid for aging and determining reproductive status in female *A. jamaicensis*, so they were used in the NZP colony and in the mark-recapture studies on BCI.

TESTES.—Fleming et al. (1972) showed that periods of testicular enlargement correspond to periods of female estrus in Panamanian and Costa Rican populations of *A. jamaicensis*, *Uroderma bilobatum*, and *Carollia perspicillata*. When our original seven males arrived at the NZP the length of the testes was 5 mm or less ($\bar{X}$ = 4.3) in the three subadults, and ranged from 6–11 mm ($\bar{X}$ = 8.0) in the four adults. Because the colony was established in July, which would have been during a period of sexual receptivity for postparturient females, the adult males should have exhibited maximum testicular length. This apparently was the case, for a decline in length was apparent after one month in captivity. This decline continued for 2.5 months until mid-October 1978, when the size of the testes in adults averaged 6.3 mm. During the same period, the size of the

testes in the three subadults increased to an average of 6.0 mm. Thereafter, testicular size increased in all the males, reaching maximum size ($\bar{X}$ = 8.2 mm; n = 5) in early February 1979 near the beginning of BG-I. A slight reduction in size (to 7.6 mm) occurred at the end of BG-I, but size increased again at the onset of BG-II. Size of the testes did not regress much after BG-II. Through the following years of observation, testes of most adult males remained above 7.0 mm and some remained as high as 10.0–11.0 mm.

During the colony's first two and a half years, the observed maximum testicular length was 11.0 mm; but later, testes up to 14.0 mm long were noted occasionally. Maximum hypertrophy continued to coincide with birth peaks, but minimum length of testes never fell below 9.0 mm during the remainder of the study.

Reproductive Cycle

ESTRUS AND COPULATION.—Copulatory behavior was first seen in August 1978, but births did not occur until January 1979. As a six-month gestation period is unlikely, delayed development of the implanted blastocyst (Fleming, 1971) probably occurred in females that had conceived. However, some copulations were noted as late as November. Subsequently, we often saw copulations by newly parturient females, indicating postpartum estrus. Newborn infants were usually attached to the mothers during the copulations.

Males showed increased interest in the females even before parturition, but successful copulation did not occur until day 2 postpartum. The greatest frequency of copulations or attempts to copulate occurred on days 3 and 4 postpartum. This activity was observed only sporadically thereafter. The latest copulations (two) occurred 25 days postpartum.

A darkening of the vaginal rim of the vulva was noted on the day of parturition. During the subsequent week, the darkened area became more extensive, extending laterally as much as 4 mm, and becoming nearly black in color. During this period, the males frequently attempted to copulate. Up to four copulations or attempted copulations per female per hour were recorded. Over the following 4–18 weeks the darkened vulval area regained normal coloration. The vulval rim was the last area to fade.

PRENATAL CHANGES.—The minimum observed interbirth interval was 112 days. This figure compares well with the four-month gestation that Fleming (1971) postulated for Panamanian *A. jamaicensis*. We could detect the presence of a fetus by palpation about six weeks before birth ($\bar{X}$ = 40.8 days, SD = 6.19, n = 12). Earlier than this, there was the possibility of mistaking the left kidney for a small embryo, although feeling both the kidney and the embryo should have been a positive clue. Later, when the female's abdomen was noticeably distended by pregnancy, the head, legs, and feet of the transversely positioned fetus could be felt. The maximum

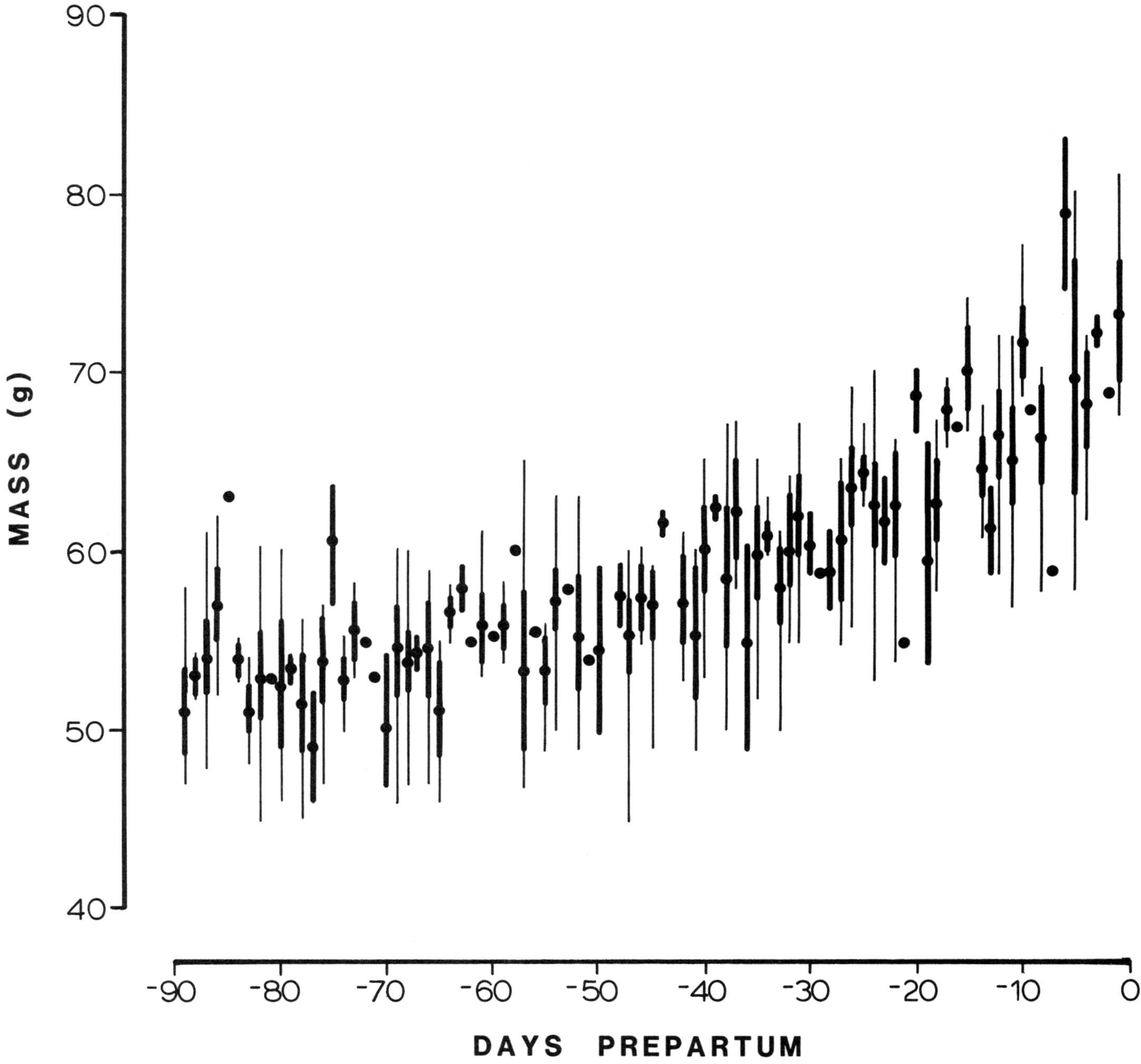

FIGURE 3-1.—Body-mass changes in pregnant *Artibeus jamaicensis* in the NZP colony. Daily means, standard errors, and ranges are shown.

transverse abdominal extension in pregnant females was ~50 mm (= crown-rump length of embryo).

When disturbed, pregnant females were more reluctant to fly than other bats, and they flew more slowly and had difficulty maneuvering. They frequently collided with obstacles or dropped to the floor and were much easier to capture with a hand net. When left alone, they returned to roosting areas. Morrison (1980c) has observed reduced flight speed in free-ranging *A. jamaicensis* carrying young.

Mass of females on the day before birth ranged from 68 to 81 g ($\bar{X}$ = 73.3 g), almost 35% more than normal nonpregnant mass (Figure 3-1). Weight changes of pregnant females have been reported (Kleiman and Davis, 1979) for only one other phyllostomid, *Carollia perspicillata*. Female *C. perspicillata* gained an equivalent of about one third of their nongravid mass during pregnancy. Their gestation length was probably 115 to 120 days. This is similar to the data we recorded for *A. jamaicensis*. Both species exhibit transverse fetal positions prior to birth. *C. perspicillata* also has a postpartum estrus (Kleiman and Davis, 1979).

PARTURITION.—We did not observe any births. Apparently they occurred during the light cycle (2000–0800 hours) or late in the dark cycle (1800–2000 hours, actual time) when we were not watching. From one to 14 hours may have elapsed between actual birth and our initial examination.

No behavioral cues were detected in females about to give birth. Adult females tended to form one large, tight group just before the light period. In the later stages of pregnancy, they frequently participated in the formation of temporary tight groups even during the dark period.

Jones (1946) watched a birth in a captive "*Artibeus planirostris trinitatis*" (= *A. jamaicensis*) from Trinidad. The female hung head-down, and presented the young head first. The baby hung by itself when only a few hours old.

LACTATION.—Milk was expressible from a mother's nipples on the day of birth. Average duration of lactation for all mothers of infants surviving past weaning age was 66.3 ± 7 days ($n = 37$; $SD = 12.758$; range, 46–95 days). Because females were examined only bimonthly for expressible milk, these figures were obtained by averaging the midpoints between the last day females were observed lactating and the first day they were noted to be dry.

Duration of lactation was analyzed by birth group, but single classification analysis of variance (ANOVA) did not reveal significant differences in duration of lactation between groups. However, mean duration of lactation for BG-II and BG-IV were both greater than the means for the other groups (Table 3-2). BG-I, BG-III, and BG-V were intervals when lactating females probably were simultaneously carrying embryos undergoing

TABLE 3-2.—Duration of lactation by birth group in the NZP colony of *Artibeus jamaicensis*.

Birth group	Mean duration of lactation		Number of females	X^2	SD
	in days	(range)			
I	65.6	(47–77)	8	35120.00	9.761
II	76.6	(57–94)	5	30078.50	13.608
III	66.7	(52–87)	7	32255.50	13.540
IV	68.5	(54–95)	7	34004.25	13.895
V	60.0	(50–87)	10	37227.50	11.679

Model II ANOVA

Source	SS	df	MS	F	P
Among	966.023	4	241.506	1.579	ns
Within	4893.754	32	152.930		
Total	5859.777	36			

TABLE 3-3.—Duration of lactation in females of the NZP colony of *Artibeus jamaicensis* in episodes of normal and delayed gestation. $t_s = 1.896$, one-tailed Student's t-test, $P < 0.05$.

Development	Mean duration of lactation in days (range)	Number of females	SD
Normal embryonic development in females lactating during BG–I, III, and V.	63.7 (47–87)	25	11.591
Delayed embryonic development in females lactating during BG–II and IV.	71.9 (54–95)	12	13.786

normal intrauterine growth, whereas arrested embryonic development characterized females gestating during BG-II and BG-IV.

When the data of BG-II and BG-IV were compared with those of BG-I, BG-III, and BG-V the two sets were significantly different at the 0.05 level (Table 3-3). Lactation in females carrying embryos during delayed embryonic development persisted approximately eight days longer than lactation during normal embryonic development.

Reproductive Rates

Information on six birth groups (BG-I to BG-VI) is given in Table 3-4. Only one wild-caught female (F14) failed to produce an infant during BG-I, but she gave birth nearly eight weeks past the last birthdate of other bats of that birth group. Two weeks later, an aborted fetus (about one-half normal neonate size) was found on the floor. One month later, the other pregnant females began having their second babies. Thus the length of BG-II was increased by almost six weeks in order to include these two birth events.

Wild-caught and captive-born females with prior birth experience (multiparous) had high conception rates, 91.7% and 85.7%, respectively (Table 3-5). However, conception rates were lower for captive-born females (~70%) during the first birth group following their transition to adulthood, as shown by the expected values obtained in an r-by-c contingency table (Choi, 1978). Some pregnancies may have gone undetected because of abortion or early fetal resorption.

Each female gave birth to only one offspring. Primary sex ratios did not differ significantly from unity for individual birth groups, or overall, or for infants born to wild-caught versus those born to captive-born females (Table 3-6). Most phyllostomids produce a litter size of one (Humphrey and Bonaccorso, 1979), although twin embryos have been found in low frequency in a few species, including *A. jamaicensis* (Barlow and Tamsitt, 1968). A female vampire produced twins in captivity, one of which was apparently stillborn (Burns, 1970).

TABLE 3-4.—Summary of reproduction and mortality in the NZP colony of *Artibeus jamaicensis*.

Birth group	Dates of birth	Origin of females	Number of adults conceiving		Sex of infants (M.F.?)	Number of deaths of infants and juveniles at various ages (in days)			
			N/total	(%)		≤1	2–30	31–60	61–110
I	01.29.79 to 03.07.79	Wild-caught	12/13	(92.3)	7.5.0	0.2.0	1.0		1.0
II	05.01.79 to 08.01.79	Wild-caught	9/13	(69.2)	4.4.1	0.0.1	1.1	0.1	
III	11.28.79 to 01.31.80	Wild-caught	12/12	(100.0)	6.6.0		3.2		1.0
		Captive-born (primiparous)	3/5	(60.0)	3.0.0	2.0.0	1.0		
IV	04.21.80 to 06.16.80	Wild-caught	12/12	(100.0)	5.7.0		0.2		1.0
		Captive-born (multiparous)	3/3	(100.0)	3.0.0				
		Captive-born (primiparous)	2/4	(50.0)	0.1.1	0.1.1			
V	11.05.80 to 12.21.80	Wild-caught	11/12	(91.7)	7.4.0				2.1
		Captive-born (multiparous)	5/6	(83.3)	1.4.0		0.1*	0.1*	
		Captive-born (primiparous)	5/5	(100.0)	1.0.4	0.0.4	1.0		
VI	03.04.81 to 04.30.81	Wild-caught	10/10	(100.0)	3.7.0	1.0.0	1.0	0.1*	0.1
		Captive-born (multiparous)	4/5	(80.0)	2.2.0	1.0.0	1.1	0.1	
		Captive-born (primiparous)	4/6	(66.7)	3.1.0	1.0.0	1.0	1.0	
Total					45.41.6	7.1.6	10.7	1.4	5.2
							23.14.6		

* Deaths occured accidentally during handling. These individuals are not included in any subsequent analysis of mortality.

However, no twinning occurred in our captive *A. jamaicensis*. The neonate of this species is so large at birth, compared with the mother, that it is difficult to imagine a female successfully carrying two young to term.

Most phyllostomids produce two young per year, one at a time, whereas temperate zone vespertilionids produce single litters of one to four young. Apparently, the differences in reproductive strategies are due to temperature fluctuations and to different temporal patterns of food abundance (Wilson, 1979).

INTERBIRTH INTERVALS.—During their first year in captivity the reproductive cycles of the NZP bats remained synchronized with the cycles of the wild population in Panamá from which they were taken. Inclusive birth dates of birth groups BG-I and BG-II of the NZP bats corresponded approximately to birth dates in the free-living Panamanian population (Table 3-4).

However, from BG-III on, births began to occur sooner than expected, based on the reproductive seasonality of the free-living Panamanian bats. Delayed embryonic development should have deferred births in BG-III until February 1980, but births in the colony began in late November 1979 and were completed by the end of January 1980. BG-IV was approximately 1.5 months ahead of the predicted schedule, and BG-V was more than two months early. If delayed embryonic

TABLE 3-5.—Frequency of conception in the NZP colony of *Artibeus jamaicensis*, tabulated by origin and reproductive experience of females in an r-by-c contingency table. Expected values are in parentheses (Choi, 1978). $X^2 = 6.62$, $df = 2$, $0.02 < P < 0.05$.

Origin and reproductive experience of females	Number of pregnancies	Number not pregnant	Percent of females conceiving
Wild-caught	66 (62.5)	6 (9.5)	91.7
Captive-born (multiparous)	12 (12.2)	2 (1.8)	85.7
Captive-born (primiparous)	14 (17.4)	6 (2.6)	70.0

TABLE 3-6.—Analysis of sex ratios of neonates in the NZP colony of *Artibeus jamaicensis* by birth group and mother's origin.

Grouping	Males (N)	Females (N)	P*
Birth group			
BG–I	7	5	0.1934
BG–II	4	4	0.2734
BG–III	9	6	0.1527
BG–IV	8	8	0.1964
BG–V	9	8	0.1855
BG–VI	8	10	0.1669
Total†	45	41	0.0783
Mother's origin			
Wild-caught	32	33	0.0978
Captive-born	13	8	0.0970

*One-tailed exact binomial probability test.

† BG–I through BG–VI, combined:
$df = 12$, -2 sum ln $P = 19.834 = X^2$, $0.10 > P > 0.05$.

TABLE 3-7.—Length of interbirth intervals in the NZP colony of *Artibeus jamaicensis*.

Interval between birth groups	Number of individual females	Mean interval in days (range)	s^2	Comparison*
I–II	7	134.6 (120–153)	198.29	X
II–III†	6	162.2 (132–184)	443.37	Y‡
III–IV	14	131.2 (113–158)	372.34	X
IV–V†	13	190.9 (159–220)	377.74	Z‡
V–VI	13	120.2 (112–137)	50.74	X

ANOVA (one-way)

Source	SS	df	MS	F	P
Between	40561.185	4	10140.296	36.354	< 0.001
Within	13388.740	48			
Total	53949.925	52			

* Student-Newman-Keuls multiple comparison test. Identical letters denote that differences were not significant at the 5% level.

† Intervals where delayed embryonic development was presumed to have occurred.

‡ $P < 0.01$.

development occurred in the captive females at approximately the same time it occurred in free-living females, then the interbirth intervals between BG-II and BG-III and between BG-IV and BG-V would be expected to be longer than interbirth intervals where development proceeded normally.

These data and interbirth intervals for individual females between consecutive birth groups (Table 3-7) suggest that delayed embryonic development did occur in the captives but was of diminished duration. The average interbirth interval for normal prenatal development would be about 122 days and the interbirth interval with delayed embryonic development would be about 213 days according to Fleming's (1971) data for free-living Panamanian *A. jamaicensis*.

Differences between intervals presumed to represent normal gestations were not significantly different from one another in the NZP bats as revealed by *a posteriori* testing (Table 3-7). The presumed delayed intervals, however, were significantly different from the normal intervals as well as from each other. The overall trend in the colony's reproductive output was toward a shortening of the interbirth intervals and earlier onsets of successive birth groups than expected if the captives had remained synchronized with the wild population. Of the two intervals where delayed development was expected, mean length of the second interval (BG-IV to BG-V) was nearly 30 days longer than that of the first, although BG-V still occurred approximately three months ahead of the Panamanian cycle. Loss of some environmental clue, such as day length, may have impacted the cycle of the captive bats.

Juvenile Mortality

Forty-three babies died before weaning (Table 3-4). Factors contributing to deaths were as follows (by age groups):

NEONATAL PERIOD (1st day; 14 deaths).—Mortality was the result of premature birth, still birth, congenital deformation, and maternal neglect.

PREVOLANT PERIOD (2–30 days of age; 17 deaths).—This period was characterized by mothers frequently leaving their infants unattended, or in the company of other young and/or adults. Mothers were in estrus during the early part of this period and were frequently pursued by males. A higher than normal density of males (never less than five adult males) may have contributed to increased levels of stress in the colony. The greatest percentage of infant mortality occurred at this time, possibly due to this stress.

NEWLY VOLANT PERIOD (31–60 days of age; 5 deaths).— Young were newly volant, but not yet weaned. Initially, they remained in the roost, exercised their wings and pectoral muscles, and accompanied their mothers on short flights. If they were not strong enough for a return flight to a roost area and could not find a substrate up which they could crawl to a roost, or a place high enough to initiate another flight, they would weaken and die unless they were retrieved by their mothers or discovered by the keepers.

WEANING PERIOD (61–110 days; 7 deaths).—Juveniles were becoming more independent, weaning was nearing completion, and the mothers were starting to reject their young by chasing them from the roost. Transition to subadulthood was beginning.

Preweaning mortality among infants of wild-caught females was 35.4% (Table 3-8). Much of this mortality was attributable to two females (F4 and F12) who consistently failed to rear offspring (seven babies between them for BG-I to BG-VI) because of congenital defects (e.g., skeletal deformities) or intestinal nematode infestations in the young. These females were able to rear young successfully beyond weaning age when they were removed from the colony and placed in a larger flight

TABLE 3-8.—Correlation of infant and juvenile mortality and level of maternal experience. Numbers of infant deaths in various age groups are tabulated in an r-by-c contingency table with expected values in parentheses (Choi, 1978). $X^2 = 20.697$, $df = 4$, $P < 0.001$.

Origin and reproductive experience of females	Number of births	Number of deaths during preweaning		Number surviving beyond weaning 110+ days	Percent preweaning mortality	Percent postweaning survival
		≤ 30 days	31–110 days			
Wild-caught	65	15 (21.91)	8 (7.30)	42 (35.79)	35.4	64.6
Captive-born						
multiparous	10	3 (3.37)	1 (1.12)	6 (5.51)	40.0	60.0
primiparous	14	12 (4.72)	1 (1.57)	1 (7.71)	92.9	7.1
Total for colony	89	30	10	49	44.9	55.1

TABLE 3-9.—Analysis of mortality among captive-born bats by sex. Numbers of infant deaths in various age groups are tabulated in an r-by-c contingency table with expected values shown in parentheses (Choi, 1978). $X^2 = 5.860$, $df = 3$, $0.25 > P > 0.10$.

Sex	Total births	Total deaths	Age in days			Survivors 110+
			≤ 1	2–30	31–110	
Males	45	23	7 (4.34)	10 (8.67)	6 (5.42)	22 (26.57)
Females	38	11	1 (3.66)	6 (7.33)	4 (4.58)	27 (22.43)
Total	83	34	8	16	10	49

cage with fewer bats. Thus, they may have been in poor nutritional condition from excess stress or social exclusion from food, factors that could have affected their offspring's growth in utero and postpartum development. Excluding the young born to females F4 and F12 would reduce mortality of juveniles of wild-caught mothers to about 27%.

Preweaning mortality among offspring of multiparous (experienced) captive-born mothers was 40%, attributable entirely to the deaths of all young born during BG-VI. We have no explanation for this catastrophe. In contrast, approximately 64% of the young born to primiparous (inexperienced) females did not live more than a day, and nearly 86% did not survive a month. Sixty percent or more of the young of experienced mothers (wild-caught females and multiparous captive-born females) survived beyond weaning, but few (7%) of the young of primiparous females survived that long.

Apparently, preweaning mortality and postweaning survivorship are dependent on the mother's level of reproductive experience (Table 3-8). Deaths of young were not evenly distributed between the sexes (Table 3-9). In our colony, males died at a rate greater than expected and females died at a rate lower than expected, but the differences were not significant.

Lower conception rates and increased infant mortality among primiparous captive-born females may be attributable to a variety of causes. In natural populations, young *A. jamaicensis* probably are forced to disperse from their natal roost before birth of the next young, and thus parenting behavior would not

be acquired through learning. Despite proximity to experienced females at the NZP, inexperience may have operated to increase mortality of infants born to our captive, primiparous females. Increased mortality due to inbreeding could also have been involved, because the inexperienced females could have mated with their male parent, male siblings, or other related males. However, the subsequent successful reproduction in these females suggests that this was not an important factor.

Kleiman (1980) discussed several physiological and behavioral means whereby reproduction may be suppressed in socially subordinate female mammals in captive situations. Abortions, stillbirths, inadequate maternal care, and one apparent case of depressed lactation were the causes of death for several of the first young born to our captive-reared females.

In Panamá and México, groups of 4–11 adult female *A. jamaicensis* accompanied by their juvenile offspring and a single adult male, occupied tree hole roosts, whereas foliage roosts (palm fronds and subcanopy trees) were occupied by solitary males or small groups of subadults (Morrison, 1979). Tree hole roosts probably are preferred over foliage roosts by reproducing females, as they provide greater protection from rain, predators, and fluctuations in temperature (see Section 8, Roosting Behavior). While foraging, female *A. jamaicensis* usually leave their young in the roosts (Fenton, 1969).

Trune and Slobodchikoff (1976) found that clustered *Antrozous pallidus* were less agitated and had less weight loss, lower oxygen consumption, smaller body temperature-ambient temperature differentials, lost less heat to the environment, and conserved more metabolic energy while inactive, than bats roosting individually. Any or all of these factors could be important to prenatal and postnatal development.

Our captive-born female *Artibeus jamaicensis* first produced young while they were still in nonclustering "subgroups." The implication is that these females were not in the appropriate social environment (harem) necessary for infant development. For some free-living populations, tree hole roosts may be critical resources for reproduction and the cluster of bats within the roosts may be vital as a source of "helpers" to maintain the appropriate thermal conditions for the young bats. These factors might explain the failure of our primiparous captive females to successfully rear offspring.

Neonatal Physiognomy

Between 29 January 1979 and 30 April 1981, at least 78 full-term bats were born at NZP. Unless otherwise noted in the descriptions that follow, observations on neonates refer to 37 of these infants on their first day of life. Neonates were first examined at the beginning of the dark period and were probably from 1 to 14 hours old when first seen. All neonates retained a segment of dried umbilical cord. Frequently, bits of dried amniotic membrane were present on their bodies as well. By the second day, the cord and all dried membranes were gone, except for two infants that still retained umbilical cords.

All neonates had their eyes open, and most had ear pinnae and noseleaf erect when they were first examined. A few had the pinnae and the noseleaf flattened against the head, but they became erect within six hours. Ear openings usually were apparent and the neonates twitched their pinnae or bodies in response to sounds. Six babies did not show ear openings or respond to sounds until as late as four days following birth. However, all infants examined had the ability to emit ultrasonic sounds in the 80 kHz range on their first day. *A. jamaicensis* emit their ultrasonic orientation calls through the nose (Griffin, 1958), and, accordingly, the mouths of infants were closed when they produced ultrasonic emissions. Their nostrils and the grooves at the base of the noseleaf on the outer edges of the nostrils quivered visibly during such vocalizations.

The neonates usually were covered sparsely with dark gray fur on the dorsum and top of the head. The ears, muzzle, wings, and venter were hairless. Nonfurred skin usually was pink. However, a few neonates had dark gray skin.

The feet of neonates were well developed and disproportionately large. The wings appeared short, narrow, and relatively less developed compared to the otherwise precocial physical condition of the neonates. Bones of fingers and joints were soft and flexible. The bones and skin of the wing tips were unpigmented and translucent, and the tip of digit III often curved in toward the body more than 180 degrees.

Average physical measurements of neonates were taken on 22 individuals, each of which survived at least three weeks. Males and females did not differ significantly in mass, forearm length, wing span length, or wing area, although average size of female neonates exceeded that of males (Table 3-10).

Phyllostomid neonates are comparatively large at birth, with neonate-to-mother mass ratios usually exceeding 0.25 and forearm ratios exceeding 0.41 (Kleiman and Davis, 1979; Figures 3-2 and 3-3; Table 3-11). *A. jamaicensis* compares well with these observations, with a neonate-to-mother mass ratio of 0.26 and a forearm ratio of 0.54 (Table 3-10).

Neonatal Behavior

Newborn young in the zoo colony typically were found hanging beside their mothers with mouth grasping one of the mother's nipples. Young thus attached kept their wings folded and did not appear to use their thumbs for clinging. Often, a mother shielded her infant, at least partially, with a wing. It was rare to find a neonate with its feet also attached to its mother while resting quietly in the roost. On the other hand, it was not unusual to find newborn infants hanging alone in the roost. They hung quietly, with both feet attached to the ceiling, their wings folded, and often had their eyes closed. They appeared to be sleeping.

Neonates were able to crawl about in a slow and wobbly

TABLE 3-10.—Body mass and measurements of wings of neonate *Artibeus jamaicensis* in the NZP colony.

Variable	Female, $N = 12$		Male, $N = 10$		Combined, $N = 22$		t_s *	P	Sex ratio (F/M)	Percent adult size
	mean	SD	mean	SD	mean	SD				
Mass (g)	14.1	2.100	13.7	1.655	13.9	1.878	0.513	ns	1.031	0.264
Forearm length (mm)	33.9	2.749	33.0	2.088	33.5	2.453	0.811	ns	1.026	0.542
Wing span (mm)	235.8	18.195	227.5	15.501	232.0	17.159	1.144	ns	1.036	0.504
Wing area (cm^2)	77.2	18.865	71.6	17.666	74.7	18.120	0.715	ns	1.078	0.276
Wing loading (g/cm^2)	0.187	0.025	0.198	0.034	0.192	0.029	0.825	ns	0.944	1.118
Aspect ratio (wing span^2/wing area)	7.41	0.869	7.49	1.252	7.45	1.034	0.168	ns	0.990	0.942

* Student's *t*-test for significant differences between the means of males and females (one-tailed).

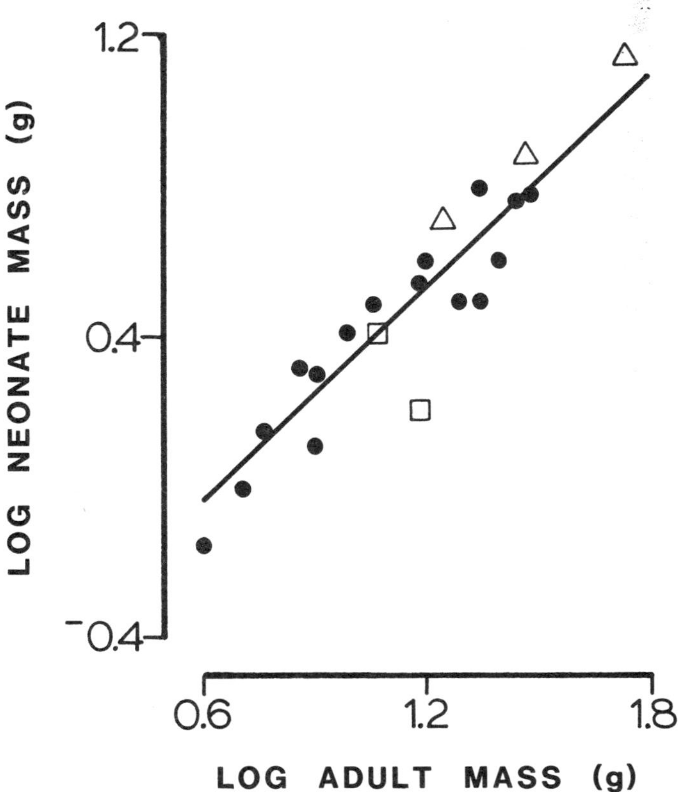

FIGURE 3-2.—The relationship between neonate mass and adult mass for 18 species of bats. Log neonate mass (g) is plotted against log adult mass (g). The regression line for vespertilionids (solid circles) is y = −0.60 + 0.93 x, (r = 0.942, P < 0.001). Phyllostomid (open triangles) and molossid (open squares) data points are shown for comparison. Data are from Table 3-11.

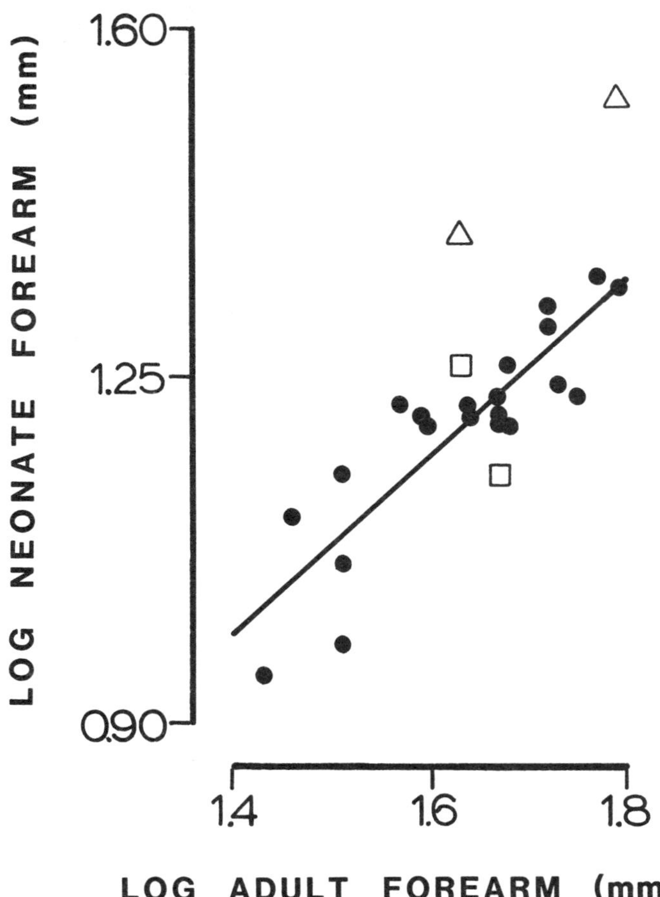

FIGURE 3-3.—The relationship between neonate and adult forearm lengths for 19 species of bats. Log neonate forearm (mm) is plotted against log adult forearm (mm). The regression line for vespertilionids (solid circles) is y = −0.24 + 0.88 x, (r = 0.879, P < 0.001). Phyllostomid (open triangles) and molossid (open squares) data points are shown for comparison. Data are from Table 3-11.

fashion and were active in the reattachment process upon reunion with their mothers. They showed a strong tendency to keep their heads downward. When a mother was held in a head-up position, the attached neonate readjusted its position with its feet until it grasped its mother's head. Turning the mother head-down caused the baby to resume the normal head-down posture. Neonates placed on a vertical wire mesh surface with their heads up immediately started to invert. They also tended to crawl upward (achieved by "walking" backward with the feet). Older babies tended to crawl upward until stopped by an obstacle or lack of a foothold, but neonates usually crawled upward only a few centimeters and then stopped.

Autogrooming in *A. jamaicensis* consists of licking the body surfaces and raking or scratching body surfaces with one foot while hanging by the other. Neonates were seen licking themselves, especially their wing surfaces, but they never hung by one foot in order to use the other foot to groom. Allogrooming is known in this species to include the grooming of infants by their mothers. Occasionally an infant extended a wing to its mother as if to solicit grooming, and the mother then licked the baby's wing. Mothers frequently sniffed young other

than their own and often huddled with them, but they licked only their own young. We never observed allogrooming between adults at the NZP, but it must occur occasionally in wild adult *A. jamaicensis*. That would best explain the chewed necklace bands we noted in some individuals that were captured on BCI.

When we separated a neonate from its mother's nipple, it appeared anxious and was quick to restore a holdfast with its mouth, but it often was not discriminatory about the substrate it happened to grasp. As they were being examined the newborns frequently succeeded in biting our fingers, and on two occasions babies remained attached and began to suck vigorously. By the second day postpartum, the babies were more discriminating about what they would grasp with their mouths. In order to avoid falling, they would bite our fingers when prevented from clasping our hands with their thumbs and hind feet, but none tried to suck the skin. Infants were never found attached to females other than their own mothers.

Disturbed mothers carried their babies with them in flight.

TABLE 3-11.—Mass, forearm length, and growth data from the literature and this study for 20 species of bats. Table headings are: A = age adult mass or forearm length attained (days); AF = adult forearm length (mm); AW = adult mass (g); NF = neonate forearm length (mm); NW = neonate mass (g); Rate = mass gain (g/day) or growth of forearm (mm/day); S = study type (C: captive, CF: captive but females pregnant when captured, F: field). Values estimated from data in the literature are in parentheses. Published growth curves were examined to determine the age when adult size was attained, which was the age when asymptotic size was first reached. To avoid problems associated with varying rates of growth, rates were derived as in Case (1978). Thus, our growth rate is the maximum rate during the linear (and relatively constant) phase of growth.

Species	Mass				Forearm				S	Reference
	AW	NW	A	Rate	AF	NF	A	Rate		
PHYLLOSTOMIDAE										
Carollia perspicillata	17.6	5.0	70	0.23	42.3	24.4	38	0.78	C	Kleiman and Davis, 1979
Desmodus rotundus	(29.5)	7.5	280	0.12					C	Schmidt and Manske, 1973
Artibeus jamaicensis	54.3	13.9	80	0.43	61.0	33.5	50	0.91	C	This study
VESPERTILIONIDAE										
Eptesicus fuscus	(16.0)	4.0	48	0.26	(47.5)	(18.0)	42	1.10	F	Davis et al., 1968
Eptesicus fuscus	19.6	3.1	60	0.28	(47.0)	16.1	50	0.90	F	Kunz, 1974
Eptesicus fuscus					(46.8)	17.0	35	0.71	CF	Gould, 1971
Eptesicus serotinus	28.3	5.8	55	0.55	(52.0)	(20.0)	28	1.50	CF	Kleiman, 1969
Myotis lucifugus	(7.3)	(2.1)	20		39.5	15.7	17	1.59	F	O'Farrell and Studier, 1973
Myotis lucifugus					(37.5)	(16.5)	30	1.18	CF	Gould, 1971
Myotis myotis	22.5	6.1	45	0.90	59.3	22.6	45	1.60	F	Kratky, 1970
Myotis adversus					(39.2)	(16.4)	42	0.77	F	Dwyer, 1970
Myotis thysanodes	(9.7)	(2.6)	21	0.24	43.8	16.3	21	1.31	F	O'Farrell and Studier, 1973
Myotis velifer	(11.6)	3.0	45	0.40	(47.1)	16.0	35	1.21	F	Kunz, 1973
Nyctalus noctula	28.9	5.7	35	0.60	51.9	20.7	35	1.07	CF	Kleiman, 1969
Nyctalus lasiopterus	(30.0)	(5.9)	45	0.75	62.0	22.1	43	1.65	C	Maeda, 1972
Nycticeius humeralis	8.4	2.0	100	0.05	(32.7)	(14.0)	46	0.61	CF	Jones, 1967
Pipistrellus pipistrellus	5.9	1.4	20	0.18	32.0	11.4	35	0.88	CF	Kleiman, 1969
Pipistrellus pipistrellus	5.1	1.0	25	0.11	(32.0)	9.5	25	0.83	F	Rakhmatulina, 1972
Plecotus townsendii					(44.0)	16.6		1.20	F	Pearson et al., 1952
Tylonycteris pachypus	(4.0)	(0.7)	40	0.05	(27.0)	(9.0)	55	0.28	F	Medway, 1972
Tylonycteris robustula	(8.0)	(1.3)	50	0.15	(29.0)	(13.0)	45	0.50	F	Medway, 1972
Miniopterus schreibersi	(15.0)	(3.5)	47	0.27	(48.0)	15.8	52	0.71	F	Dwyer, 1963
Antrozous pallidus	22.2	(3.1)	35	0.22	53.9	17.4	38	1.15	F	Davis, 1969b
Antrozous pallidus	25.0	(4.0)	82	0.55	(56.5)	(17.0)	28	1.20	C	Brown, 1976
MOLOSSIDAE										
Tadarida brasiliensis	(15.0)	(1.5)	30	0.45	(47.0)	(14.0)	46	0.73	F	Pagels and Jones, 1974
Tadarida brasiliensis	11.8	(2.5)	20	0.38	43.0	(18.0)	32	0.98	F	Short, 1961

Apparently the babies instantaneously detached their feet from the ceiling as the mother spread her wings to take off. When a mother with a baby attached to one of her nipples was netted in flight, the infant usually was found in a position parallel to its mothers's body, its feet gripping her femur or inguinal region. Rarely, an infant was found in a crosswise posture such as that noted by Kleiman and Davis (1979) for *Carollia perspicillata*. In this case, the infant's feet were attached to the opposite nipple region and the baby was carried across the mother's chest just behind her throat.

Growth and Development of Young

GROWTH OF HAIR.—Follicular activity preceding growth of fur caused the color of the skin to change from the neonatal pink to gray. Skin of the underparts darkened around seven days of age, and hair growth began at 12 days. At this time, young also began to groom themselves with their feet. The sparse, appressed fur on the head and dorsum was erect and starting to thicken on day 15. The muzzle area darkened around day 17, and facial fur growth started about day 20. By day 22

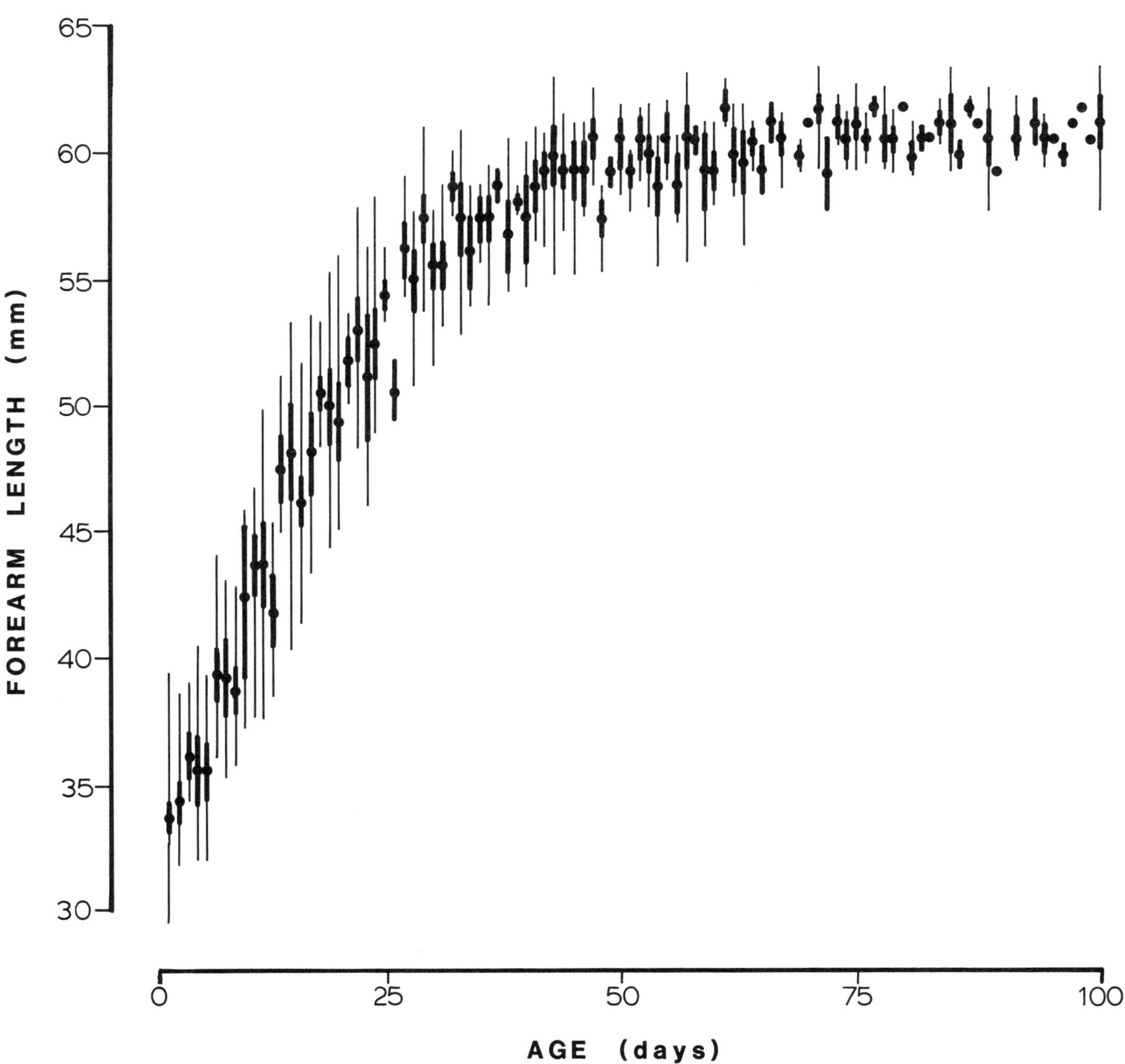

FIGURE 3-4.—Growth of forearm in the NZP colony of *Artibeus jamaicensis*. Daily means, standard errors, and ranges are shown.

the ventrum was covered with sparse fur. The general and superciliary vibrissae were only red spots about the corners of the mouth and eyes until day 25, when they began to protrude. Thereafter, short, soft vibrissae were also evident. Young had full pelage by 30 days of age and whitish facial stripes were apparent on some individuals.

BODY SIZE.—Growth curves for the four measured characters (forearm length, wing span, wing area, and mass) are presented in Figures 3-4 to 3-7. Daily means, standard errors, and ranges are shown for 22 young from BG-I and II. Data derived from deformed or markedly underdeveloped young that died within a few days of birth were not included in the computations.

Forearm length showed the greatest development at birth relative to other measured characters and the fastest growth rate ($\bar{X} = 0.9$ mm/day during the maximum linear growth phase; Figure 3-4). Forearm growth stabilized around 50 days of age at $\bar{X} = 61$ mm, 1 mm less than the average forearm length of bats of the original adult colony (Tables 3-1 and 3-12).

Wing span increased by $\bar{X} = 5.8$ mm per day and wing area by $\bar{X} = 3.8$ cm^2 per day during the maximum growth phase, and both reached adult proportions in $\bar{X} = 70$ days (Figures 3-5 and 3-6). Body mass showed the slowest rate of increase (~ 0.5 g per day; Figure 3-7) and took nearly 80 days to stabilize at $\bar{X} = 48$ g, which was 5 g less than the average adult body mass of

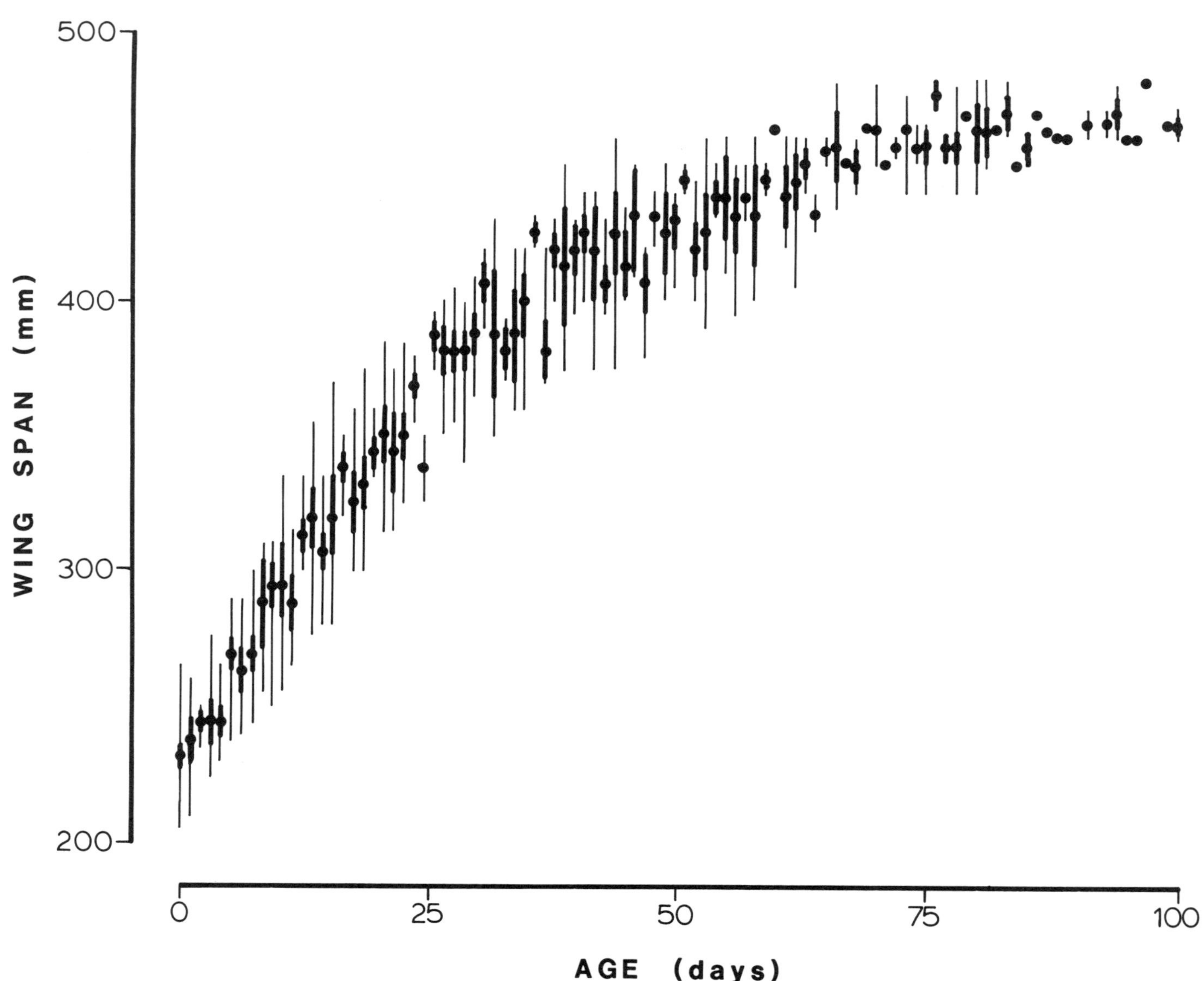

FIGURE 3-5.—Growth of wing span in the NZP colony of *Artibeus jamaicensis*. Means, standard errors, and ranges are shown.

TABLE 3-12.—Average asymptotic measurements by sex of *Artibeus jamaicensis* in the NZP colony.

Variable	Females			Males			t_s*	P†	Sex ratio (F/M)
	mean	SD	n	mean	SD	n			
Mass (g)	48.4	3.462	20	47.6	2.923	18	0.077	ns	1.017
Forearm (mm)	61.7	1.377	19	60.7	1.635	18	2.058	<0.025	1.016
Wing span (mm)	466.7	10.630	15	457.1	91.400	14	2.600	<0.01	1.021
Wing area (cm²)	264.73	18.155	7	257.77	8.789	6	0.853‡	ns	1.027

* Student's *t*-test for significant differences between the means.

† One-tailed.

‡ The variances for measurements of male and female wing area were not homoscedastic, so a Wilcoxon two-sample test was also performed ($U_s = 25$, ns).

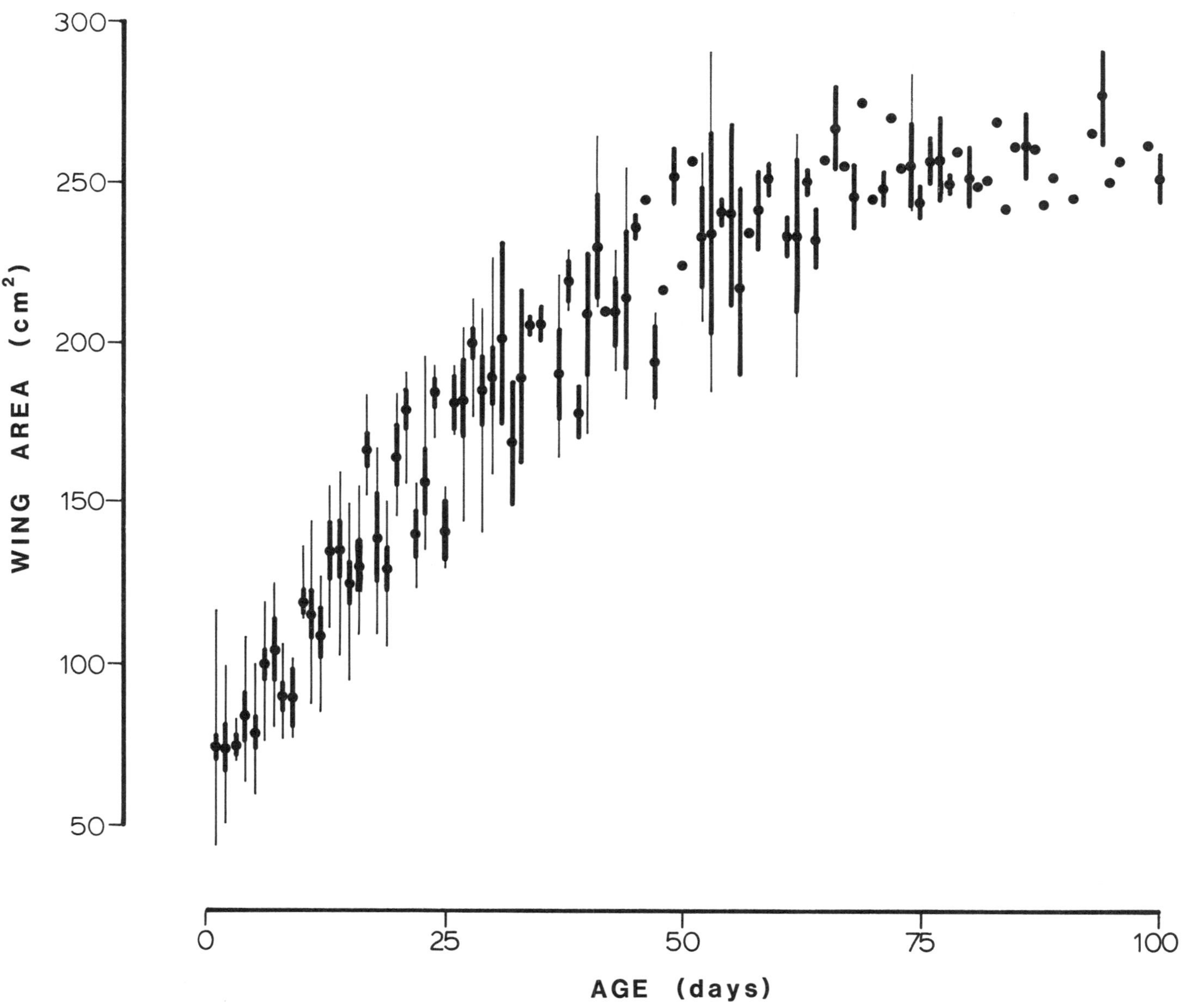

FIGURE 3-6.—Growth of wing area in the NZP colony of *Artibeus jamaicensis*. Means, standard errors, and ranges are shown.

the colony's original members (Tables 3-1 and 3-11).

Sexual dimorphism was not as apparent in the asymptotic size of the captive-reared young as it was in the wild-caught bats. Again, females exceeded males in all measured variables, but the differences were significant only for forearm length and wing span. The magnitude of the differences between the sexes was also reduced, and females exceeded males in size by only 2%–3% (Table 3-12). The variable best correlated with age was mass ($r = 0.9387$, $P < 0.0001$; where r is the correlation coefficient and P is probability). All size variables were highly correlated with one another. The average asymptotic masses of 19 captive-born young were compared with their neonatal masses, and a significant relationship was found ($r = 0.7097$, $t_s = 4.1459$, $P < 0.001$; Figure 3-8). Large infants tended to become large adults.

INFANT-MOTHER SIZE RATIO.—There was a tendency for larger females to produce large infants (Figure 3-9), but the relationship was not significant ($n = 12$; $r = 0.552$; $0.10 > P >$

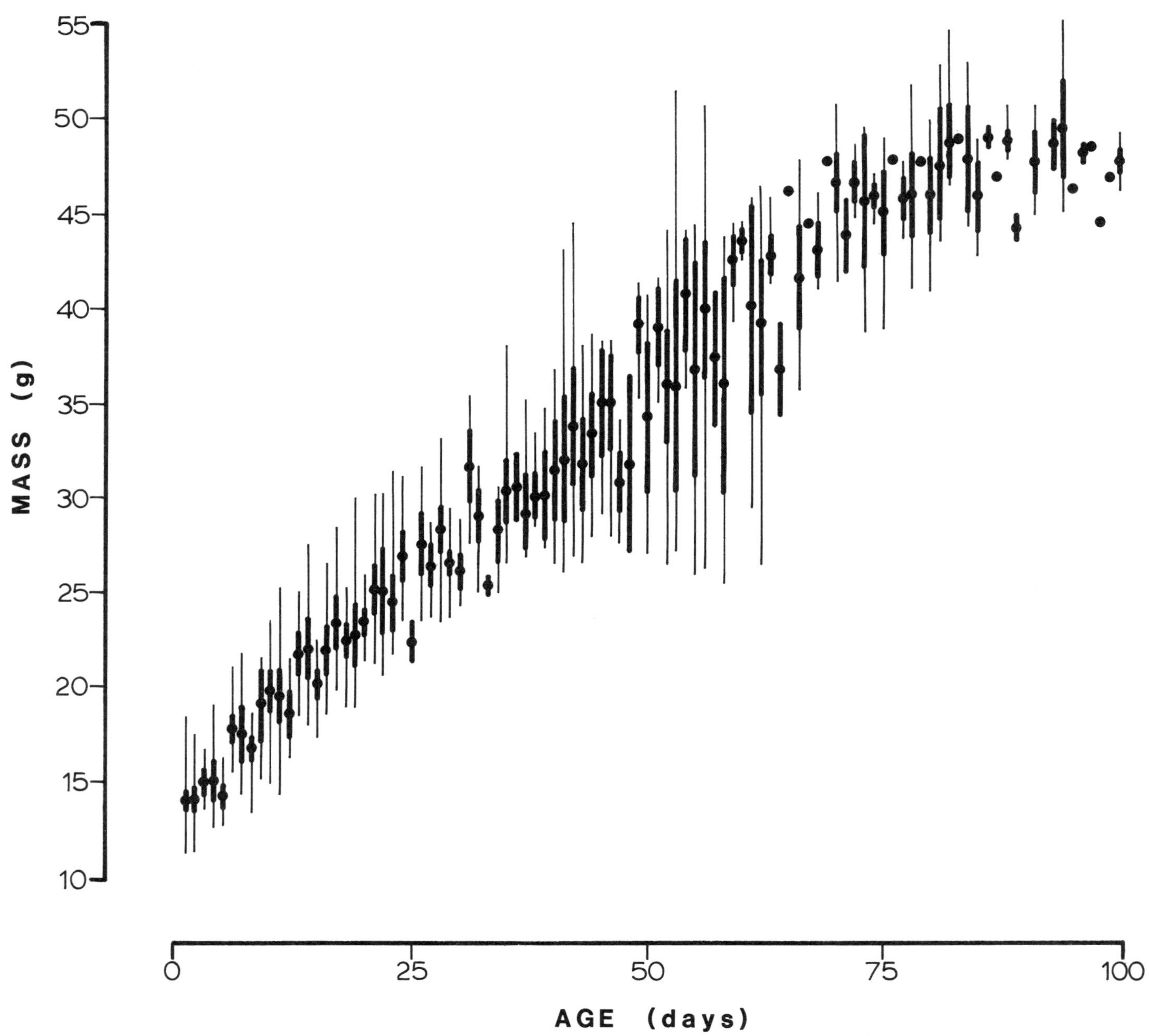

FIGURE 3-7.—Growth of mass in the NZP colony of *Artibeus jamaicensis*. Means, standard errors, and ranges are shown.

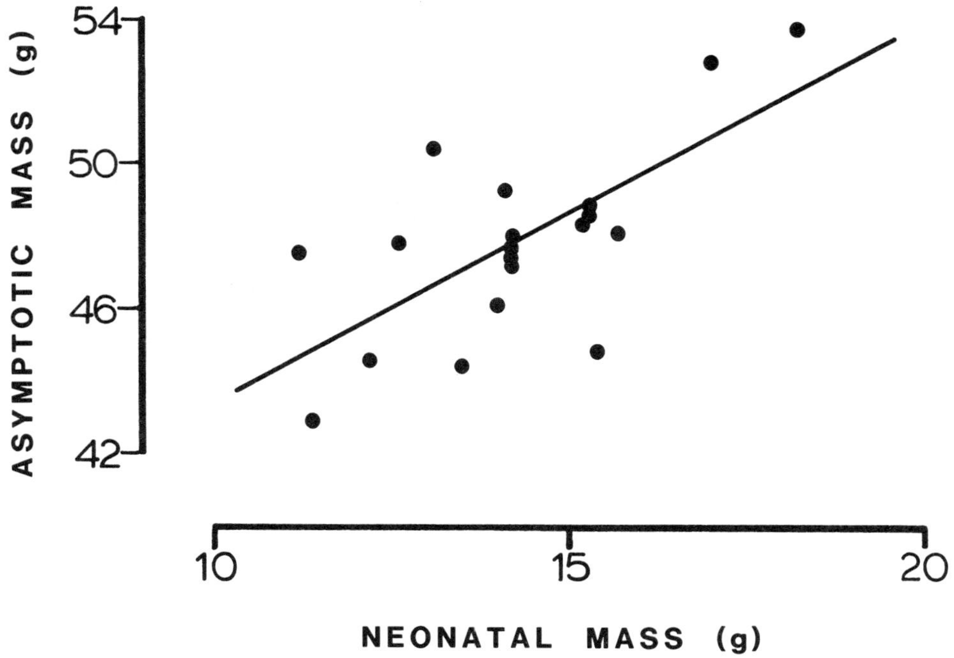

FIGURE 3-8.—The relationship between the asymptotic mass of captive-reared young and their neonatal mass. The equation for the regression line is y = 32.94 + 1.05 x, (r = 0.710, P < 0.001).

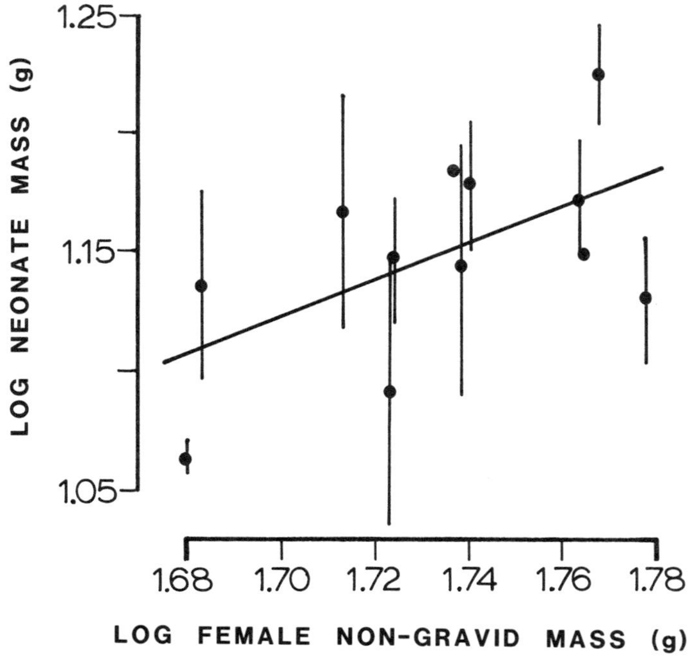

FIGURE 3-9.—The relationship between neonate and mother mass in the NZP colony of *Artibeus jamaicensis*. A mean nongravid mass was determined for each adult female. Means (solid circles) and ranges (bars) of infant mass are shown. The equation for the regression line is y = −0.14 + 0.74 x, (r = 0.552, 0.10 < P < 0.05).

0.05). However, relative infant size (as a percentage of the mother's nonpregnant mass) was correlated with the mother's potential weight-bearing capacity (wing area/nongravid mass).

A positive relationship between relative neonatal mass and the mother's wing weight-bearing capacity has not been recorded for any other species of bat. However, Kunz (1974) noted that female wing size was larger than male wing size in *Eptesicus fuscus* in Kansas where the normal litter is one. He speculated that if there is any selective advantage gained by large wing size, the degree of sexual dimorphism in *E. fuscus* should be even greater in the more eastern states where two is the usual litter size.

More attention should be paid to species' wing-loading values and other variables of wing shape and aerodynamic ability and their relationships to a variety of developmental, behavioral, physiological, and ecological factors. Wings are essential for the location, pursuit, and transport of food. Other potential functions of wings that also could be wing-size dependent are heat dissipation, evaporative water loss, transport or shielding of young, and behavioral displays.

The mass of a developing fetus must be of particular significance to those bats that forage while flying. However, beyond a certain point in relative size, the larger wings themselves would represent a significant weight load with much increased drag and would probably be quite difficult to move, given a limited muscle mass (Davis, 1969a).

Oddly, in our colony the female with the greatest proportional wing size (F12) produced infants that were among the smallest in size (both proportionately and absolutely). Of course, there may be many other reasons why a female would produce small young.

DENTITION.—Fifteen young from BG-I and BG-VI were monitored to determine ages when deciduous teeth were lost and permanent teeth erupted. The deciduous dental formula of *A. jamaicensis* is 2/2 1/1 2/2 0/0 × 2 = 20. At birth, both deciduous upper incisors, the upper canine, upper premolar 4, both lower incisors, and the lower canine projected from the gingivum. The deciduous incisors were tiny, rounded nubs barely visible between the canines. Upper and lower deciduous canines and upper deciduous premolar 4 were sharp, slender, and recurved. The six functional milk teeth served to grasp the flesh surrounding the mother's nipple. Great care had to be taken when removing attached young from nipples as the skin was easily torn by the teeth.

Patterns of replacement of deciduous teeth were variable. The upper deciduous first incisors were shed at about 7 days of age, followed by the eruption of the upper and lower permanent first incisors. The upper permanent canines began eruption around 17 days, followed by the lower canines at 21 days. The deciduous canines were retained about ten more days until the secondary canines had fully erupted.

The permanent premolars and molars were visible through the gum from birth and began erupting before canine eruption was completed. Eruption of teeth in the upper jaw slightly preceded that of the lower. The unerupted deciduous premolars sometimes broke through the gum along with their respective permanent teeth and were lost as eruption was completed. The second deciduous incisors were the last milk teeth to be shed, followed by the eruption of the permanent second incisors, which are small in the adult and probably nonfunctional. The adult dental formula, 2/2 1/1 2/2 2/3 × 2 = 30, was achieved by 40 ± 6 days.

Compared with vespertilionids, phyllostomids generally have a milk dentition that is simpler—teeth smaller, reduced in number, and not so strongly recurved. Kleiman and Davis (1979) thought this correlated with the phyllostomid's tendency when foraging to carry attached young rather than depositing them in crèches. However, *Artibeus jamaicensis* does not often transport its young, and our data from BCI do not support the idea that phyllostomids in general carry their young while foraging. Thus, in that regard they resemble many temperate zone vespertilionids (Davis, 1970; Fenton, 1969).

PATTERNS OF DEVELOPMENT.—Although there are a number of reports on growth and development in vespertilionids, there is not enough detailed information on phyllostomids to permit meaningful comparisons of the two families. It is possible, however, in search of patterns, to describe trends among the vespertilionids and then to compare them with the meager data for the phyllostomids.

1. Development at Birth: Attempts to rank neonatal bats as altricial, precocial, or intermediate have resulted in classifying phyllostomids as generally precocial and vespertilionids as comparatively altricial (Gould, 1975; Kleiman and Davis, 1979).

2. Development of Sight: Whether or not the opened eyes of phyllostomid neonates are functional has not been established. If they are, they could be useful in orienting the young away from illuminated and exposed substrates during day roosting. A number of phyllostomids roost in exposed locations such as the undersides of foliage and branches, in rock crevices, under overhanging roots, under stream banks, and in hollow logs and abandoned burrows (Tuttle, 1976).

3. Hairiness: Because most tropical phyllostomids, including *A. jamaicensis,* have small nursery colonies (Bradbury, 1977) or have nursery roosts in exposed locations, it may be necessary for their neonates to be at least partially furred. However, two phyllostomids, *Vampyrum spectrum* (Ditmars, 1936) and *Phyllostomus hastatus* (Gould, 1975), have hairless neonates. *Vampyrum spectrum,* the largest bat in the New World (180 g), roosts in small groups (<10), in tree holes (Vehrencamp et al., 1977). *Phyllostomus hastatus* is the third largest New World bat (80–120 g). Harems of five to ten females and a single male roost in hollow logs, hollow trees, houses, and caves. McCracken and Bradbury (1977, 1981) found thousands occupying large caves in Trinidad. Because thermal conductance (heat loss) is negatively correlated with body size in mammals (Bradley and Deavers, 1980), temperature stress may be less of a problem for these larger species (their babies are probably larger as well).

Positive relationships between colony size, ambient cave temperatures, and enhanced growth rates have been reported in *Myotis grisescens* by Tuttle (1975). The large numbers probably help maintain homeothermy and a high roost temperature and thus alleviate the need for the newborn to be furred.

4. Metabolic Rates: Bats show some life history characteristics that depart from the norm for small mammals. *Artibeus jamaicensis* and bats in general are unusually long-lived, have low intrinsic rates of natural increase, usually have a litter size of one (maximum recorded litter is five in *Lasiurus cinereus*), and show relatively slow prenatal and postnatal development (see review in Eisenberg, 1981).

5. Brain Size: Most bats that pursue insects on the wing (e.g., many vespertilionids and molossids) have an Encephalization Quotient (EQ; see Jerison, 1973) that barely exceeds those shown by many of the Insectivora (Eisenberg, 1981). Conversely, phyllostomids and pteropodids have brains that are larger than would be predicted from their body masses (EQ > 1.0). Eisenberg and Wilson (1978) noted that most species in these families are foraging for rich food resources isolated in small patches, an activity requiring relatively larger brains to process and store information.

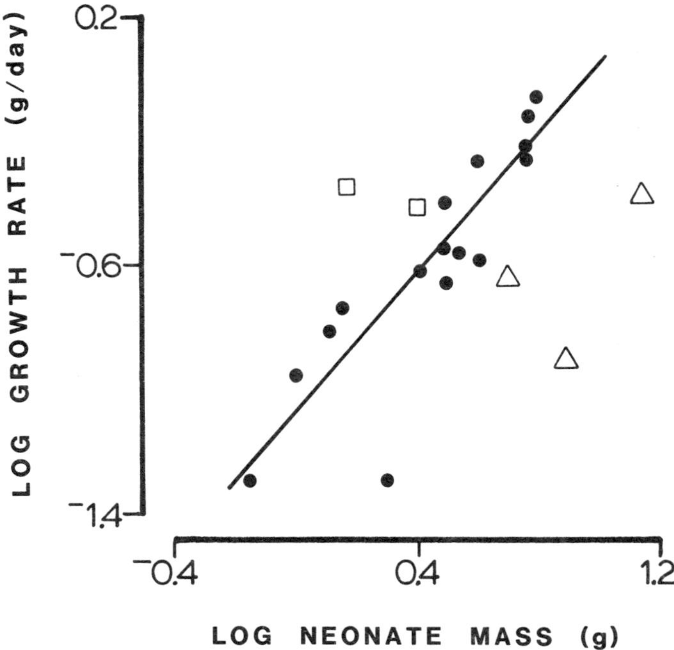

FIGURE 3-10.—Rates of mass increase versus neonate mass for 18 species of bats. Log growth rate (g/day) is plotted against log neonate mass (g). The regression line for vespertilionids (solid circles) is y = −1.09 + 1.14 x, (r = 0.873, P < 0.001). Phyllostomid (open triangles) and molossid (open circles) data points are shown for comparison. Data are from Table 3-11.

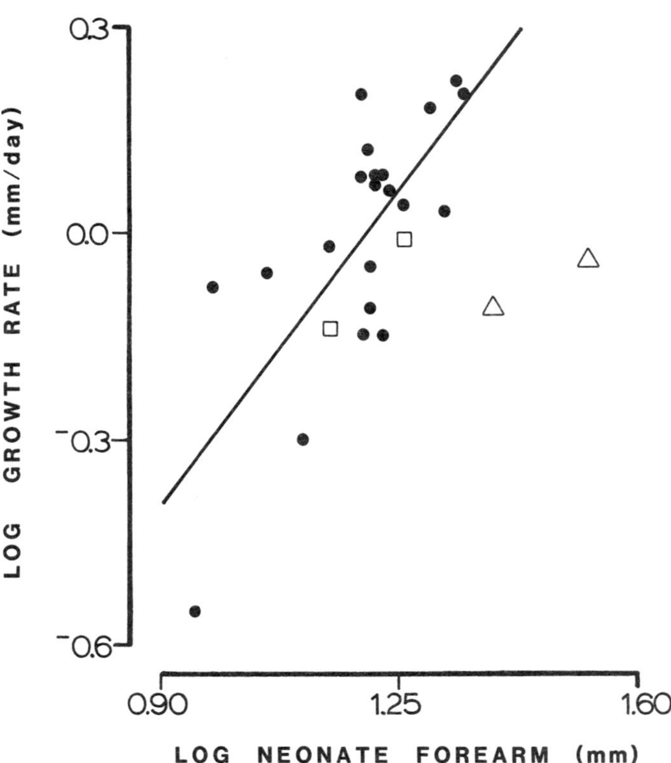

FIGURE 3-11.—Rate of forearm growth versus neonate forearm length for 19 species of bats. Log growth rate (mm/day) is plotted against log neonate forearm length (mm). The regression line for vespertilionids (solid circles) is y = −1.56 + 1.29 x, (r = 0.736, P < 0.001). Phyllostomid (open triangles) and molossid (open squares) data points are shown for comparison. Data are from Table 3-11.

6. Gestation: Eisenberg (1981) showed that among species of similar body size, gestation is longer in the species with a larger adult EQ. Longer gestations and larger brains in phyllostomids contrast with brief gestations and smaller brains of most vespertilionids and of other bats that pursue flying insects. Our data on *A. jamaicensis* support Eisenberg's hypothesis.

Duration of undelayed gestation varies in bats from as few as 40 days in insectivores to over 100 days in frugivores. The estimate of 112 days gestation for *A. jamaicensis* accords with the trend suggested by Eisenberg (1981) of extended gestations in tropical frugivores.

7. Size and Growth Rates: These relationships are best presented as double logarithmic plots. We used least-squares regression to calculate the y-intercept (a) and the slope (b) for the equation: log y = log a + b log x, where x is body mass and y is the life history variable under investigation.

We examined pairs of variables separately for mass and forearm length in the vespertilionids (Figures 3-10 and 3-11). Correlations for both measures between neonate size and adult size, and between growth rate and either neonate or adult size are significant. Comparable data for molossids are not sufficient to show trends, but the larger, slower-growing phyllostomid young appear to depart from the vespertilionid tendencies.

Development of Wings and Flight

MORPHOMETRICS.—Findley et al. (1972) defined and described the probable functional significance of three wing variables as follows:

1. Wing-loading is the mass (in grams) supported by the surface area of the wings (in cm^2). If the mass-to-wing area ratio is high, a high speed is necessary to stay airborne. Thus, rapid fliers will tend to have high wing-loading values, and slow flight probably requires low wing-loading.

2. Aspect ratio is a measure of relative wing width, and as calculated by Findley et al. (1972) is wing length divided by wing width. We calculated it according to Vaughn (1959) the conventional way as wing span (mm)^2/wing area (cm^2). High aspect ratios denote narrow wings and low aspect ratios indicate broad ones. With high aspect ratios drag is decreased and greater speed is possible, but lift is reduced. Therefore, more rapid flight is required to remain aloft. On the other hand, with low aspect ratios greater lift is possible at low speeds, but there is more drag at higher speeds.

3. Tip index, a measure of propulsion, is calculated as wing tip length (length of third digit, mm) divided by length of

forearm (mm). The wing tip is the main propulsive part of the wing. Consequently, relatively longer tips are correlated with greater speed. Hovering species tend to have relatively long wing tips (hovering is very rapid flight with increased drag to lessen forward motion).

During the first 40 days of life of our captive bats wing-loading diminished 26% from near 0.195 to 0.145 g/cm², meaning that growth in wing area was proceeding much faster than growth in body mass. After day 40, wing-loading

increased again to an asymptote of approximately 0.185 g/cm² (Figure 3-12). Note that the initiation of flight occurred during the time when wing-loading was minimal. The other aerodynamic variables (aspect ratio and tip index) were at intermediate values when young first became volant (Figure 3-12). Changes in wing shape are shown for a young of known age (Figure 3-13).

DEVELOPMENT OF FLIGHT.—At first, during drop tests, infants fell directly to the ground, although most spread their

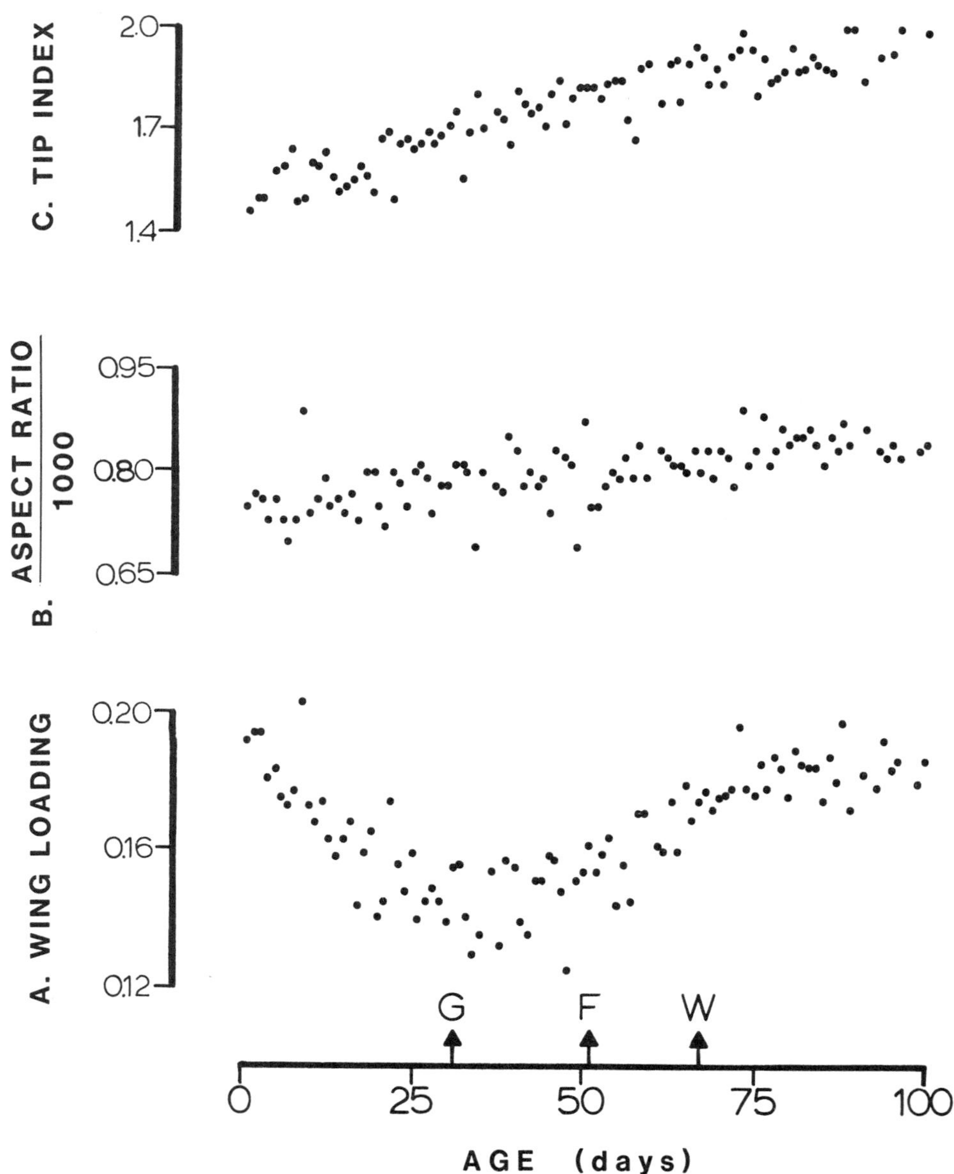

FIGURE 3-12.—Developmental changes in three aerodynamic parameters of the wings of *Artibeus jamaicensis* in the NZP colony. Dots represent daily mean values. Tip index is tip length (mm)/forearm length (mm). Aspect ratio = wing span (mm)²/wing area (cm²). Wing loading = body mass (g)/wing area (cm²). G = mean age when young are first capable of gaining altitude and negotiating clumsy landings during drop tests. F = mean age when young regularly initiate flights. W = mean age of weaning.

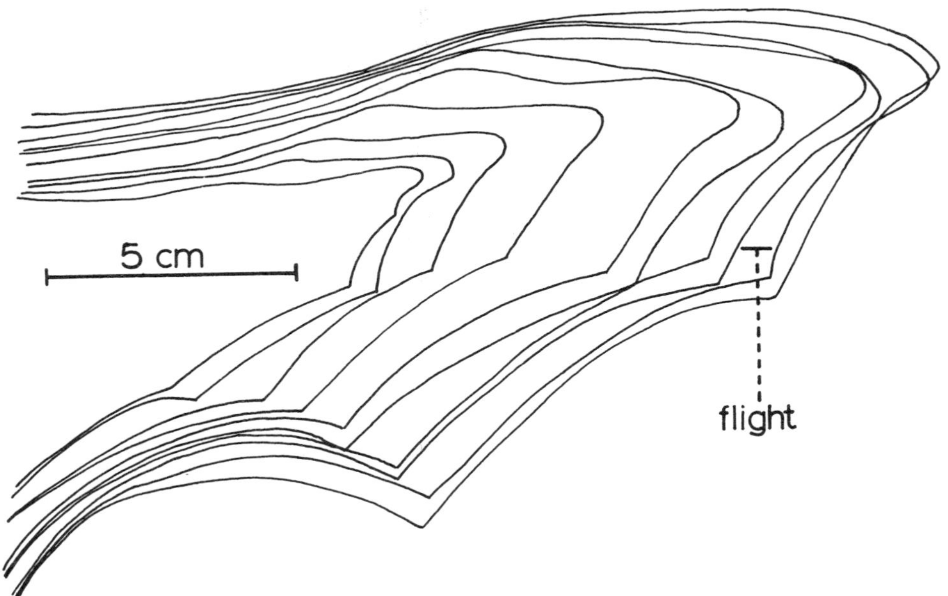

FIGURE 3-13.—Wing outlines from a captive-born *Artibeus jamaicensis* in the NZP colony. Outlines are actual tracings of the wing, made when the bat was 2, 5, 8, 15, 23, 30, 39, 42, 53, and 58 days of age. Flight was initiated between the 42nd and 53rd days.

wings partially while falling. From day 4 to day 19, the infants flapped their wings, but still dropped straight down. Around day 19, the young began beating their wings vigorously, with full up and down strokes in the roost, but they did not fly. These stationary wing-beating sessions lasted about 1.5 minutes each, and there were as many as eight sessions per infant per hour. Presumably, the function of this behavior was to develop and condition the muscles necessary for flight. This behavior was correlated with the last observed transport of offspring by mothers between different roosting areas. At the same time, the young also began vigorous scanning in the manner of the adults, i.e., hanging by one or both feet, with the head upturned, facing into the center of the flight room, and rapidly moving the ear pinnae and noseleaf.

By day 24, infants began to achieve some forward motion when dropped, and they were able to gain some altitude around day 29. By day 31 they could make clumsy landings on obstacles. Day 48 was approximately the last day that young were seen attached to the mother's nipple during the dark period. We do not know whether infants were attached to the mother during the light part of the cycle because at that time all the bats came together into a large clump, in which individuals were not distinguishable. At this time the young also began to show interest in the nectar ration that was always available. They lapped it from food cups and licked droplets off our fingers. By day 51, young were regularly initiating flights, and they followed their mothers on flights. They were able to maneuver and land proficiently by day 65 and thus were distinguished as "juveniles." The flight skills of juveniles and

subadults were visibly (if not quantifiably) distinct from those of the adults, and at these ages they were much easier to capture during hand-netting procedures.

Because wings are not essential in the earliest stages of a young bat's life, it is not unusual to find that the wings are relatively underdeveloped in newborns. Wing-loading in young bats and the development of flight abilities have been investigated in *Nycticeius humeralis* (Jones, 1967), *Antrozous pallidus* (Davis, 1969a), and *Myotis thysanodes* and *M. lucifugus* (O'Farrell and Studier, 1973). As in *A. jamaicensis,* initial wing growth was faster than growth in body mass and resulted in diminishing wing-loading up to the time of flight initiation. Subsequently, there followed an increase in wing-loading as body mass increased and the flight abilities were mastered. We have found no previous reports of vigorous stationary wing beating for any other bat even though it probably occurs in most species.

McManus and Nellis (1972) provided an anomalous example of ontogeny of wing-loading for *A. jamaicensis* from the Virgin Islands. They found an exponential relationship between log wing-loading (g/cm^2) and body mass (g), with larger bats having progressively higher wing-loading values.

Transition to Subadulthood

The age at which the cartilaginous epiphyseal gap characteristic of juvenile bats could no longer be detected seemed to be highly variable in our colony, but this may reflect our limitations in recognizing precisely the point of closing.

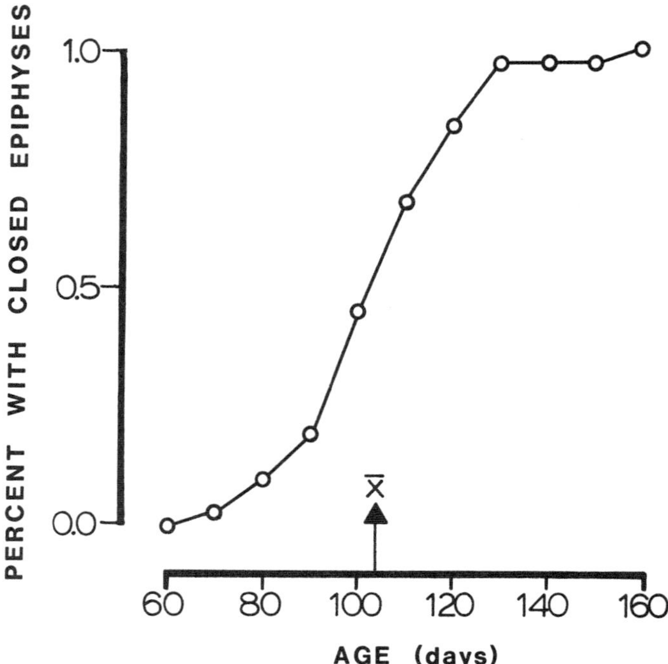

FIGURE 3-14.—Age of epiphyseal closure of *Artibeus jamaicensis* in the NZP colony. The percentage of young bats (n = 31) showing closed epiphyses is plotted against age. Mean age of closure was 104 days.

Closure times thought to be reasonably accurate for 31 individuals were averaged. Ages of closure ranged from 70 to 160 days with a mean age of 104.4 days ($SD = 19.038$, Figure 3-14). More than 80% of the young made the transition from juvenile to subadulthood between three and four months of age.

Transition to Adulthood

MALES.—In young males, testes were discernable by palpation at least as early as weaning, when they were 3–4 mm long. Subsequently, the testes gradually increased to their maximum size (9–14 mm), which in all cases corresponded to a time of female postpartum sexual receptivity. In eight of 14 males, sexual activity was initiated during the second birth group following their own when they were nearly one year old. The remaining six apparently were capable of mating in the first group following their own. However, the majority of these males were born in birth groups where the subsequent interval to the next birth group was extended due to the occurrence of delayed development of embryos in the gestating females. These males were approximately eight months old at the time of their transition from subadult to adult. Following initial maximal enlargement, regression of testis size to less than 6.0 mm was not seen.

FEMALES.—Twelve captive females were nearly one year old at first parturition ($\bar{X} = 332.3$ days, $SD = 55.178$). Assuming a gestation of almost four months, they must have reached sexual maturity by eight months. Two more females gave birth at around 18 months, but they may have had undetected abortions previously.

Preparturient nipple enlargement in captive-born females was similar to that seen in the original wild-caught females. The nipples enlarged slightly (from 0.5 to 2.0 mm diameter) at 251 days of age ($n = 8$, $SD = 16.999$), which was 84 days prepartum ($n = 8$, $SD = 20.197$). None of the first infants born to the 14 captive-born females survived more than five days, and at least six of these infants were probably premature. After this initial reproductive failure, the nipples in eight females had regressed to small by 17 days ($SD = 10.379$) after their infant's death. Apparently, the nipples of one female had not enlarged by the time she had given birth.

Five of these females gave birth again, and four of their young completed weaning. Two females did not show complete regression of nipples, and the two others had regressed by 125 and 127 days postpartum. The nipples of the female whose infant died regressed to small in 23 days.

Summary

Bimodal reproduction and reproductive synchrony occurred in the captive females as it did in the free-living Panamanian population from which they came. Reproductive cycles of the captive bats remained in synchrony with those of the wild population for the first year, then tended toward shorter interbirth intervals. Every second birth group showed an extended gestation period, which we believe was due to delayed development of the implanted blastocyst, as in the wild. Duration of normal undelayed gestation was 3.5–4 months. The minimum observed interbirth interval was 112 days (mean 129 days). Mean interval with delayed embryonic development was 177 days. Maximum weight gain by pregnant females was 32%. The weight gain adversely affected the flight abilities of females in late pregnancy. Estrus followed parturition and peak receptivity was three to four days postpartum.

Fecundity was depressed in reproductively inexperienced females but averaged 91% in multiparous females, whether wild-caught or captive-born. Preweaning mortality was high (almost 100%) for reproductively inexperienced females and low for experienced females (35%–40%).

Maximum nipple enlargement, hair loss, and pigmentation occurred 1.5 weeks prepartum. An increase in the nipple size of pregnant subadult females was detected four weeks before fetuses became palpable (ten weeks prepartum). Mean duration of lactation was 66 days, although it was significantly longer in females that were gestating dormant embryos (~72 days) than in those females gestating normally developing embryos (~64 days). Postparturient nipple condition of parous females was always distinctly different from that of subadults and is, therefore, useful in assigning females to adult or subadult age classes.

The neonates were physically and behaviorally precocial.

Their eyes were open, they had fur on the dorsum, and they were vocal, responsive to sounds, and very active. They were large (more than 25% of nongravid female mass, and forearm lengths exceeded 50% of adult size). The permanent dentition was complete at 40 days of age.

Developmental rates of captive-born young were quite variable. Body mass was the measure most highly correlated with age and stabilized around 80 days of age. Forearm length increased most rapidly and growth was completed in 50 days. Wing span and wing area reached adult size in 70 days. Beyond 104 days of age most individuals showed complete epiphyseal fusion. Initiation of flight in young bats occurred during a developmental stage when wing-loading values were low (31–51 days of age).

Reproductive maturity in males, as indicated by testicular hypertrophy, occurred as early as eight months of age, but more often by one year of age. Females usually first gave birth when they were one year old, indicating that they had been sexually active at about eight months of age.

Significant sexual dimorphism in four wing measurements (span, area, width, and tip length) was detected in wild-caught adult captives. Body mass and forearm length were greater in females than in males, but significance was not demonstrated. Males had higher wing-loading values than females.

4.   Reproduction on Barro Colorado Island

Don E. Wilson, Charles O. Handley, Jr.,
and Alfred L. Gardner

Data on reproduction in bats have accumulated at an accelerating rate during the past 25 years (Racey, 1982; Wilson, 1979). Because *Artibeus jamaicensis* is common and widespread its reproductive cycle is better known than those of most other Neotropical bats. However, in-depth studies at a single locality over long periods, which are essential for interpreting isolated pieces of information and for elucidating details of a reproductive cycle, have been lacking.

We augmented our understanding of reproduction in *A. jamaicensis,* gained from our colony at the National Zoological Park (NZP), by gathering information from populations on Barro Colorado Island (BCI) and vicinity. Our data on reproduction, accumulated over several annual cycles from wild-caught bats, demonstrate a high degree of reproductive synchrony in this species.

In 1971 Fleming presented evidence that embryos of *A. jamaicensis* undergo retarded development during part of the year, and Fleming et al. (1972) demonstrated bimodal polyestry to be the basic reproductive pattern in this species. Our large data set from BCI and the adjacent mainland, together with the results of the study of our NZP colony (see Section 3, Reproduction in a Captive Colony) provide us the opportunity to examine this cycle in some detail. We base our outline of the reproductive cycle of *A. jamaicensis* on 4447 individual females captured between 1972 and 1980 on BCI and vicinity.

The Basic Pattern

We confirm the previously postulated pattern of bimodal polyestry (Fleming et al., 1972; Wilson, 1979), and Fleming's (1971) statements on delayed embryonic development (Figure 4-1, Table 4-1). In general, females are palpably pregnant in January and begin to give birth by late February or early March. They are so synchronized that normally more than half of the

Don E. Wilson, and Charles O. Handley, Jr., National Museum of Natural History, Smithsonian Institution, Washington, D.C. 20560. Alfred L. Gardner, NERC, U.S. Fish and Wildlife Service, National Museum of Natural History, Washington, D.C. 20560.

population gives birth in those two months. During a postpartum estrus most of them become pregnant again. This results in a brief period when individuals are both pregnant and lactating. Palpably pregnant, but still lactating individuals are most likely in April, May, or June, between the February–March and July–August birth peaks.

Young born during the initial birth peak (February and March) are weaned in April and May. The second cycle of births commences in July and August with the same pattern of postpartum estrus and lactation. The young from the second birth peak are then weaned in September and October. The difference between the two periods of reproductive activity occurs when the blastocyst from the second postpartum estrus implants, but does not develop at the same rate as the preceding one. The period of delayed embryonic growth covers approximately three months, and is followed by a period of essentially normal development during the remaining four.

We were able to recognize individuals that were pregnant (terminal six weeks), lactating, or postlactating, and as these events tracked one another over the years of our study, we recorded peak times for each event. Some variation from year to year is obvious (Figure 4-1), but the amount of congruence from one year to the next is striking.

Another way we summarized the data was to combine the information from all years into a single "average" year (Figure 4-2). Thus summarized and divided into 3-week intervals, these averages show the bimodal cycle and emphasize the 12-week diapause between July and November.

Individual Variation

In spite of the great number of marked bats in our sample, and even though we sampled almost continuously for three years, we were unable to accumulate a complete record of an individual female's reproductive history throughout her lifetime. To do so, we would have to capture the same female at least twice per year during the birth periods. We did catch some individuals two or more times each year, but not always at appropriate stages of reproduction. Although our data are not based on complete individual reproductive histories, it seems

TABLE 4-1.—Reproductive condition (by week) of adult female *Artibeus jamaicensis* on BCI, 1975 through 1980.

Week		No reproduction		Pregnant		Lactating		Young on teat		Pregnant & lactating		Post-lactating		Total
		N	%	N	%	N	%	N	%	N	%	N	%	
1975	9	12	75.0	4	25.0	0	0.0	0	0.0	0	0.0	0	0.0	16
	10	14	87.5	2	12.5	0	0.0	0	0.0	0	0.0	0	0.0	16
	11	8	34.8	11	47.8	4	17.4	0	0.0	0	0.0	0	0.0	23
	12	1	33.3	2	66.7	0	0.0	0	0.0	0	0.0	0	0.0	3
Total		35		19		4		0		0		0		58
1976	10	27	51.9	25	48.1	0	0.0	0	0.0	0	0.0	0	0.0	52
	11	3	50.0	3	50.0	0	0.0	0	0.0	0	0.0	0	0.0	6
	12	1	2.8	35	97.2	0	0.0	0	0.0	0	0.0	0	0.0	36
	13	2	5.9	32	94.1	0	0.0	0	0.0	0	0.0	0	0.0	34
	41	16	84.2	0	0.0	2	10.5	0	0.0	0	0.0	1	5.3	19
	42	38	97.4	0	0.0	1	2.6	0	0.0	0	0.0	0	0.0	39
	43	51	89.5	0	0.0	6	10.5	0	0.0	0	0.0	0	0.0	57
	44	9	64.3	0	0.0	3	21.4	0	0.0	0	0.0	2	14.3	14
	45	27	96.4	0	0.0	0	0.0	0	0.0	0	0.0	1	3.6	28
	46	35	92.1	0	0.0	1	2.6	0	0.0	0	0.0	2	5.3	38
	47	28	100.0	0	0.0	0	0.0	0	0.0	0	0.0	0	0.0	28
	48	37	100.0	0	0.0	0	0.0	0	0.0	0	0.0	0	0.0	37
Total		274		95		13		0		0		6		388
1977	27	0	0.0	1	100.0	0	0.0	0	0.0	0	0.0	0	0.0	1
	28	3	9.4	19	59.4	5	15.6	0	0.0	0	0.0	5	15.6	32
	29	3	6.3	13	27.1	14	29.2	0	0.0	0	0.0	18	37.5	48
	30	2	15.4	2	15.4	4	30.8	0	0.0	0	0.0	5	38.5	13
	31	0	0.0	0	0.0	3	42.9	0	0.0	0	0.0	4	57.1	7
	32	0	0.0	0	0.0	1	100.0	0	0.0	0	0.0	0	0.0	1
	33	1	14.3	2	28.6	0	0.0	0	0.0	0	0.0	4	57.1	7
	34	3	27.3	0	0.0	1	9.1	0	0.0	0	0.0	7	63.6	11
	35	10	43.5	0	0.0	6	26.1	0	0.0	0	0.0	7	30.4	23
	36	5	45.5	0	0.0	2	18.2	0	0.0	0	0.0	4	36.4	11
	37	15	65.2	0	0.0	2	8.7	0	0.0	0	0.0	6	26.1	23
	38	5	63.3	0	0.0	0	0.0	0	0.0	0	0.0	1	16.7	6
	39	1	25.0	0	0.0	0	0.0	0	0.0	0	0.0	3	75.0	4
	40	1	100.0	0	0.0	0	0.0	0	0.0	0	0.0	0	0.0	1
	41	8	28.6	0	0.0	3	10.7	0	0.0	0	0.0	17	60.7	28
	42	33	51.6	0	0.0	1	1.6	0	0.0	0	0.0	30	46.9	64
	43	13	40.6	0	0.0	2	6.3	0	0.0	0	0.0	17	53.1	32
	44	9	58.3	0	0.0	0	0.0	0	0.0	0	0.0	7	43.8	16
	45	7	53.8	0	0.0	0	0.0	0	0.0	0	0.0	6	46.2	13
	46	16	80.0	0	0.0	0	0.0	0	0.0	0	0.0	4	20.0	20
	47	2	100.0	0	0.0	0	0.0	0	0.0	0	0.0	0	0.0	2
	48	1	50.0	0	0.0	0	0.0	0	0.0	0	0.0	1	50.0	2
	49	10	100.0	0	0.0	0	0.0	0	0.0	0	0.0	0	0.0	10
	50	53	84.1	7	11.1	1	1.6	0	0.0	0	0.0	2	3.2	63
	51	21	55.3	16	42.1	1	2.6	0	0.0	0	0.0	0	0.0	38
	52	69	94.5	2	2.7	1	1.4	0	0.0	0	0.0	1	1.4	73
Total		291		62		47		0		0		149		549
1978	1	101	99.0	1	1.0	0	0.0	0	0.0	0	0.0	0	0.0	102
	2	24	70.6	9	26.5	0	0.0	0	0.0	0	0.0	1	2.9	34
	4	2	100.0	0	0.0	0	0.0	0	0.0	0	0.0	0	0.0	2
	5	4	80.0	1	20.0	0	0.0	0	0.0	0	0.0	0	0.0	5
	6	16	66.7	8	33.3	0	0.0	0	0.0	0	0.0	0	0.0	24
	7	6	40.0	9	60.0	0	0.0	0	0.0	0	0.0	0	0.0	15
	8	4	6.9	54	93.1	0	0.0	0	0.0	0	0.0	0	0.0	58
	9	1	4.3	21	91.3	0	0.0	0	0.0	0	0.0	1	4.3	23
	10	0	0.0	25	78.1	7	21.9	0	0.0	0	0.0	0	0.0	32
	11	5	8.6	16	27.6	35	60.3	0	0.0	0	0.0	2	3.4	58
	12	1	1.8	20	35.7	30	53.6	0	0.0	0	0.0	5	8.9	56
	13	3	13.6	3	13.6	14	63.6	0	0.0	0	0.0	2	9.1	22
	14	5	9.6	2	3.8	31	59.6	0	0.0	0	0.0	14	26.9	52
	15	5	17.9	1	3.6	14	50.0	0	0.0	0	0.0	8	28.6	28
	16	0	0.0	0	0.0	10	55.6	0	0.0	0	0.0	8	44.4	18
	17	0	0.0	0	0.0	1	50.0	0	0.0	0	0.0	1	50.0	2
	18	0	0.0	0	0.0	3	30.0	0	0.0	0	0.0	7	70.0	10
	19	0	0.0	1	12.5	0	0.0	0	0.0	0	0.0	7	87.5	8
	20	0	0.0	0	0.0	0	0.0	0	0.0	0	0.0	2	100.0	2
	21	2	40.0	0	0.0	1	20.0	0	0.0	0	0.0	2	40.0	5
	22	0	0.0	0	0.0	0	0.0	0	0.0	0	0.0	1	100.0	1
	24	1	33.3	2	66.7	0	0.0	0	0.0	0	0.0	0	0.0	3
	26	4	18.2	7	31.8	4	18.2	1	4.5	2	9.1	4	18.2	22
	27	1	20.0	1	20.0	0	0.0	0	0.0	0	0.0	3	60.0	5

TABLE 4-1.—Continued.

Week	No reproduction		Pregnant		Lactating		Young on teat		Pregnant & lactating		Post-lactating		Total
	N	%	N	%	N	%	N	%	N	%	N	%	
28	0	0.0	3	75.0	1	25.0	0	0.0	0	0.0	0	0.0	4
29	1	100.0	0	0.0	0	0.0	0	0.0	0	0.0	0	0.0	1
30	3	9.7	1	3.2	16	51.6	0	0.0	1	3.2	10	32.3	31
31	2	100.0	0	0.0	0	0.0	0	0.0	0	0.0	0	0.0	2
32	0	0.0	0	0.0	3	50.0	0	0.0	0	0.0	3	50.0	6
33	1	14.3	0	0.0	4	57.1	0	0.0	0	0.0	2	28.6	7
43	11	91.7	0	0.0	0	0.0	0	0.0	0	0.0	1	8.3	12
44	4	80.0	0	0.0	1	20.0	0	0.0	0	0.0	0	0.0	5
45	14	82.4	0	0.0	0	0.0	0	0.0	0	0.0	3	17.6	17
46	13	86.7	0	0.0	0	0.0	0	0.0	0	0.0	2	13.3	15
47	35	92.1	0	0.0	0	0.0	0	0.0	0	0.0	3	7.9	38
48	16	94.1	0	0.0	0	0.0	0	0.0	0	0.0	1	5.9	17
49	45	95.7	0	0.0	0	0.0	0	0.0	0	0.0	2	4.3	47
50	7	100.0	0	0.0	0	0.0	0	0.0	0	0.0	0	0.0	7
51	21	95.5	0	0.0	0	0.0	0	0.0	0	0.0	1	4.5	22
52	59	98.3	0	0.0	0	0.0	0	0.0	0	0.0	1	1.7	60
Total	417		185		175		1		3		97		878
1979 1	28	96.6	0	0.0	0	0.0	0	0.0	0	0.0	1	3.4	29
2	13	100.0	0	0.0	0	0.0	0	0.0	0	0.0	0	0.0	13
3	8	29.6	19	70.4	0	0.0	0	0.0	0	0.0	0	0.0	27
4	2	22.2	7	77.6	0	0.0	0	0.0	0	0.0	0	0.0	9
5	2	33.3	4	66.7	0	0.0	0	0.0	0	0.0	0	0.0	6
7	1	33.3	2	66.7	0	0.0	0	0.0	0	0.0	0	0.0	3
9	8	47.1	2	11.8	6	35.3	0	0.0	0	0.0	1	5.9	17
10	4	20.0	3	15.0	10	50.0	0	0.0	0	0.0	3	15.0	20
11	3	3.0	4	4.0	89	89.9	0	0.0	0	0.0	3	3.0	99
12	3	18.8	0	0.0	12	75.0	0	0.0	0	0.0	1	6.3	16
13	7	15.6	1	2.2	31	68.9	0	0.0	0	0.0	6	13.3	45
14	0	0.0	0	0.0	2	100.0	0	0.0	0	0.0	0	0.0	2
15	4	3.8	5	4.8	63	60.0	1	1.0	0	0.0	32	30.5	105
16	9	7.4	11	9.1	71	58.7	0	0.0	0	0.0	30	24.8	121
17	4	6.9	7	12.1	26	44.8	0	0.0	0	0.0	21	36.2	58
18	1	100.0	0	0.0	0	0.0	0	0.0	0	0.0	0	0.0	1
19	3	25.0	1	8.3	0	0.0	0	0.0	0	0.0	8	66.7	12
20	3	25.0	5	41.7	2	16.7	0	0.0	0	0.0	2	16.7	12
21	4	10.5	24	63.2	7	18.4	0	0.0	3	7.9	0	0.0	38
22	2	3.9	38	74.5	2	3.9	0	0.0	0	0.0	9	17.6	51
23	1	3.7	18	66.7	7	25.9	0	0.0	0	0.0	1	3.7	27
24	2	3.7	39	72.2	13	24.1	0	0.0	0	0.0	0	0.0	54
26	0	0.0	2	66.7	1	33.3	0	0.0	0	0.0	0	0.0	3
29	1	9.1	1	9.1	9	81.8	0	0.0	0	0.0	0	0.0	11
36	12	52.2	0	0.0	6	26.1	0	0.0	0	0.0	5	21.7	23
37	36	56.3	0	0.0	6	9.4	0	0.0	0	0.0	22	34.4	64
38	34	72.3	0	0.0	4	8.5	0	0.0	0	0.0	9	19.1	47
39	26	78.8	0	0.0	3	9.1	0	0.0	0	0.0	4	12.1	33
40	9	69.2	0	0.0	0	0.0	0	0.0	0	0.0	4	30.8	13
41	43	74.1	0	0.0	1	1.7	0	0.0	0	0.0	14	24.1	58
42	36	92.7	1	2.4	0	0.0	0	0.0	0	0.0	2	4.9	41
43	19	95.0	0	0.0	0	0.0	0	0.0	0	0.0	1	5.0	20
44	11	100.0	0	0.0	0	0.0	0	0.0	0	0.0	0	0.0	11
45	90	100.0	0	0.0	0	0.0	0	0.0	0	0.0	0	0.0	90
46	95	94.1	4	4.0	0	0.0	0	0.0	0	0.0	2	2.0	101
47	48	98.0	1	2.0	0	0.0	0	0.0	0	0.0	0	0.0	49
48	5	83.3	0	0.0	1	16.7	0	0.0	0	0.0	0	0.0	6
49	27	93.1	2	6.9	0	0.0	0	0.0	0	0.0	0	0.0	29
50	70	98.6	0	0.0	0	0.0	0	0.0	0	0.0	1	1.4	71
51	1	100.0	0	0.0	0	0.0	0	0.0	0	0.0	0	0.0	1
Total	677		201		372		1		3		182		1436
1980 1	8	88.9	1	11.1	0	0.0	0	0.0	0	0.0	0	0.0	9
2	59	96.7	2	3.3	0	0.0	0	0.0	0	0.0	0	0.0	61
3	33	97.1	1	2.9	0	0.0	0	0.0	0	0.0	0	0.0	34
4	6	85.7	1	14.3	0	0.0	0	0.0	0	0.0	0	0.0	7
5	3	75.0	1	25.0	0	0.0	0	0.0	0	0.0	0	0.0	4
6	17	94.4	1	5.6	0	0.0	0	0.0	0	0.0	0	0.0	18
7	2	50.0	2	50.0	0	0.0	0	0.0	0	0.0	0	0.0	4
8	17	63.0	10	37.0	0	0.0	0	0.0	0	0.0	0	0.0	27
9	1	100.0	0	0.0	0	0.0	0	0.0	0	0.0	0	0.0	1
13	1	2.4	1	2.4	35	83.3	0	0.0	0	0.0	5	11.9	42
14	10	8.9	6	5.4	82	73.2	0	0.0	0	0.0	14	12.5	112
15	3	2.2	18	13.2	84	61.8	0	0.0	0	0.0	31	22.8	136

TABLE 4-1.—Continued.

Week	No reproduction		Pregnant		Lactating		Young on teat		Pregnant & lactating		Post-lactating		Total
	N	%	N	%	N	%	N	%	N	%	N	%	
16	0	0.0	0	0.0	2	50.0	0	0.0	0	0.0	2	50.0	4
18	0	0.0	2	66.7	0	0.0	0	0.0	0	0.0	1	33.3	3
19	9	32.1	1	3.6	3	10.7	0	0.0	0	0.0	15	53.6	28
20	2	8.3	13	54.2	7	29.2	0	0.0	0	0.0	2	8.3	24
21	7	38.9	7	38.9	0	0.0	0	0.0	0	0.0	4	22.2	18
32	8	61.5	0	0.0	4	30.8	0	0.0	0	0.0	1	7.7	13
33	15	27.3	0	0.0	36	65.5	0	0.0	0	0.0	4	7.3	55
34	9	13.8	0	0.0	20	30.8	0	0.0	0	0.0	36	55.4	65
35	13	23.2	0	0.0	9	16.1	0	0.0	0	0.0	34	60.7	56
36	5	14.3	0	0.0	8	22.9	0	0.0	0	0.0	22	62.9	35
37	26	40.0	0	0.0	13	20.0	0	0.0	0	0.0	26	40.0	65
38	36	64.3	0	0.0	3	5.4	0	0.0	0	0.0	17	30.4	56
39	14	66.7	0	0.0	1	4.8	0	0.0	0	0.0	6	28.6	21
40	33	64.7	0	0.0	1	2.0	0	0.0	0	0.0	17	33.3	51
41	59	73.8	0	0.0	2	2.5	0	0.0	0	0.0	19	23.8	80
42	38	88.4	0	0.0	0	0.0	0	0.0	0	0.0	5	11.6	43
43	6	100.0	0	0.0	0	0.0	0	0.0	0	0.0	0	0.0	6
Total	440		67		310		0		0		261		1078

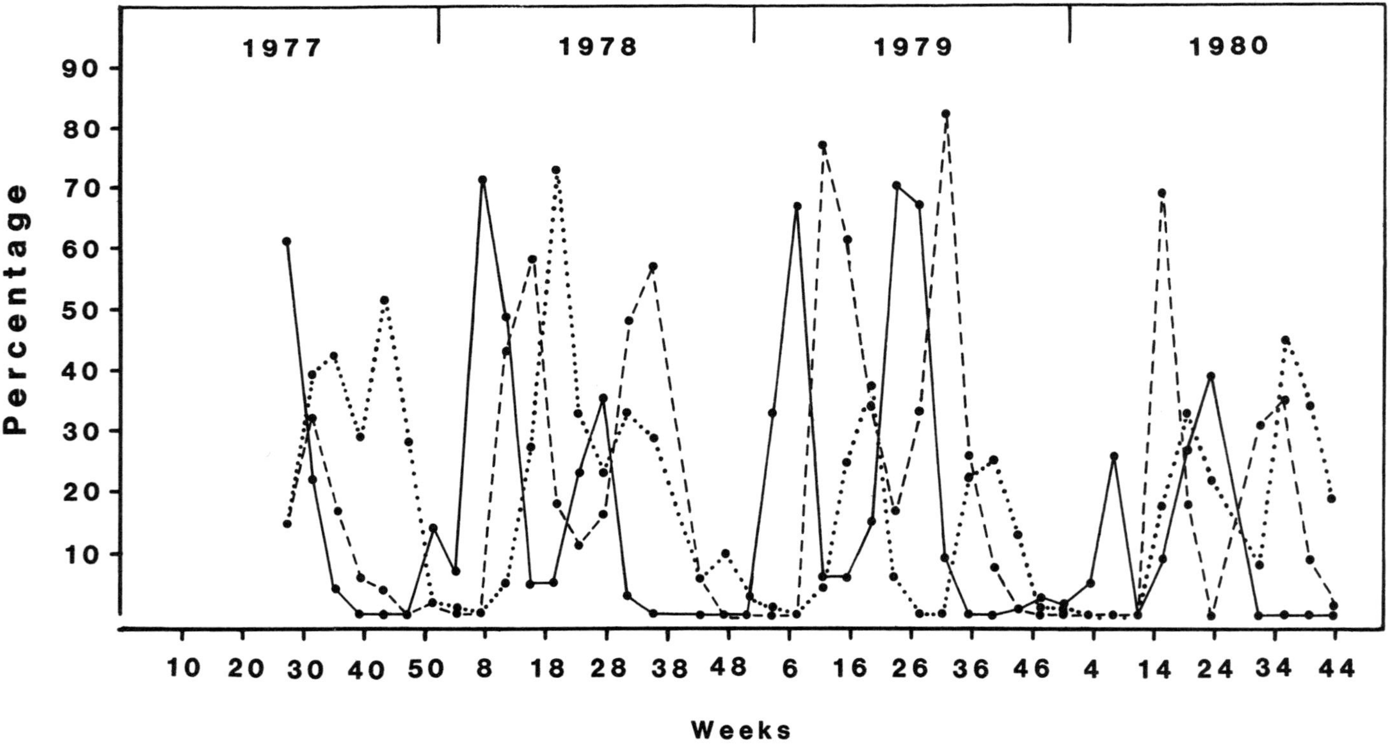

FIGURE 4-1.—Percentages of pregnant (solid line), lactating (dashed line), and postlactating (dotted line) female *Artibeus jamaicensis* on BCI. Each record represents four weeks combined.

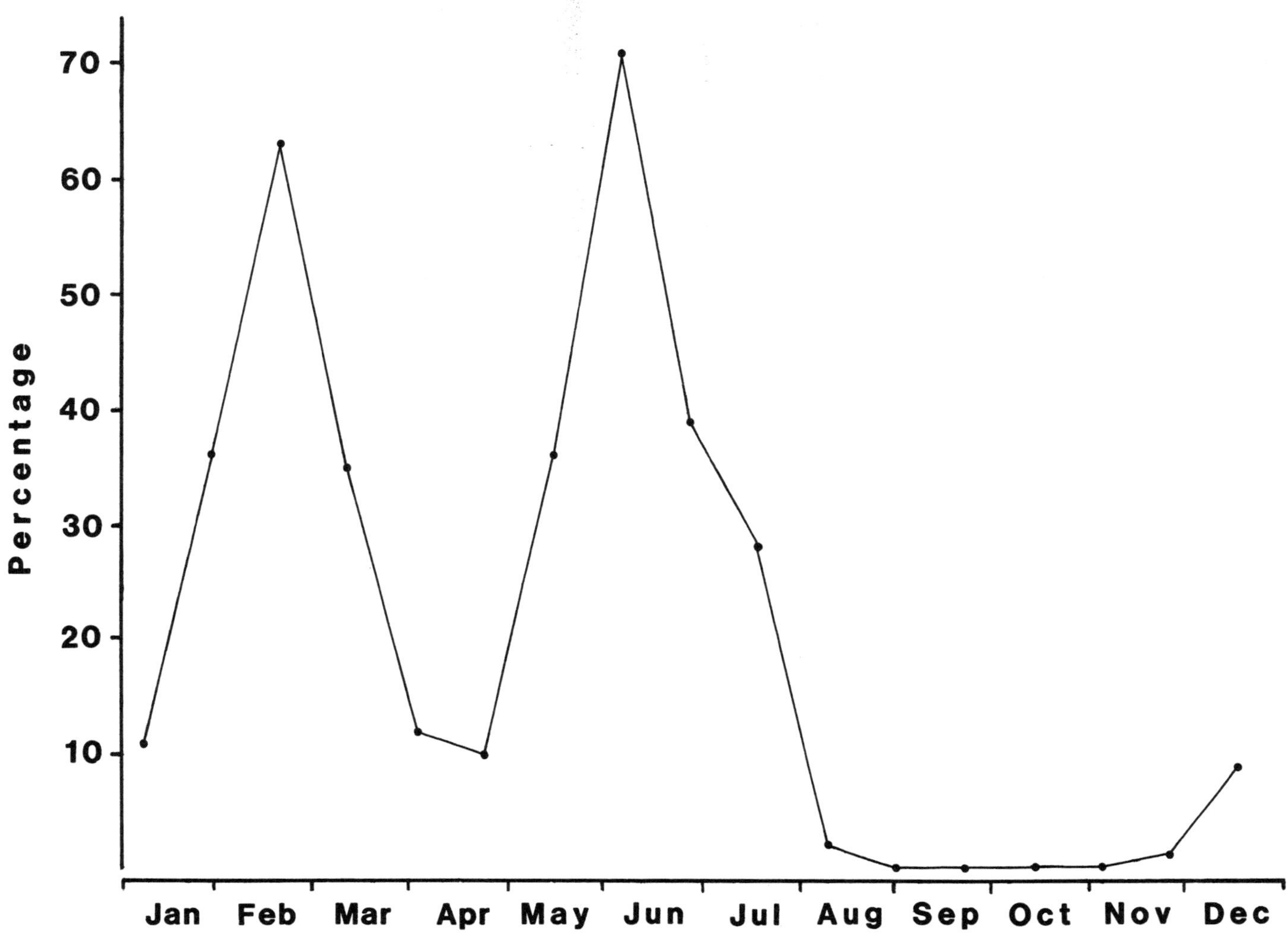

FIGURE 4-2.—Average percentage of pregnant *Artibeus jamaicensis* on BCI (based on data from 1972 through 1980, pooled in three week increments).

clear that not every female takes part in every breeding period. There are records from each week of the year of females showing no evidence of reproductive activity (Table 4-1). At the peak of each reproductive episode, the number of pregnant females averages about 70% (Figure 4-2).

For females originally banded as juveniles or subadults, we have 894 capture records in the year following their birth. Of these, 547 or slightly more than 60% were recorded as nonreproductive (neither palpably pregnant, lactating, nor postlactating). From a sample of 1121 capture records of adult females (individuals that had experienced an earlier pregnancy), 449 or 40% were nonreproductive. During any given sampling period, there are almost always more reproductively active adults than there are yearlings (Figures 4-3 and 4-4). A summary graph (Figure 4-5) combining data from 1977 through 1980 shows that yearlings not only breed less frequently, they are more likely to be out of synchrony with the reproductively experienced adults.

The Reproductive Cycle

ESTRUS.—Each female has two periods of postpartum estrus per year. The first occurs immediately following the birth in February-March; the second immediately following the birth in July-August. Thus, the estrous cycle is easily controlled by the timing of parturition except in those females that, for whatever reason, fail to give birth during a particular reproductive period. We do not know what physiological and environmental factors cue estrus at the appropriate time. Perhaps estrus in females that failed to become pregnant is resynchronized by contact with

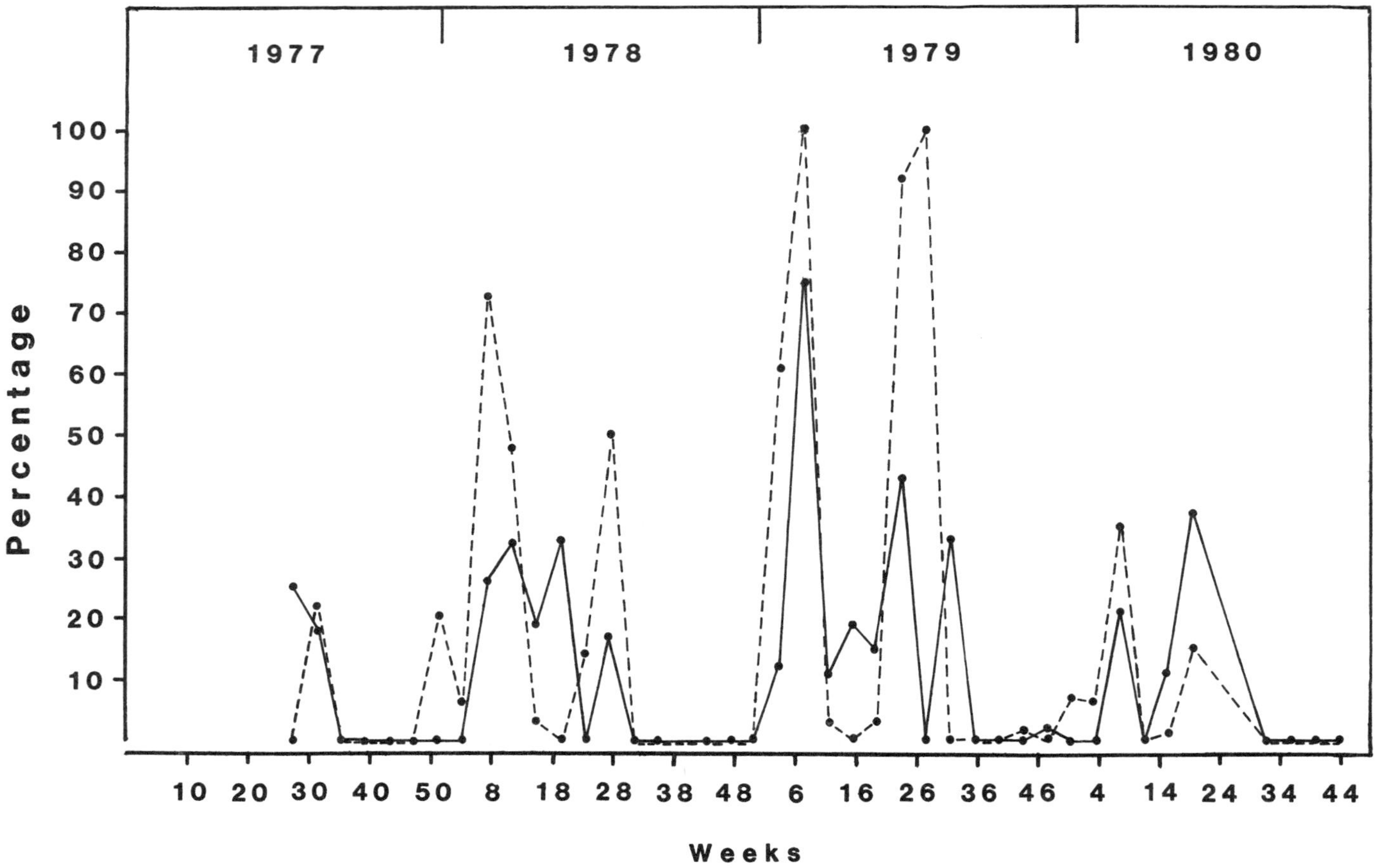

1977
1978
1979
1980
Percentage
100
90
80
70
60
50
40
30
20
10
10 20 30 40 50 8 18 28 38 48 6 16 26 36 46 4 14 24 34 44
Weeks

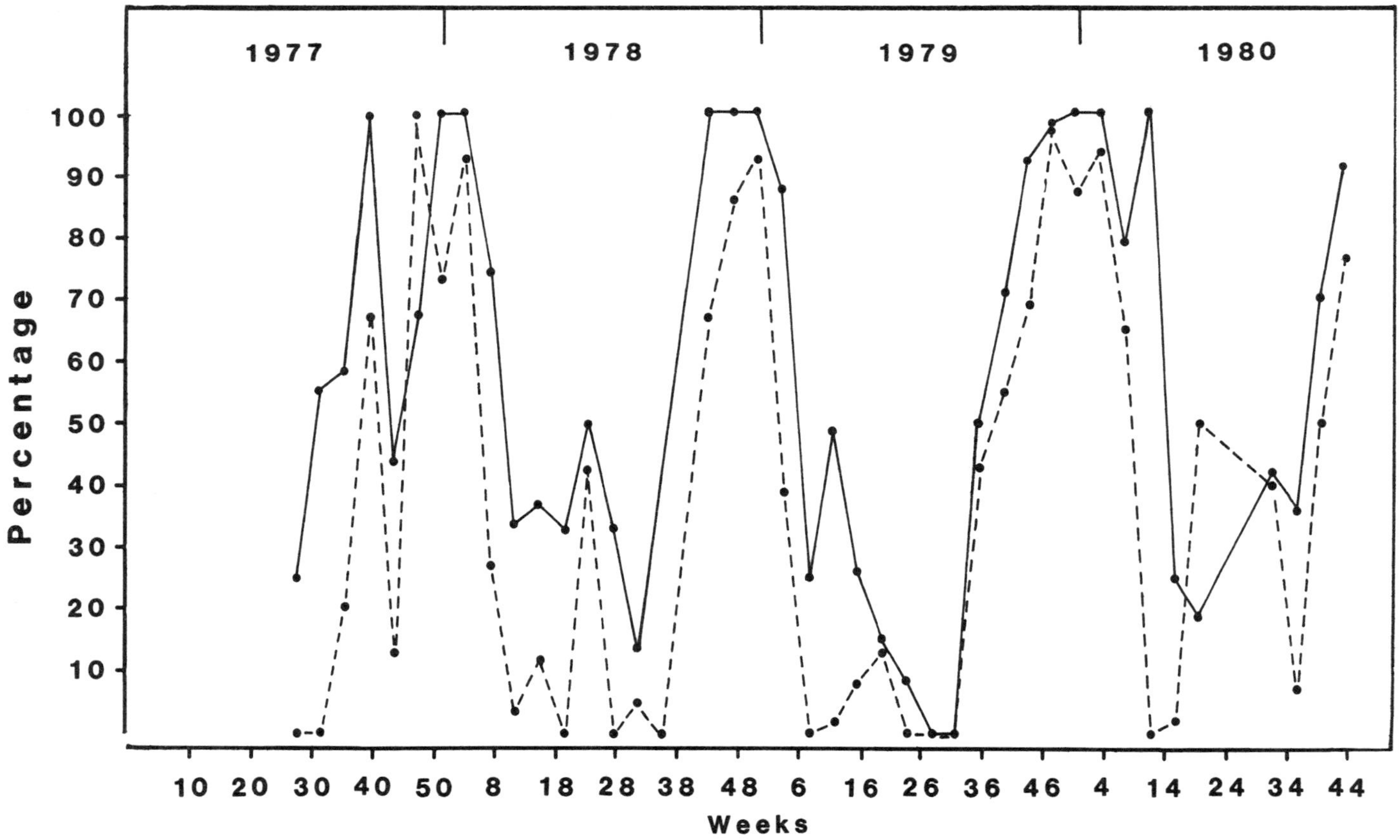

1977
1978
1979
1980
Percentage
100
90
80
70
60
50
40
30
20
10
10 20 30 40 50 8 18 28 38 48 6 16 26 36 46 4 14 24 34 44
Weeks

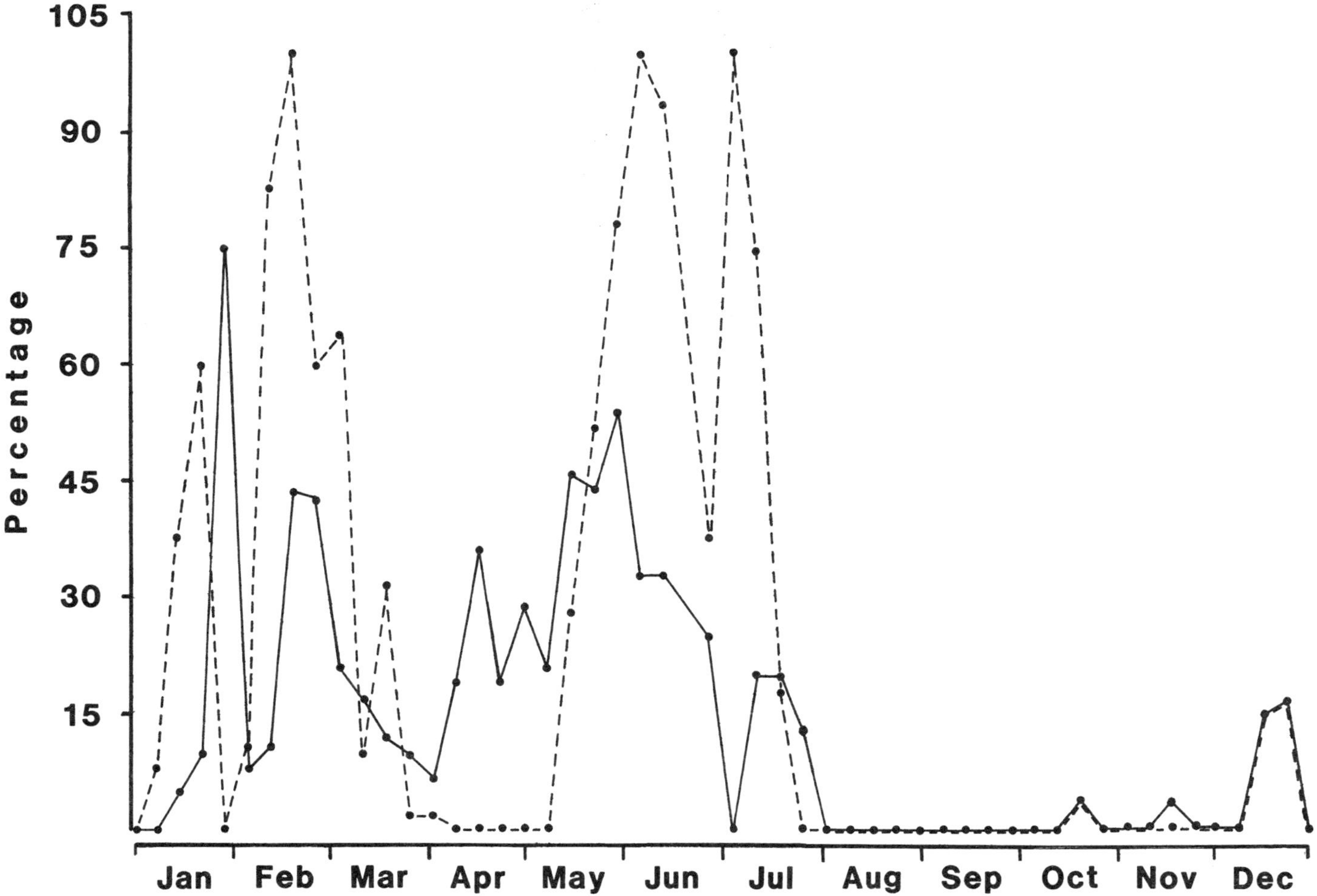

FIGURE 4-5.—Average percentages of pregnant yearling (solid line) and older (dashed line) female *Artibeus jamaicensis* on BCI, based on data from 1977–1980 combined. Each record represents one week.

adult female roost mates who are following a normal cycle.

COPULATION.—Males have enlarged testes during the periods when females are undergoing postpartum estrus (see Section 3, Reproduction in a Captive Colony). The degree of asynchrony in the system indicates that males may be capable of inseminating females for relatively long periods overlapping both birth peaks.

EMBRYONIC DEVELOPMENT.—Fertilization and implantation follow a normal sequence during each of the reproductive episodes, but subsequent development differs between the two. The normal gestation period following implantation during the first reproductive episode of the year is 3.5–4 months.

FIGURE 4-3 (opposite, top).—Percentages of pregnant yearling (solid line) and older (dashed line) female *Artibeus jamaicensis* on BCI. Each record represents four weeks combined.

FIGURE 4-4 (opposite, bottom).—Percentages of nonreproductive yearling (solid line) and nonyearling (dashed line) female *Artibeus jamaicensis* on BCI. Each record represents four weeks combined.

Implantation during the second episode is followed by delayed embryonic development, which results in a gestation period of about seven months (Fleming, 1971). This is a unique feature of the reproductive cycle of *A. jamaicensis*. The cues are unknown, but because the cycle persists when animals are moved into captivity under environmental conditions unlike the natural ones (see Section 3, Reproduction in a Captive Colony), it must be under some genetic control. Eventually, the synchronization breaks down under constant conditions of captivity, suggesting that one or more environmental cues are necessary to reset the system.

PARTURITION.—The two peaks of parturition are in February–March and July–August. The neonates are well-enough developed to be able to hang by themselves within a few hours of birth (see Section 3, Reproduction in a Captive Colony). Only on two occasions did we capture a female with an attached young. Apparently, the young are left behind shortly after birth while the females forage.

LACTATION.—Lactation is easily detected and lasts for about two months (see Section 3, Reproduction in a Captive Colony).

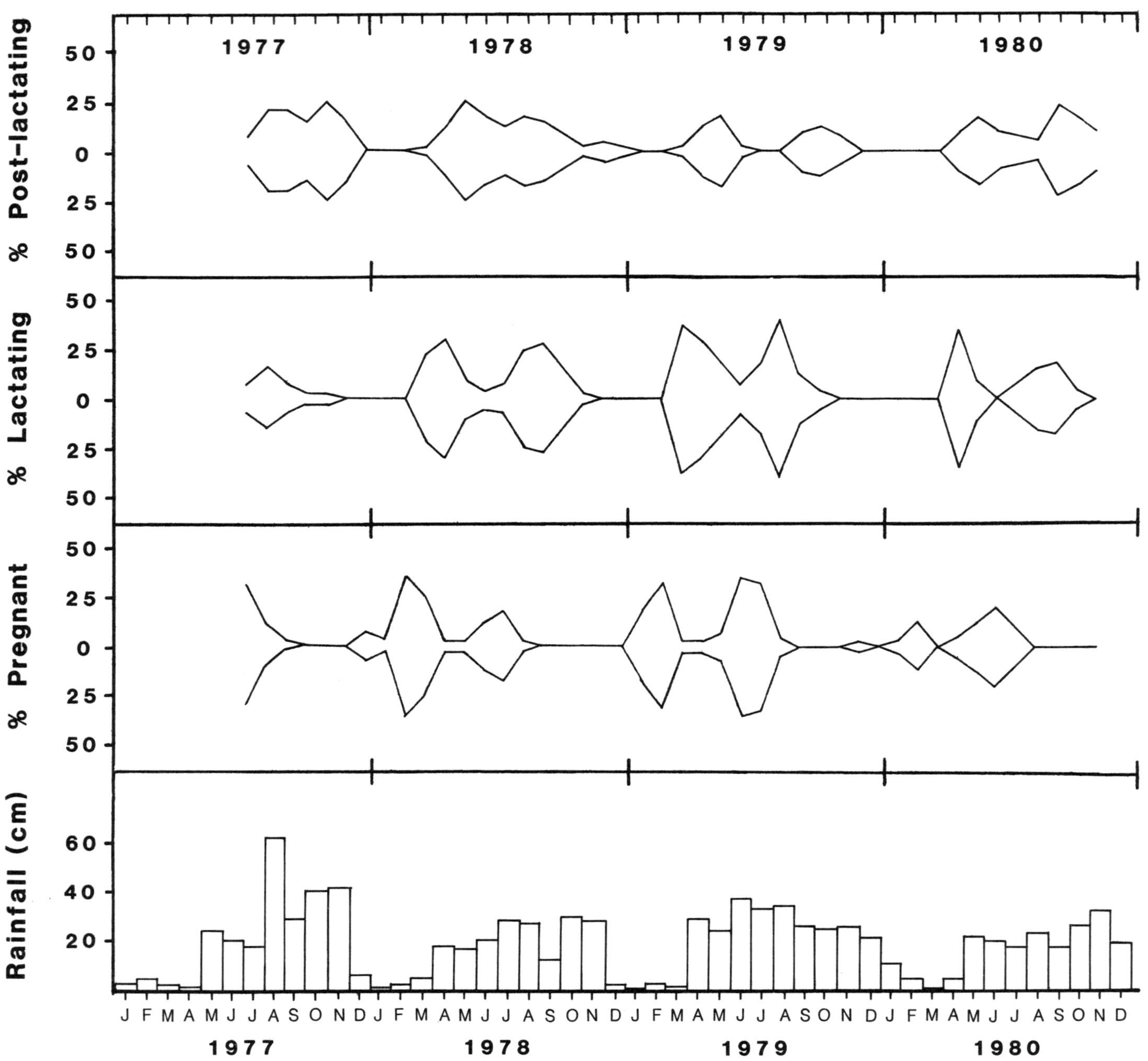

FIGURE 4-6.—Schematic representation of the reproductive cycle of *Artibeus jamaicensis* on BCI, showing temporal variation in percentages of pregnant, lactating, and postlactating females. Mean rainfall is shown in the bar graph at the bottom. Time is in months and years.

Peaks in percentages of lactating females follow about a month behind the peaks for pregnant females (Figures 4-1 and 4-6). Lactation in captivity lasts a bit longer in females that are carrying dormant embryos following the second breeding episode. This phenomenon can be inferred in our wild population (Figure 4-6).

POSTLACTATION.—We were able to glean additional information about the cycle from individuals we recognized as postlactating. Apparently there is more individual variation in the length of time that nipples are classified as postlactating than in other reproductive criteria (Figures 4-1 and 4-6). Part of this variation may be an artifact of stages of refinement of our definition of postlaction. Early in the Bat Project, we presumed a bat to be postlactating if no milk could be expressed from nipples although the nipples were large, flabby, and surrounded by naked skin. Later, the category was expanded to include females showing new hair growing in around the regressing nipple, regardless of its size.

Seasonality, Rainfall, and Abundance of Food

Previous hypotheses (Wilson, 1979) have focused on energetics as the primary control on reproduction, and this indeed seems the most logical explanation for the basic pattern of the cycle in *A. jamaicensis*. Other factors being equal, one would expect animals to achieve maximum reproductive output without sacrificing survivorship of the young.

If we assume that the physiological constraints operating on an *A. jamaicensis* result in a litter size of one and a gestation period of 3.5–4 months, then in a world of abundant energetic resources, females simply should reproduce continually, producing up to three young per year. That they do not suggests that there must be a time when energy sources are in short supply.

The time of maximum stress might occur at any of several critical points in the reproductive cycle. Pregnancy, as does lactation, carries with it an increase in energy demands to individual females (see Section 2, Physiology; Studier et al., 1972). Reproductive success should be enhanced if these stressful periods coincide with periods of peak resource availability (Bradbury and Vehrencamp, 1977b).

The critical time for most bats is when the young are weaned. Not only do the growing young have high energy demands, they are forced to learn to forage at the same time, and presumably it would be easier to do so when food is readily available. Readily available food resources also are critical at the time of weaning in a species showing synchronized reproductive cycles, because the population size is markedly increased at that time. There is now a considerable body of evidence supporting the position that reproductive events in bats are timed to synchronize with seasonally abundant food resources (Dinerstein, 1983; Heithaus et al., 1975; Racey, 1982).

Young born in the first reproductive episode (February–March) are weaned in April–May, and those born in the second period (July–August) are weaned in September–October. This pattern is confirmed by our capture records, as the number of juveniles and subadults increases dramatically during these periods. On BCI this results in a general pattern in which the young are weaned during the rainy season (Figure 4-6). If the bats conceived during the second episode were to develop normally (in 3.5–4 months), they would be born in November and weaned at the beginning of the dry season in January.

Foster (1982) has shown that the peaks in fruit availability on BCI occur in April–May and September–October, with a low in January at the beginning of the dry season (Figure 4-7). Therefore, it seems that delayed development has evolved as a mechanism to avoid weaning young during the dry season when fruiting trees are scarce. Thus, the reproductive cycle of *A. jamaicensis* seems to be an adaptation to the seasonal cycle of food abundance. Evidence from other localities supports this hypothesis (Wilson, 1979).

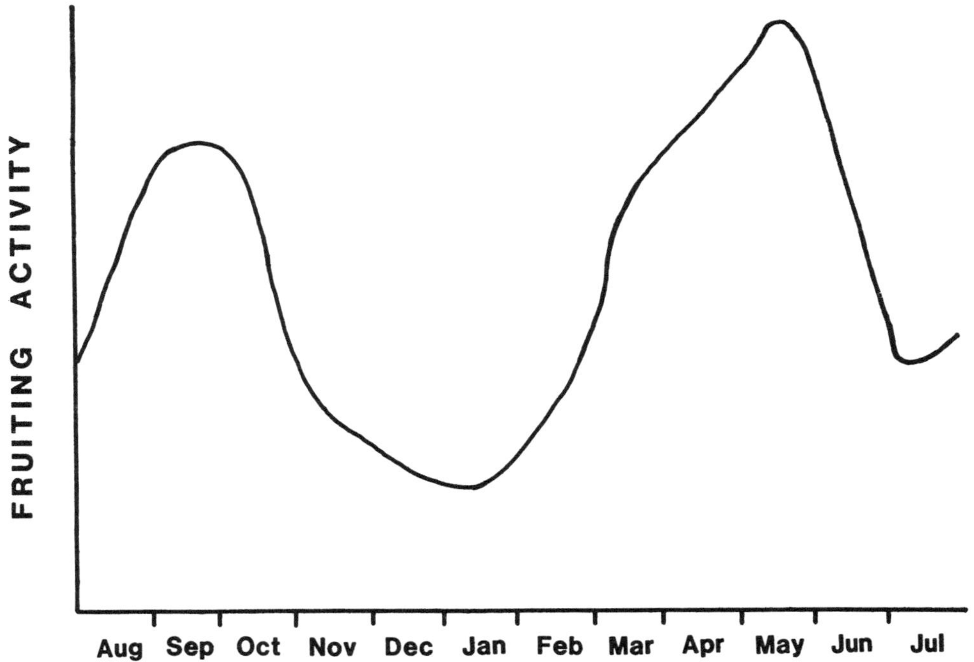

FIGURE 4-7.—Schematic representation of seasonality of fruiting activities of canopy trees on BCI. "Activity" represents the proportion of species that are fruiting (determined by seed traps). Redrawn from Foster (1982, fig. 12).

Environmental Cues

This unique cycle persisted after the *A. jamaicensis* were brought into the laboratory, although it soon began to drift toward shorter periods of delayed development. This suggests that in addition to being under genetic control, the bats are influenced by some environmental cues that reset and maintain the periodicity of the reproductive cycle each year. Not only must an environmental cue be detectable by the animals, it also has to occur at the appropriate time for them to react to the upcoming period of limited resources. Although the rainfall pattern probably is responsible for the pattern of fruit availability, thus providing the evolutionary drive for the system, it is difficult to see how either rainfall or fruit availability could serve as proximal cues to trigger the onset of reproductive episodes.

If the cycle depended entirely on environmental controls other than photoperiod, we would expect to see slight variations from year to year, extreme variation in rare years of unusual environmental fluctuations, and an abrupt change in the cycle under constant conditions. If the cycle were under complete genetic control, we would expect to see no annual variation, persistence of the cycle in unusual climactic years, and persistence under constant conditions. There are slight annual variations (Figure 4-1), no data for unusual years (Figure 4-6), and no abrupt change under constant conditions of captivity.

In the Old World, an ecologically equivalent species, *Haplonycteris fischeri* of the Philippines, has a monestrous cycle with delayed development (Heideman, 1988). It shows no annual variation, persistence in unusual climatic years, and a curious pattern of geographic variation. The persistence lends support to the hypothesis of genetic control; geographic variation argues for an environmental component as well (Heideman, 1988).

Perhaps genetic control is effected through the postpartum estrus, which essentially sets the time of copulation and fertilization. In *A. jamaicensis,* postpartum estrus persists in captivity, and the period of delayed development diminishes over time. This would suggest that an appropriate proximal environmental cue to fine-tune the cycle should be one that determines the end of the period of delayed development and the beginning of normal rate of growth of the embryo. Assuming the 3.5–4 month gestation time seen in the alternate reproductive period, the critical time would be during November–December, or the end of the rainy season.

The end of the rainy season is a time of declining food resources (Figure 4-7) making it unlikely that the proximal cue to begin normal development of an embryo would be a sudden change in the amount or kind of energy resource available. Rainfall itself is an equally poor proximal cue, because the rainy season does not end abruptly and the amount of annual variation in daily rainfall at that time of year is greater than the amount of variation shown by the timing of the first parturition period 3.5–4 months later.

What, then, is the proximal environmental cue that resets the cycle on an annual basis? Perhaps variation in night length? This is one of the more intriguing questions that remain to be answered about the reproductive cycle of *A. jamaicensis* and perhaps other related species that show a similar bimodal pattern.

Fleming (1988) found a similar bimodal pattern in *Carollia perspicillata* in Costa Rica. Although there is no delayed development, the cycle also persists in captivity. As with *A. jamaicensis* on BCI, the cycle is correlated with rainfall seasonality, and presumably with differential availability of food resources.

Summary

Artibeus jamaicensis has a unique reproductive pattern consisting of alternating episodes of normal embryonic development and delayed development. Individual females usually produce two young per year, although occasionally they produce only one. Young females do not give birth in the calendar year of their birth, but may do so before they are one year old. Yearlings produce young at a lower rate than do nonyearlings. The unusual reproductive pattern seems adapted to take advantage of periods of maximum food availability by insuring that young are weaned at such energetically favorable times. The environmental cues responsible for maintaining the periodicity remain enigmatic, although the persistence of the pattern in captive animals suggests some measure of genetic control.

5.   Survival and Relative Abundance

Alfred L. Gardner, Charles O. Handley, Jr.,
and Don E. Wilson

Relative Abundance

Artibeus jamaicensis was consistently the most common bat netted on BCI and adjacent mainland from 1976 to 1980. Its abundance, however, whether measured on the basis of absolute numbers or relative to all other species netted did not appear to follow either a monthly or seasonal pattern. Average nightly catch of *A. jamaicensis* per month ranged from less than five to nearly 80 per netting session and was inconsistent from month to month throughout the study (Figure 5-1). Because part of that inconsistency may have resulted from uneven netting effort from year to year, we examined the capture record for 1979, the year for which our efforts were the most intense and for which July and August were the only months we did not sample. Numbers of *A. jamaicensis* still showed great fluctuation from month to month whether measured directly (averaged on a per night basis) or calculated as a percentage of total catch (Figures 5-2, 5-3).

We also compared the 1979 record for *A. jamaicensis* with that for the five other most common species of frugivorous bats (*Artibeus lituratus, A. phaeotis, Uroderma bilobatum, Chiroderma villosum,* and *Phyllostomus discolor*) netted the same year (Table 5-1; Figure 5-3). If records for February are ignored as being too few to be useful, *A. jamaicensis* made up 44%–76% of the nightly catch in 1979 and were caught on 150 of the 151 nights netted. With the exception of *A. phaeotis,* the numbers of the other species also fluctuated greatly (Table 5-1): *A. lituratus,* 2%–15% (120 nights); *U. bilobatum,* 1%–15% (60 nights); *C. villosum,* 0.5%–7% (39 nights); and *P. discolor,* 0.2%–7% (40 nights). *A. phaeotis* (2%–3%; 107 nights) appears to be relatively evenly dispersed at low density and most of those caught probably were resident in the vicinity of the netting station.

We tested the monthly differences in captures of these bats (Table 5-1) to see if the differences were more apparent than

Alfred L. Gardner, NERC, U.S. Fish and Wildlife Service, National Museum of Natural History, Washington, D.C. 20560.
Charles O. Handley, Jr., and Don E. Wilson, National Museum of Natural History, Smithsonian Institution, Washington, D.C. 20560.

real. The differences proved to be highly significant. Even when the data for February are removed, overall Chi-square values are insignificant only for *A. phaeotis* ($X^2 = 11.790$, $df = 9$, ns; where df is degrees of freedom and ns means not significant). The overall Chi-square values are astronomical for the other five species, even when based on a reduced data set (minus February; range, 150.926–495.984). For the full set of data (Table 5-1), numbers of *A. jamaicensis* are significantly low ($P < 0.001$) for January and November, and high ($P < 0.001$) for April, June, September, and December.

Unusually high numbers of *C. villosum* and *U. bilobatum* were taken in October and November, and greater numbers of *A. lituratus* were caught in November, December, and January (Table 5-1) than at other times of the year. We reduced the data set to see what influence the monthly variation in numbers of the others species may have had on these tests. When the data for February and all *A. lituratus, U. bilobatum, C. villosum,* and *P. discolor* are removed from the data set (Table 5-1), the Chi-square values for *A. jamaicensis* are significantly low for March ($P < 0.01$) and May ($P < 0.05$), and high ($P < 0.01$) only for April and December.

The causes of year-to-year variation in population size are difficult to assess. Annual variations in numbers caught are at least partly the result of differences in personnel and their relative experience as well as the evolution of techniques and capture strategies used from year to year as the study progressed. Duration of netting episodes, selection of netting sites, and frequency of nights of netting per month along with learned net avoidance, undoubtedly were additional contributing factors.

Numbers of *A. jamaicensis,* best seen in the October, November, and December records (Figure 5-1), generally increased from 1976 to 1980. Captures of *A. jamaicensis* tended to be highest when young are first volant (April–June and September–October) and lowest when the population is composed mostly of subadults and adults. We do not know how much of this variation resulted from adverse conditioning due to having been caught before.

Occasional unusual concentrations of bats at certain fruiting trees also distorted the monthly averages and percentages of

TABLE 5-1.—Seasonal variation in abundance of six species of bats on Barro Colorado Island during 1979.

Character	JAN	FEB	MAR	APR	MAY	JUN	SEP	OCT	NOV	DEC	TOTAL
Total Bats (N)	595	35	804	1066	762	438	1418	1594	1976	430	9118
Number of netting nights	17	6	17	10	17	5	20	21	34	10	157
Mean number per netting night	35.0	5.8	47.3	106.6	44.8	87.6	70.9	75.9	58.1	43.0	58.1
Artibeus jamaicensis	261	8	456	790	474	334	1024	884	937	316	5484
Percentage of catch	43.9	22.9	56.7	74.1	62.2	76.3	72.2	55.5	47.4	73.5	60.1
Number of nights caught	16	1	17	10	17	5	20	21	34	10	151
Percentage of nights caught	94	17	100	100	100	100	100	100	100	100	96
Mean number per netting night	15.4	1.3	26.8	79.0	27.9	66.8	51.2	42.1	27.6	31.6	34.9
Mean number per night caught	16.3	8.0	26.8	79.0	27.9	66.8	51.2	42.1	27.6	31.6	36.3
Artibeus lituratus	87	0	55	64	32	7	41	63	305	63	717
Percentage of catch	14.6	0	6.8	6.0	4.2	1.6	2.9	4.0	15.4	14.7	7.9
Number of nights caught	15	0	11	9	12	3	18	18	25	9	120
Percentage of nights caught	88	0	65	90	71	60	90	86	74	90	76
Mean number per netting night	5.1	0	3.2	6.4	1.9	1.4	2.1	3.0	9.0	6.3	4.6
Mean number per night caught	5.8	0	5.0	7.1	2.7	2.3	2.3	3.5	12.2	7.0	6.0
Artibeus phaeotis	19	1	24	23	30	7	28	46	47	10	235
Percentage of catch	3.2	2.9	3.0	2.2	3.9	1.6	2.0	2.9	2.4	2.3	2.6
Number of nights caught	11	1	13	8	12	3	14	18	22	6	108
Percentage of nights caught	65	17	76	80	71	60	70	86	65	60	69
Mean number per netting night	1.1	0.2	1.4	2.3	1.8	1.4	1.4	2.2	1.4	1.0	1.5
Mean number per night caught	1.7	1.0	1.8	2.9	2.5	2.3	2.0	2.6	2.1	1.7	2.2
Uroderma bilobatum	30	0	15	25	7	4	2	160	307	1	551
Percentage of catch	5.0	0	1.9	2.3	0.9	0.9	0.1	10.0	15.5	0.2	6.0
Number of nights caught	8	0	4	4	5	2	2	9	25	1	60
Percentage of nights caught	47	0	24	40	29	40	10	43	74	10	38
Mean number per netting night	1.8	0	0.9	2.5	0.4	0.8	0.1	7.6	9.0	0.1	3.5
Mean number per night caught	3.5	0	3.8	6.3	1.4	2.0	1.0	17.8	12.3	1.0	9.2
Chiroderma villosum	3	0	21	2	9	0	16	111	6	4	172
Percentage of catch	0.5	0	2.6	0.2	1.2	0	1.1	7.0	0.3	0.9	1.9
Number of nights caught	3	0	6	2	4	0	8	9	4	3	39
Percentage of nights caught	18	0	35	20	24	0	40	43	12	30	25
Mean number per netting night	0.2	0	1.2	0.2	0.5	0	0.8	5.3	0.2	0.4	1.1
Mean number per night caught	1.0	0	3.5	1.0	2.3	0	2.0	12.3	1.5	1.3	4.4
Phyllostomus discolor	44	0	10	2	8	0	24	13	24	5	130
Percentage of catch	7.4	0	1.2	0.2	1.1	0	1.7	0.8	1.2	1.2	1.4
Number of nights caught	6	0	6	2	3	0	7	6	7	3	40
Percentage of nights caught	35	0	35	20	18	0	35	29	21	30	25
Mean number per netting night	2.6	0	0.6	0.2	0.5	0	1.2	0.6	0.7	0.5	0.8
Mean number per night caught	7.3	0	1.7	1.0	2.7	0	3.4	2.2	3.4	1.7	3.3
All Other Species	151	26	223	183	202	86	283	317	350	31	1852
Percentage of catch	25.4	74.3	27.7	17.2	26.5	19.6	20.0	19.9	17.7	7.2	20.3
Mean number per netting night	8.9	4.3	13.1	18.3	11.9	17.2	14.2	15.1	10.3	3.1	11.8

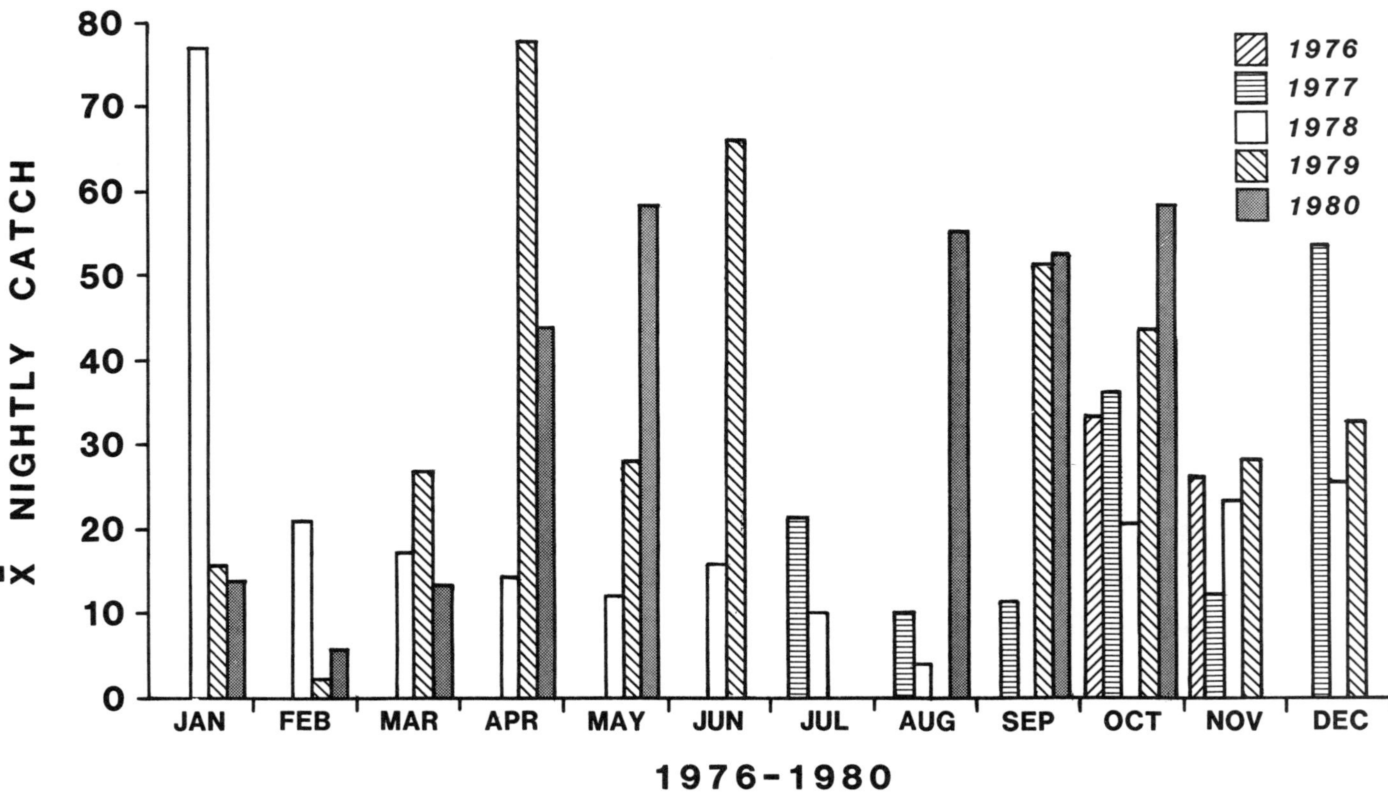

FIGURE 5-1.—Mean nightly catch of *Artibeus jamaicensis* on BCI and adjacent mainland per month from October 1976 through October 1980.

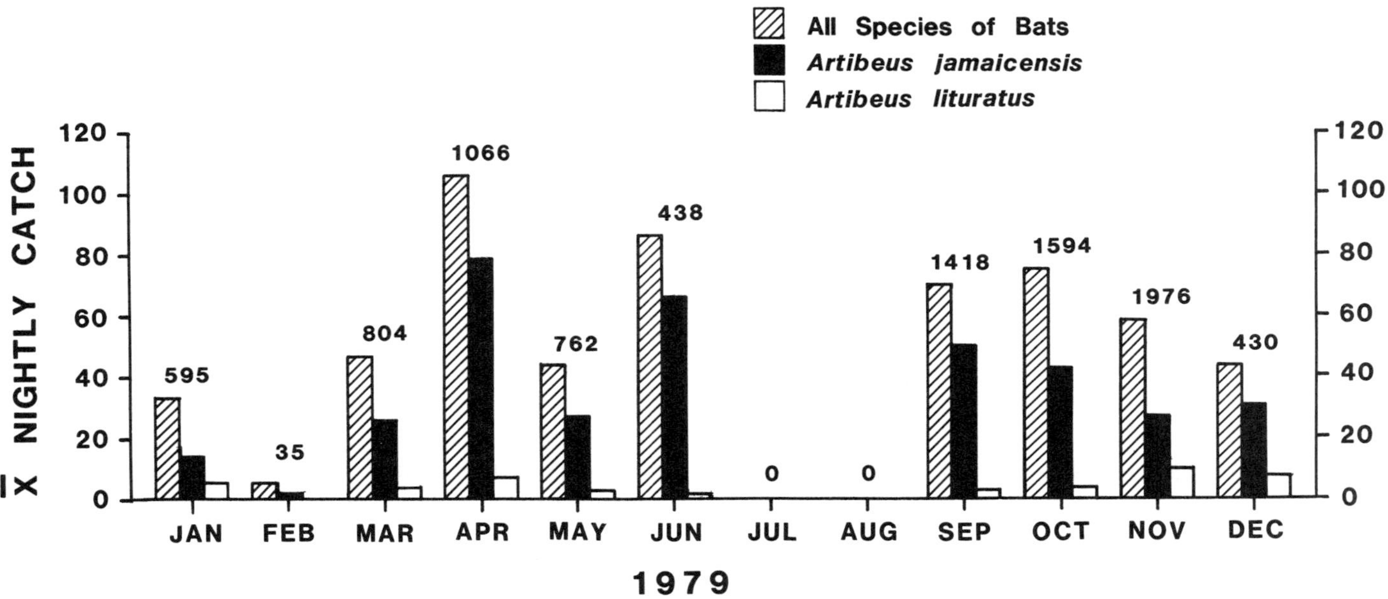

FIGURE 5-2.—Mean nightly catch of all bats caught on BCI and adjacent mainland per month during 1979 and the mean nightly catch of *Artibeus jamaicensis* and *A. lituratus* during the same period. Figures above bars are monthly catch of all bats.

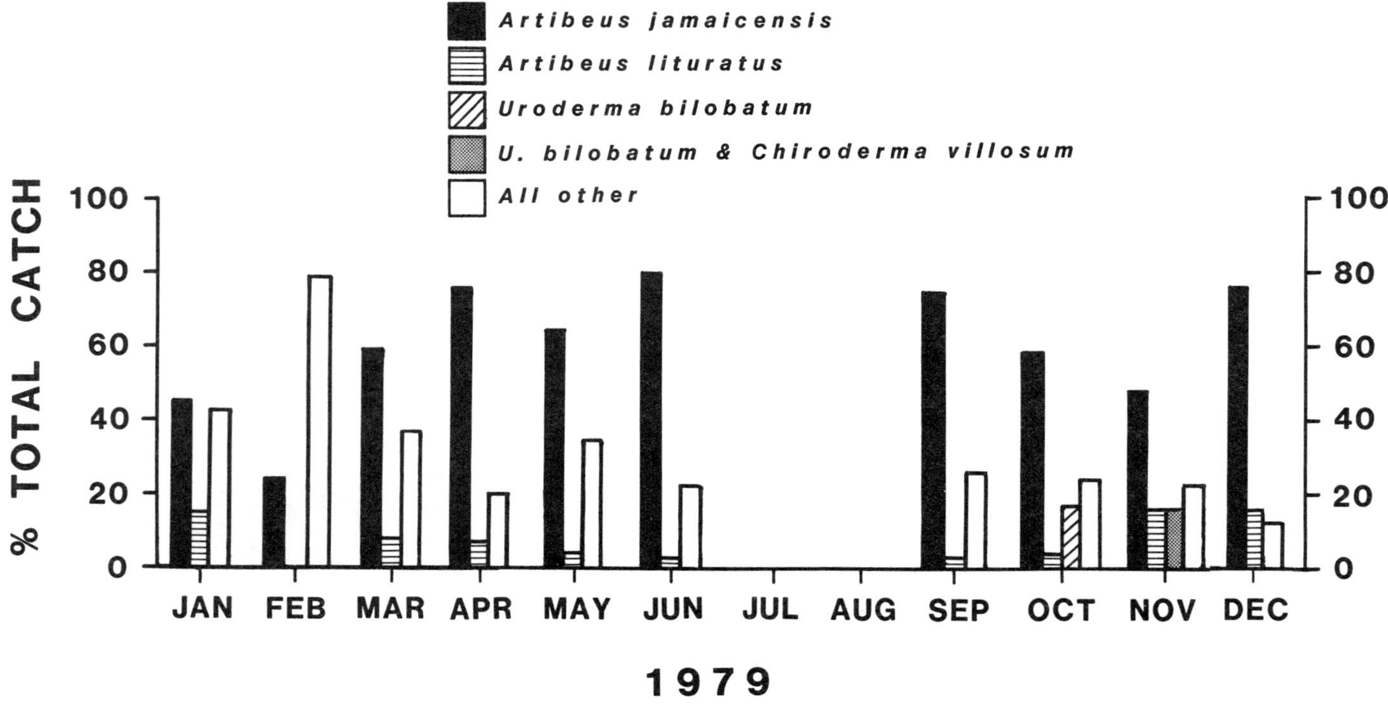

FIGURE 5-3.—Monthly composition of the bat fauna caught on BCI and the adjacent mainland during 1979. Bars represent percentages of total catch.

bats captured. The night of 25–26 October 1979 produced 282 bats of 16 species. Only 86 were *A. jamaicensis*, however; *Uroderma bilobatum* (115) was the most numerous species and *Chiroderma villosum* (50) was third. Another night with an unusually high catch was that of 23–24 April 1979. We handled 255 bats comprising 13 species of which *A. jamaicensis* numbered 215, including 100 previously unmarked juveniles and 53 adult females. Only 40 of the adult females could have been the mothers of these juveniles (the other adult females were either pregnant or nonreproductive). This suggests the possibility of net avoidance, because if first-flying juveniles are flying with their mothers, as seems a reasonable yet unproven assumption, then the mothers of 60 were active in the vicinity, but not caught.

Recapture Frequency

The entire sample of *A. jamaicensis* marked between October 1976 and June 1980, consisted of 8907 bats captured a total of 15,728 times (Figure 5-4, Table 5-2). Fifty-seven percent (5061 individuals) were caught only once, leaving 3846 captured two or more times (6821 records; Table 5-2, line B). The number of subsequent capture records per bat diminished until we were left with a single individual captured 11 times.

From these data we calculated the mean number of captures per *A. jamaicensis* (Table 5-2, line C) in each capture cohort–for the whole sample, for two or more captures, for three or more captures, and so on. The 8907 bats, captured a total of 15,728 times, averaged 1.77 captures per individual (Table 5-2). The next cohort, those captured two or more times, also has a mean of 1.77 captures per individual. The individual capture rates of the first five groups (one or more captures through five or more captures) vary little (1.76–1.77). The capture rates, however, for each of the remaining capture cohorts are lower (1.43–1.65, Table 5-2) and more variable. When the data for the first five capture cohorts are summed and the derived mean capture rate (1.77) is contrasted with the mean capture rate for the remaining six cohorts (1.573, data from Table 5-2), the differences are significant ($t = 6.752$, $df = 8$, $P < 0.001$).

We used the derived mean number of captures per individual for one to five captures (1.77) and for six or more captures (1.57) to estimate the expected numbers (Table 5-2, line D) in each capture cohort. In most cohorts the actual capture rate is close to the expected rate. Individuals disappear and recapture records diminished in our sample at a nearly constant rate through the first five times a bat is captured. The probability of recapturing a bat is roughly the same regardless of the number of previous captures. Therefore, emigration, necklace loss, mortality rates, and other factors contributing to the disappearance of bats must be nearly constant.

Another illustration of the relationship between successive

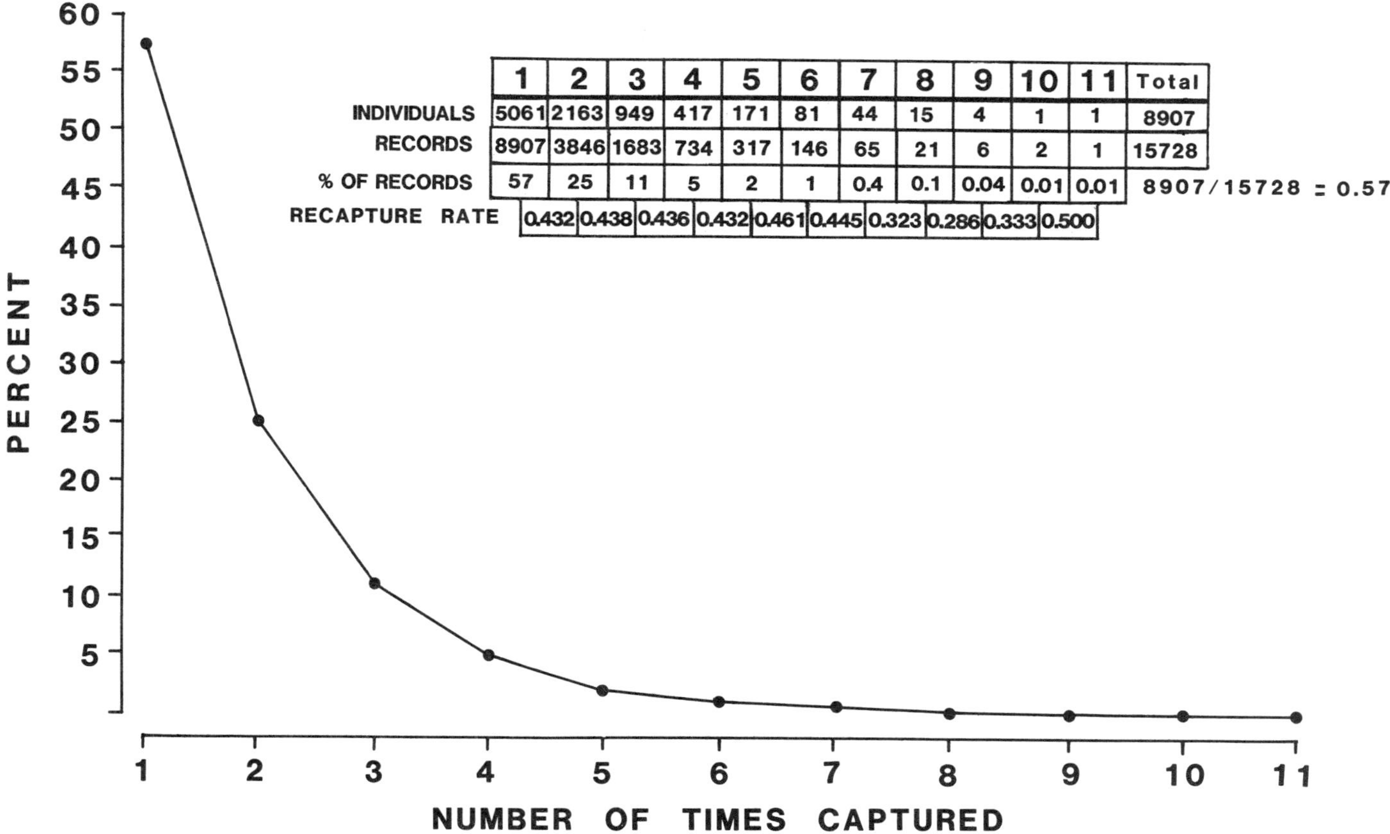

	1	2	3	4	5	6	7	8	9	10	11	Total
INDIVIDUALS	5061	2163	949	417	171	81	44	15	4	1	1	8907
RECORDS	8907	3846	1683	734	317	146	65	21	6	2	1	15728
% OF RECORDS	57	25	11	5	2	1	0.4	0.1	0.04	0.01	0.01	
RECAPTURE RATE	0.432	0.438	0.436	0.432	0.461	0.445	0.323	0.286	0.333	0.500		

FIGURE 5-4.—Recapture curve, recapture rate (rate of decrease), and frequency of capture of *Artibeus jamaicensis* on BCI and adjacent mainland between 1976 and 1980.

capture groups is the rate (percentage) of decrease between adjacent groups (Figure 5-4). This we calculated by dividing the number of records in a capture group by the number of records in the preceding group. The rate of decrease (Figure 5-4) is relatively constant up to seven captures (range, 0.432–0.461), but more variable thereafter (range, 0.286–0.500).

Observed Survivorship

To study survivorship of *A. jamaicensis* marked from 1976 through 1980, we divided calendar years into half years: January–June = "spring" and July–December = "fall." Bats caught and marked as juveniles were assumed to have been born in the half year in which they were caught. On BCI and vicinity most births occur between the 8th and 16th week and between the 25th and 33rd week of each year. Bats first caught and marked as subadults were assumed to have been born in the previous half year. We derived a conservative estimate of survivorship by dividing the number of juveniles, subadults, or adults marked in a half year into the number in that birth-cohort

sample known to be alive based on the total recapture record up to October 1980. Bats marked later in the study (especially adults and subadults marked in 1979 and 1980) had depressed survivorship rates because of the lack of sufficient time to accumulate a more complete recapture record. The data reported on here only cover a span of 4.5 years, but we know from Wilson and Tyson (1970) of a 7-year-old *A. jamaicensis,* and Handley caught two bats after 1980 that were at least nine years old.

Survival to the first half year after being marked varied among year-and-age cohorts from a low of 8% for adult females marked in the spring of 1980 to a high of 58% for juvenile females marked in the spring of 1978 (Figure 5-5, Tables 5-3 to 5-8). Other factors in addition to the lack of sufficient time to accumulate recapture records influenced survival estimates. One of the more important was the lower probability of recapturing bats from the channel markers (Buena Vista and Peña Blanca) and from the four mainland sites (Bohio, Frijoles Road, Gigante, and Mona Grita Point) because these bats normally ranged outside of the area we routinely sampled. These sites were not netted on a regular

TABLE 5-2.—Captures of *Artibeus jamaicensis* marked on BCI and adjacent mainland between October 1976 and May 1980, and recaptured through October 1980. D is the expected number of individuals in each capture category (A -[A/C] = D, where C = 1.7675 for categories 1 through 5, and C = 1.5726 for categories 6 through 11: e.g., 8907 -[8907/1.7675] = 3868).

Captures	1 or more	2 or more	3 or more	4 or more	5 or more	6 or more	7 or more	8 or more	9 or more	10 or more	11
A. Number of individuals	8907	3846	1683	734	317	146	65	21	6	2	1
Percentage of original sample remaining	100	43.180	18.895	8.241	3.559	1.639	0.730	0.236	0.067	0.022	0.011
B. Number of records of captures	15728	6821	2975	1292	558	241	95	30	9	3	1
C. Mean number of captures (B)/individuals remaining (A)	1.7658	1.7735	1.7677	1.7602	1.7603	1.6506	1.4615	1.4286	1.5000	1.5000	1.0000
D. Number expected = A −(A/C)		3868	1670	731	319	138	53	24	8	2	1

TABLE 5-3.—Survivorship of juvenile female *Artibeus jamaicensis* on BCI and vicinity in successive half years of life. N = number of individuals surviving. Survivorship of all juveniles on last line.

Period	Number juveniles marked	Survival in successive half years of life															
		2		3		4		5		6		7		8		9	
		N	%	N	%	N	%	N	%	N	%	N	%	N	%	N	%
Fall 1976	74			28	38	21	28	11	15	11	15	10	14	7	9	3	4
Fall 1977	236	58	25	40	17	29	12	19	8	7	3	3	1				
Spring 1978	54	31	57	23	43	13	24	8	15	5	9						
Fall 1978	83	32	39	24	29	11	13	8	10								
Spring 1979	272	108	40	52	19	27	10										
Fall 1979	331	91	27	50	15												
Spring 1980	368	107	29														
Spring 1978–80	694	246/694	35	75/326	23	40/326	12	8/54	15	5/54	9						
Fall 1976–79	724	181/650	28	142/724	20	61/393	16	38/393	10	18/310	6	13/310	4	7/74	9	3/74	4
All periods	1418	427/1344	32	217/1050	21	101/719	14	46/447	10	23/364	6	13/310	4	7/74	9	3/74	4
Sexes combined	2934	851/2772	31	384/2225	17	168/1484	11	79/940	8	36/760	5	20/659	3	10/162	6	5/162	3

TABLE 5-4.—Survivorship of juvenile male *Artibeus jamaicensis* on BCI and vicinity in successive half years of life. N = number of individuals surviving.

Period	Number juveniles marked	2		3		4		5		6		7		8		9	
		N	%	N	%	N	%	N	%	N	%	N	%	N	%	N	%
Fall 1976	88			18	20	9	10	5	6	5	6	4	5	3	3	2	2
Fall 1977	261	79	30	33	13	27	10	22	8	7	3	3	1				
Spring 1978	47	17	36	7	15	5	11	2	4	1	2						
Fall 1978	97	32	33	21	22	9	9	4	4								
Spring 1979	272	102	37	40	15	17	6										
Fall 1979	410	103	25	48	12												
Spring 1980	341	91	27														
Spring 1978–80	660	210/660	32	47/319	14	22/319	7	2/47	4	1/47	2						
Fall 1976–79	856	214/768	28	120/856	14	45/446	10	31/446	7	12/349	3	7/349	2	3/88	3	2/88	2
All periods	1516	424/1428	30	167/1175	14	67/765	9	33/493	7	13/396	3	7/344	2	3/83	4	2/83	2

TABLE 5-5.—Survivorship of subadult female *Artibeus jamaicensis* on BCI and vicinity in successive half years of life. N = number of individuals surviving. Survivorship of all subadults on last line.

Period	Number subadults marked	3		4		5		6		7		8		9		10	
		N	%	N	%	N	%	N	%	N	%	N	%	N	%	N	%
Fall 1976	227			83	17	60	26	43	19	36	16	27	12	14	6	7	3
Fall 1977	358	120	34	94	26	76	21	55	15	28	8	14	4				
Spring 1978	171	54	32	47	27	34	20	23	13	12	7						
Fall 1978	289	131	45	96	33	43	15	27	9								
Spring 1979	141	54	38	32	23	20	14										
Fall 1979	257	68	26	39	15												
Spring 1980	19	4	21														
Spring 1978–80	331	112/331	34	79/312	25	54/312	17	23/171	13	12/171	7						
Fall 1976–79	1131	319/904	35	312/1131	28	179/874	20	125/874	14	64/585	11	41/585	7	14/227	6	7/227	3
All periods	1462	431/1235	35	391/1443	27	233/1186	20	148/1045	14	76/756	10	41/585	7	14/227	6	7/227	3
Sexes combined	2636	751/2236	34	599/2552	23	371/2099	18	209/1804	12	109/1356	8	53/1058	5	18/400	4	8/400	2

TABLE 5-6.—Survivorship of subadult male *Artibeus jamaicensis* on BCI and vicinity in successive half years of life. N = number of individuals surviving.

Period	Number subadults marked	Survival in successive half years of life														
		3		4		5		6		7		8		9		10	
		N	%	N	%	N	%	N	%	N	%	N	%	N	%	N	%
Fall 1976	173			49	28	38	22	18	10	15	9	9	5	4	2	1	1
Fall 1977	300	108	36	53	18	42	14	28	9	12	4	3	1				
Spring 1978	127	35	28	28	22	22	17	11	9	5	4						
Fall 1978	159	62	39	31	19	17	11	4	3								
Spring 1979	154	60	39	28	18	19	12										
Fall 1979	196	45	23	19	10												
Spring 1980	65	10	15														
Spring 1978–80	346	105/346	30	56/281	20	41/281	15	11/127	9	6/127	5						
Fall 1976–79	828	215/655	33	152/828	18	97/632	15	50/632	8	27/473	6	12/473	3	4/173	2	1/173	1
All periods	1174	320/1001	32	208/1109	19	138/913	15	61/759	8	33/600	5	12/473	3	4/173	2	1/173	1

TABLE 5-7.—Survivorship of adult female *Artibeus jamaicensis* on BCI and vicinity in successive half years after mark. N = number of individuals surviving. Survivorship of all adults on last line.

Period	Number adults marked	Survival in successive half years after mark														
		1		2		3		4		5		6		7		8	
		N	%	N	%	N	%	N	%	N	%	N	%	N	%	N	%
Fall 1976	178			87	49	65	37	40	22	33	19	23	13	14	8	7	4
Fall 1977	366	164	45	130	36	108	30	82	22	46	13	28	8				
Spring 1978	287	86	30	78	27	55	19	31	11	19	7						
Fall 1978	103	37	36	20	19	11	11	7	8								
Spring 1979	368	104	28	52	14	26	7										
Fall 1979	155	38	25	23	15												
Spring 1980	194	15	8														
Spring 1978–80	849	205/849	24	130/655	20	81/655	12	31/287	11	19/287	7						
Fall 1976–79	802	239/624	38	260/802	18	184/647	28	129/647	20	79/544	15	51/544	9	14/178	8	7/178	4
All periods	1651	444/1473	30	390/1457	27	265/1302	20	160/934	17	98/831	12	51/544	9	14/178	8	7/178	4
Sexes combined	2905	761/2576	30	610/2638	23	387/2286	17	224/1676	13	123/1466	8	60/994	6	18/329	5	7/178	6

TABLE 5-8.—Survivorship of adult male *Artibeus jamaicensis* on BCI and vicinity in successive half years of life.
N = number of individuals surviving.

Period	Number adults marked	Survival in successive half years after mark															
		1		2		3		4		5		6		7		8	
		N	%	N	%	N	%	N	%	N	%	N	%	N	%	N	%
Fall 1976	151			48	32	26	17	15	10	12	8	5	3	4	3	0	0
Fall 1977	299	123	41	61	20	43	14	27	9	9	3	4	1				
Spring 1978	185	57	31	44	24	30	16	16	9	4	2						
Fall 1978	107	42	39	26	24	13	12	6	6								
Spring 1979	242	45	19	20	8	10	4										
Fall 1979	197	39	20	21	11												
Spring 1980	73	11	15														
Spring 1978–80	500	113/500	23	64/427	15	40/427	9	16/185	9	4/185	2						
Fall 1976–79	754	204/603	34	156/754	21	82/557	9	48/557	9	21/450	5	9/450	2	4/151	3		
All periods	1254	317/1103	29	220/1181	19	122/984	12	64/742	9	25/635	4	9/450	2	4/151	3		

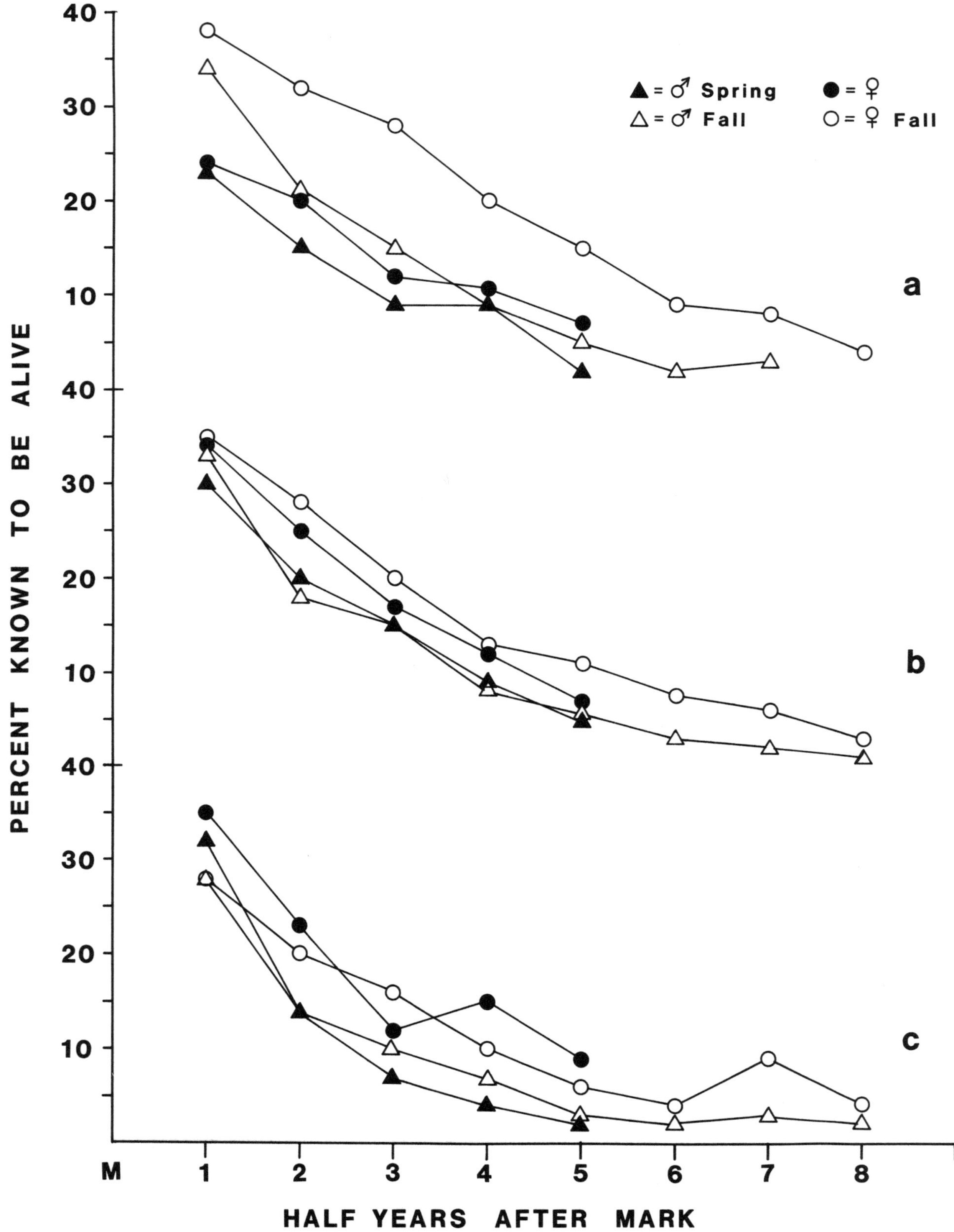

FIGURE 5-5.—Comparison of survivorship of seasonal cohorts of *Artibeus jamaicensis* marked on BCI and adjacent mainland from 1976 to 1980; a = adults, b = subadults, and c = juveniles.

basis and, although producing important information on movements, these bats did not form a major part of the main population we were studying. When the *A. jamaicensis* from these off-island sites and their recapture records are removed from the data (presented in Table 5-9), thereby restricting the analysis to bats marked on BCI and Orchid Island, survival estimates are higher and vary from a low of 17% for adult males marked in the first half 1980 to a high of 71% for juvenile females marked in the spring of 1978 (data from Table 5-10).

JUVENILES.—Overall survivorship into the second half year of life by juvenile *A. jamaicensis* was 31% (Figure 5-6, Tables 5-3 and 5-4). Survivorship of juveniles ranged from 25% for males marked in the falls of 1977 and 1979 to 57% for females marked in the spring of 1978 (Table 5-3). Survivorship from combined data from the fall of 1977 to the spring of 1980 suggests that approximately 68% of the juvenile cohort is lost by the time juveniles enter their second half year (Tables 5-3, and 5-4). We had anticipated high mortality during the first six months of life based on studies of temperate-zone bats (Brenner, 1968; Davis, 1967; Foster et al., 1978; Humphrey and Cope, 1970, 1976, 1977; Keen and Hitchcock, 1980; Mills et al., 1975; Pearson et al., 1952; Stevenson and Tuttle, 1981; Tuttle and Stevenson, 1982). Survivorship from the second to sixth half year after marking among bats marked as juveniles averages lower than that of adults and subadults (Figure 5-6).

Dispersal during the first year of life is the most likely explanation for this higher observed "mortality" or disappearance rate. Nevertheless, factors such as death, learned net avoidance, and loss of necklace (which renders the individual unrecognizable), as well as dispersal away from the area of our study, contribute to the disappearance of these bats. Survivorship by the end of the second half year of life (second half year after mark) averaged 17% (Figure 5-6, Tables 5-3 and 5-4). Thereafter, the rate of decline parallels that of subadults and adults.

Necklace loss may be greater in juveniles because we had to fit bats of this age class with adult-size necklaces. However, necklace loss cannot explain the more rapid decline in the second half year of life by which time the bats have attained near adult size. Survivorship by the end of the second half year of life (second half year after mark) averaged 17% (Figure 5-6, Tables 5-3 and 5-4). Thereafter, the rate of decline parallels that of subadults and adults.

We examined the number of bats caught in each half year versus the number known to be alive in that half year (Figure 5-7, Table 5-11) to understand the probable causes for the more rapid decline from the first to second half year of life in juveniles. The consistently higher percentage of males caught of those known to be alive in the first half year after being marked reflects an initially higher recapture rate among males

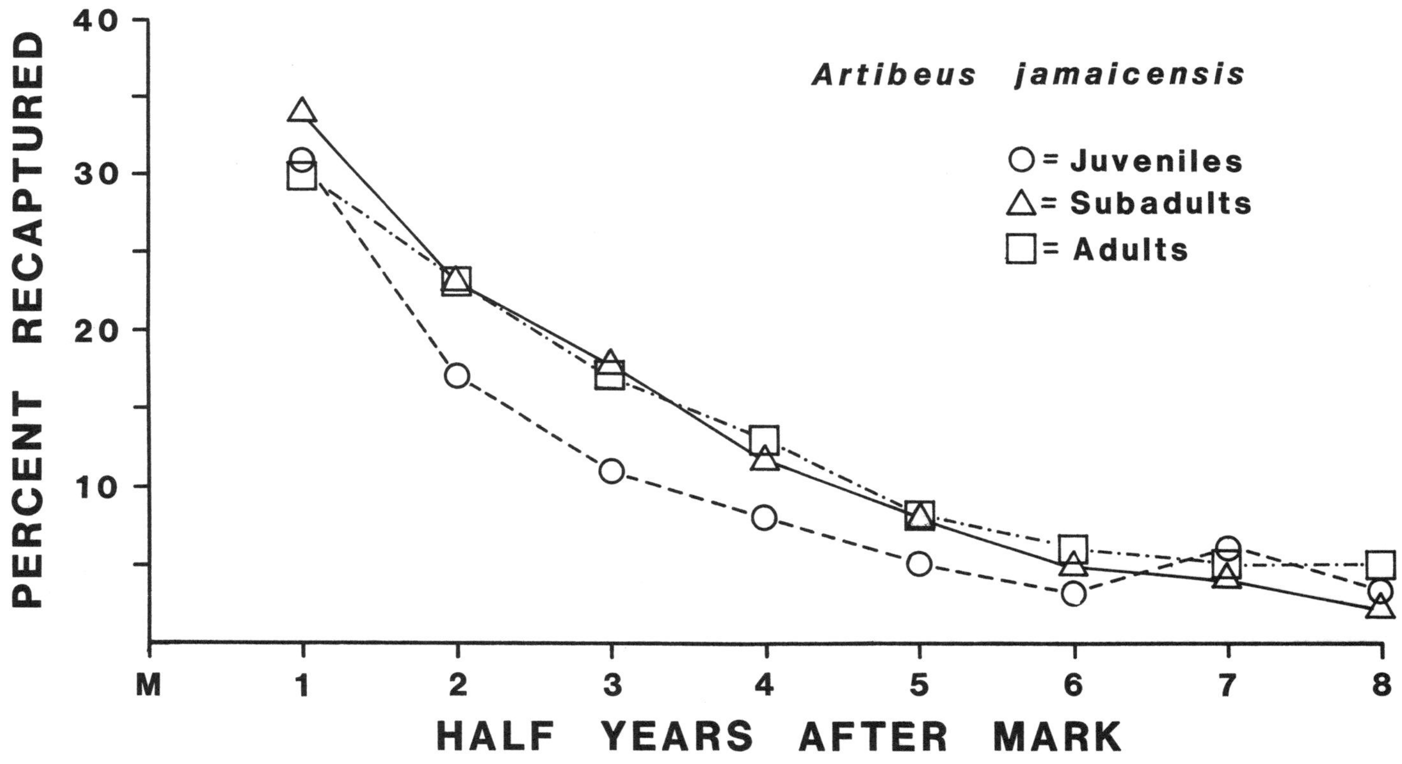

FIGURE 5-6.—Survivorship curves of the three age classes of *Artibeus jamaicensis* marked on BCI and adjacent mainland from 1976 to 1980.

TABLE 5-9.—Summary by sex and age class of the *Artibeus jamaicensis* marked on BCI and adjacent mainland in each half year and the number subsequently recaptured (in parentheses); number of adults of each sex marked in previous half years recaptured during successive half years.

Sex/Age	Fall 1976	Fall 1977	Spring 1978	Fall 1978	Spring 1979	Fall 1979	Spring 1980	Fall 1980
FEMALES								
Juveniles	74 (28)	236 (58)	54 (31)	83 (32)	272 (108)	331 (91)	368 (107)	500
Subadults	227 (83)	358 (120)	171 (54)	289 (131)	141 (54)	257 (68)	19 (4)	286
Adults	178 (87)	366 (164)	287 (86)	103 (37)	368 (104)	155 (38)	194 (15)	135
Recaptures of bats marked in previous half years	0	91	249	169	326	471	315	489
MALES								
Juveniles	88 (18)	261 (79)	47 (17)	97 (32)	272 (102)	410 (103)	341 (91)	496
Subadults	173 (49)	300 (108)	127 (35)	159 (62)	154 (60)	196 (45)	65 (10)	198
Adults	151 (48)	299 (123)	185 (57)	107 (42)	242 (45)	197 (39)	73 (11)	101
Recaptures of bats marked in previous half years	0	79	290	124	242	334	256	291

TABLE 5-10.—Summary by sex and age class of the number of *Artibeus jamaicensis* marked on BCI (sample restricted to those marked on BCI and Orchid Island) in each half year and (in parentheses) the number subsequently recaptured.

Sex/Age	Fall 1976	Fall 1977	Spring 1978	Fall 1978	Spring 1979	Fall 1979	Spring 1980
FEMALES							
Juveniles	74 (28)	229 (57)	42 (30)	75 (32)	257 (105)	323 (90)	356 (107)
Subadults	227 (83)	354 (118)	142 (46)	258 (126)	114 (47)	247 (63)	14 (3)
Adults	178 (87)	326 (138)	227 (67)	65 (26)	222 (90)	149 (36)	84 (13)
MALES							
Juveniles	88 (18)	261 (79)	41 (16)	95 (32)	261 (102)	405 (103)	329 (90)
Subadults	173 (49)	297 (107)	110 (33)	144 (56)	126 (60)	193 (45)	48 (9)
Adults	151 (48)	297 (122)	162 (54)	100 (40)	153 (38)	196 (39)	42 (7)
SEXES COMBINED							
Juveniles	162 (46)	490 (136)	83 (46)	170 (64)	518 (207)	728 (193)	685 (197)
Subadults	400 (132)	651 (225)	252 (79)	402 (182)	240 (107)	440 (108)	62 (12)
Adults	329 (135)	623 (286)	389 (121)	165 (66)	375 (128)	345 (77)	126 (20)

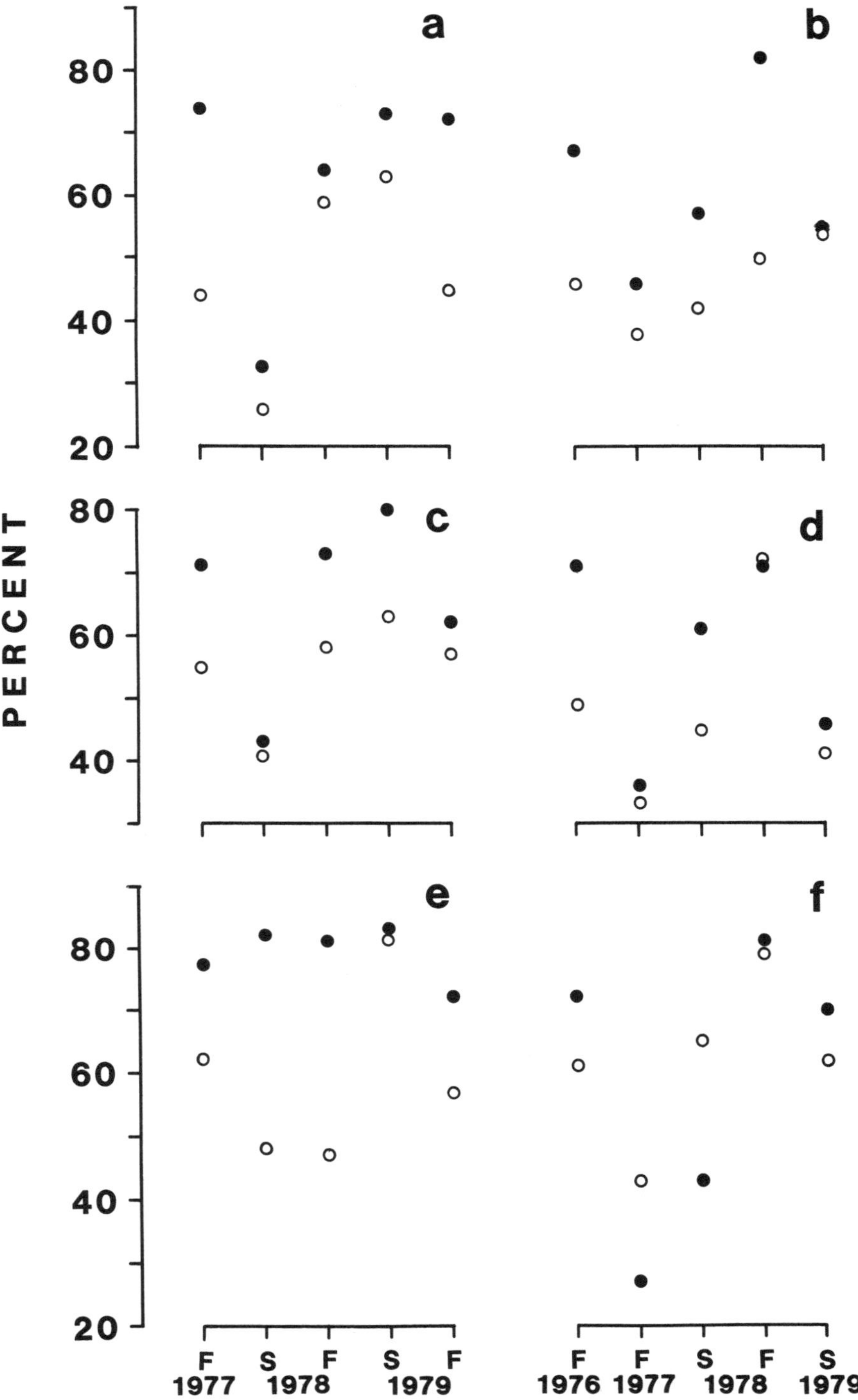

FIGURE 5-7.—Comparison of numbers of *Artibeus jamaicensis* marked on BCI and adjacent mainland and recaptured in the first and second half years after mark. Values plotted are recaptures expresssed as percentages of the numbers known to be alive in the same half year; open circles = females, closed circles = males. Adults, (a) first half year and (b) second half year of life; subadults, (c) first half year and (d) second half year of life; and juveniles, (e) first half year and (f) second half year of life.

(more recaptures within a shorter time). The same pattern among subadults (Table 5-12) and adults (Table 5-13) for the half year following their being marked suggests behavioral differences between the sexes that in turn increases susceptibility to recapture among males. The pattern is similar among the three age classes for the second half year after mark, except that the relationship is reversed for juveniles in the fall of 1977 and the spring of 1978, and for subadults in the fall of 1978. Thereafter, there is no pattern emphasizing greater catchability of males.

Juveniles of either sex marked in their first half year of life are known to be alive in the second half year in about equal numbers (Tables 5-3 and 5-4; 30% for males versus 32% for females). Nevertheless, by the third and fourth half years, the survivorship rate is higher in females (21% versus 14%, X^2 = 13.383, $P < 0.01$, for third half year; 14% versus 9%, X^2 = 9.159, $P < 0.02$), suggesting either greater mortality in males, differential emigration, or both, along with possible learned net avoidance resulting from higher actual recapture rate during the first year after mark.

Chi-square analysis of bats marked as juveniles showed some significant differences in survivorship correlated with sex and season of birth. In general, females have higher survival rates than males. Spring newborn of both sexes survive better into the second half year of life than do fall newborn with the difference between season of birth for females significant at the 0.05 level (35% versus 28%, respectively; X^2 = 6.103). In subsequent half years, the survival rate is similar regardless of birth season. Spring-born young may have the advantages of long gestation (seven months versus four), birth in the dry season, weaning at the beginning of the wet season, and attainment of subadult age (and near adult size) by the time of fig scarcity at the height of the rains. Fall-born young, the product of a short (4-month) gestation period, are born in the wet season, weaned at the height of the rains when ripe figs are beginning to be scarce, and become subadults in the dry season.

Preweaning events such as length of gestation and coincidence of birth with wet or dry seasons, may have little effect on survival of baby bats. Postweaning conditions, however, when young bats are on their own for the first time, may be critical. Fall-born bats are weaned in the season of heaviest rains when there is a greater probability of getting wet while foraging. The likelihood of getting wet is increased by the scarcity of figs because there is more reason to fly in the rain if quality food is harder to find. Wetting and consequent chilling could be fatal for weanling *A. jamaicensis* because of their restrictive, relatively inflexible energy budget (see Section 2, Physiology).

Survivorship to the second half year (first half year after mark) is not significantly different for males and females marked as juveniles. However, females appear to survive at a higher rate by the third and fourth year ($P < 0.01$ and $P < 0.02$, respectively; based on data in Tables 5-3 and 5-4). Survival in the third half year (second half year after mark) favored fall-born females over fall-born males ($P < 0.05$). Reasons for

these differential survival rates are unclear. The higher initial recapture rate characteristic of males and their higher subsequent attenuation rate in the population are both reflected in these rates. Nevertheless, we can not be sure how much of the apparent attenuation rate actually reflects greater differential emigration by males (higher probability of continued residency by females) instead of mortality.

SUBADULTS.—Speculation on survival and recapture rates of bats marked as subadults must include the assumption that these bats are already in the second half year when first captured. Bats marked as subadults in the fall were born the previous spring, and those marked in the spring were born the previous fall. Therefore, the first half year after being marked is the third half year after birth. Any comparisons of these records with those of juveniles are on this basis.

There is some unavoidable "slop" in aging subadults. Subadults caught late in the fall include all individuals born in the previous spring and some born early in the fall. In the first weeks of spring the pool of subadults includes all young of the previous year, and should be at its largest. Midway through spring the pool should have diminished to its lowest level because most have become adults (reproductive) regardless of age. Later in the spring, juveniles of that spring begin to cross the threshold to subadulthood. However, numbers of subadults marked in the fall greatly exceed the number marked in the spring (Tables 5-5, 5-6, and 5-9). This is simply because in a year-round capture program, most of the subadults are captured and marked in the fall, leaving fewer unmarked bats to be captured in the spring. Bats are aged as juveniles only on the basis of open epiphyses of the metacarpals and phalanges of the wings. Transition from juvenile to subadult status (ossification of the epiphyses in the wing) is relatively rapid (see Section 3, Reproduction in a Captive Colony).

Assuming that subadults already are in their second half year when first captured, survival rate is slightly higher than in juveniles (Figure 5-6). Nevertheless, the proportional loss from the population of the cohort marked as juveniles through the second half year after marking must have already been absorbed by the cohort marked as subadults when these bats complete the first half year after having been marked. Reasons for that loss are the same as those outlined for losses among bats marked as juveniles, except that necklace loss may not be as important a factor.

Observed overall survivorship through the first half year after mark (34%) for subadults was not significantly different from that for juveniles (32%, X^2 = 3.225), whereas recapture rate for the second half year after mark was significantly lower (17% versus 23%, X^2 = 22.300, $P < 0.001$) for juveniles (Tables 5-3 and 5-5). The highest survival rate for bats marked as subadults was 43% in the fall of 1978 (males, 45%; females, 39%; Table 5-9). As we suggested for our survivorship data on juveniles, the survival rate for subadults is undoubtedly higher than the 43% we recorded during the seven half-year periods in which we accumulated survivorship data. If we restrict our data

TABLE 5-11.—Comparison of recaptures with numbers known to be alive in each half year after mark of *Artibeus jamaicensis* marked on BCI and adjacent mainland as juveniles. N = number of individuals surviving.

Period	Survival in successive half years after mark															
	1		2		3		4		5		6		7		8	
	N	%	N	%	N	%	N	%	N	%	N	%	N	%	N	
FEMALES																
Fall 1976			17/28	61	13/21	62	2/11	18	3/11	27	4/10	40	4/7	57	3	
Fall 1977	36/57	62	17/40	43	17/29	59	13/19	68	6/7	86	3					
Spring 1978	15/31	48	15/23	65	8/13	62	4/8	50	5							
Fall 1978	15/32	47	19/24	79	5/11	45	8									
Spring 1979	87/108	81	35/52	62	27											
Fall 1979	52/90	57	50													
Spring 1980	107															
MALES																
Fall 1976			13/18	72	5/9	56	1/5	20	1/5	20	1/4	25	1/3	33	2	
Fall 1977	61/79	77	9/33	27	15/27	56	17/22	77	4/7	57	3					
Spring 1978	14/17	82	3/7	43	3/5	60	2		1							
Fall 1978	26/32	81	17/21	81	8/9	89	4									
Spring 1979	85/102	83	28/40	70	17											
Fall 1979	74/103	72	48													
Spring 1980	91															

TABLE 5-12.—Comparison of recaptures with numbers known to be alive in each half year after mark of *Artibeus jamaicensis* marked on BCI and adjacent mainland as subadults. N = number of individuals surviving.

Period	Survival in successive half years after mark															
	1		2		3		4		5		6		7		8	
	N	%	N	%	N	%	N	%	N	%	N	%	N	%	N	
FEMALES																
Fall 1976			41/83	49	21/60	35	14/43	33	19/36	53	17/27	63	8/14	57	7	
Fall 1977	66/120	55	31/94	33	40/76	53	32/35	58	16/28	57	14					
Spring 1978	22/54	41	21/47	45	18/34	53	11/23	48	12							
Fall 1978	76/131	58	69/96	72	22/43	51	27									
Spring 1979	34/54	63	13/32	41	20											
Fall 1979	39/68	57	39													
Spring 1980	4															
MALES																
Fall 1976			35/49	71	27/38	71	7/18	39	9/15	60	6/9	67	3/4	75	1	
Fall 1977	77/108	71	19/53	36	23/42	55	20/28	71	9/12	75	3					
Spring 1978	15/35	43	17/28	61	14/22	44	6/11	55	6							
Fall 1978	45/62	73	22/31	71	14/17	82	4									
Spring 1979	48/60	80	13/28	46	19											
Fall 1979	28/45	62	19													
Spring 1980	10															

TABLE 5-13.—Comparison of recaptures with numbers known to be alive in each half year after mark of *Artibeus jamaicensis* marked on BCI and adjacent mainland as adults. N = number of individuals surviving.

Period	Survival in successive half years after mark															
	1		2		3		4		5		6		7		8
	N	%	N	%	N	%	N	%	N	%	N	%	N	%	N
FEMALES															
Fall 1976			40/87	46	35/65	54	11/40	28	15/33	45	12/23	51	9/14	64	7
Fall 1977	72/164	44	49/130	38	48/108	53	47/82	57	23/46	50	28				
Spring 1978	22/86	26	33/78	42	34/55	62	14/31	45	19						
Fall 1978	22/37	59	10/20	50	5/11	45	7								
Spring 1979	65/104	63	28/52	54	26										
Fall 1979	17/38	45	23												
Spring 1980	15														
MALES															
Fall 1976			32/48	67	21/26	81	6/15	40	8/12	67	3/5	60	4		
Fall 1977	91/123	74	28/61	46	28/43	65	20/27	74	5/9	56	3/4	75	1		
Spring 1978	19/57	33	25/44	57	16/30	53	13/16	81	4						
Fall 1978	27/42	64	21/26	81	9/13	69	6								
Spring 1979	33/45	73	11/20	55	10										
Fall 1979	74/103	72	21												
Spring 1980	11														

to bats marked on BCI (including Orchid Island), highest survival rates for *A. jamaicensis* marked as subadults are for bats marked in the spring of 1979 (males, 48%; females, 41%; Table 5-10). Conservative cumulative survivorship values also reflect the progressively shorter time we had from which to record a survivor as this phase of the project neared its end. The proportion of implied "mortality" resulting from emigration is unknown, but may be offset by immigration from other populations.

ADULTS.—Survival among adults parallelled that of subadults (Figure 5-6, Tables 5-7 and 5-8). Observed overall survivorship through the first half year after mark (30%) for adults was not significantly different from that for juveniles (32%) or subadults (34%, $X^2 = 3.225$), whereas survival for the second half year after mark was the same as that for subadults, but significantly higher (23% versus 17%, $X^2 = 22.300$, $P < 0.001$) than that for juveniles (Figure 5-6; Tables 5-3, 5-5, and 5-8). The highest survival rate for bats marked as adults was 43% in the fall of 1977 (males, 41%; females, 45%; Tables 5-7, 5-8, and 5-9). As suggested for our survivorship data on juveniles and subadults, the survival rate for adults is undoubtedly higher than the 43% we recorded. If we restrict our data to bats marked on BCI (including Orchid Island), highest survival rates for *A. jamaicensis* marked as adults are for bats marked in the fall of 1976 (males, 39%; females, 49%; Table 5-10). However, survival estimates based only on bats marked in the fall of 1976 are not comparable to estimates from other half years because the numbers of bats known to be alive accrued from the second half year after mark (fall 1977) instead of from the first half year as in all other periods.

Adults as an age group are much more heterogeneous in actual age than are either subadults or juveniles. Nevertheless, the attenuation rate of adults closely traced the survival curve for subadults (Figure 5-6) and indicates that the factors contributing to losses act similarly in kind and degree on *A. jamaicensis* marked as either subadults or adults. The only factor distinguishing the survival rates of adults from those of the other age cohorts is the higher rate of loss of males from the marked sample.

GENERAL OBSERVATIONS.—Thus far in this section we have concentrated on examining our mark and recapture record to extract information of possible demographic interest, except for estimating the size of the bat population on BCI (see Section 6, Population Estimates). Our survival estimates are based on the numbers known to be alive as determined from the entire capture-recapture record. To better estimate actual survival would require a longer mark and recapture record, and would require that estimates for necklace loss, net avoidance, and emigration be factored into the results.

Although we have used the term survival when discussing recaptures and relative abundance, we acknowledge our sometimes loose interpretation of the word. There are other terms, such as "residency," that we could have used. However, even that seemingly appropriate term has its drawbacks, because we have been unable to adequately measure site fidelity and can not distinguish between and among the possible resident and visitor (or vagrant) categories of bats. Also, we do not know how far an *A. jamaicensis,* belonging to any of these residency categories, will travel to visit a fruiting tree (see Section 7, Movements).

Our information on necklace loss is based on 5 of 73 (6.8%) *A. jamaicensis* that were double banded (necklace and wingband) and subsequently recaptured without a necklace. These bats had been wing-banded before 1975 and, therefore, were all adults when necklaced. Some loss occurs when the bats are being removed from the net and the necklace slips over the head after becoming caught in the mesh. We do not know what other circumstances contribute to necklace loss. Although lacking supportive evidence, we suspect that loss is higher for bats marked as juveniles because we must apply necklaces that fit adults. The looser necklaces could be more easily slipped over the head of juveniles if this is the usual means for necklace loss. Fleming (1988) reported a low rate of necklace loss (6.5%) in *Carollia perspicillata* and said that losses sometimes occurred several years after marking.

Frequency of capture seems to have less to do with age than with sex. The two *A. jamaicensis* captured most often were males. Out of 8907 individuals marked between October 1976 and June 1980, only one was captured 11 times, and another was captured ten times. The male with ten captures was marked as a juvenile and recaptured three times in the next four months as a subadult, and six times in the next 14 months as an adult. Its age when last caught was just under two years. The bat with 11 captures was marked as a subadult, recaptured four months later as a subadult, and then nine times in the next 14 months as an adult. At last capture it also was about two years old. Of the four captured nine times each (Figure 5-4, Table 5-2), one was a female and three were males. Although approximately equal numbers of males and females were caught five or more times, males had consistantly shorter overall capture records. Only 13 males with five or more captures had records of 30 or more months between mark and last recapture, and only one of these exceeded 40 months. However, 35 females had full capture records of 30 or more months and ten had records exceeding 40 months.

We believe that most bats are easier to capture the first time than subsequently, regardless of age. The frequency-of-capture curve (Figure 5-4) is steep at first: 57% caught once, 25% caught only twice, and thereafter levels out. The remaining 18% (1683 bats) were recaptured two or more times (2975 records; recapture rate averaged 1.77). The survival curves for each age cohort (Figure 5-6) is similarly steep at first and thereafter levels out.

Sex Ratio

Our sample of almost 17,000 records of *A. jamaicensis* shows that both adult and subadult females outnumber adult and subadult males 55 : 45, while juvenile females are outnumbered by juvenile males 48 : 52 (Table 5-14). The recapture histories of bats marked as juveniles also support the contention that females have higher survival (or residency) potential than males (Tables 5-3 and 5-4). The differences, however, are not significantly different from an even sex ratio ($X^2 = 5.25$; $P = 0.02$).

To determine if there was seasonal variation in sex ratios we organized our data (Table 5-14) into half years with January–June representing the dry season (three half years) and July–December representing the wet season (five half years). In the adult sample ($n = 7594$), females were in the majority in every half year. In the subadult sample ($n = 4856$) females predominated in six of the eight half years. Males predominated in two of the three dry seasons in the sample. Among juveniles ($n = 4516$) more males than females were caught during the wet season. Although not statistically significant ($X^2 = 1.13$; $P = 0.29$), ratios favored juvenile females during two of the dry seasons and were equal in the third.

We obtained a different subset of sex-ratio data from roostlings in the two canal-marker roosts we sampled at Buena Vista and Peña Blanca. Young large enough to mark, but probably still nursing, totaled 45. Females outnumbered males 26 : 19 (58% : 42%), not far off of the 55 : 45 ratio seen in mist-netted adults and subadults, but the near converse of the 48 : 52 ratio found among mist-netted juveniles. A possible explanation is that juvenile males may leave or be ejected from the maternity roost earlier than female offspring. Judging from the observation that we seldom recaptured bats in the roost where they were marked as juveniles, we believe that all young of both sexes must disperse before they reach adulthood. Subadult sex ratio in the roosts was 27 : 1. None of the 27 subadult females had a previous history in either canal-marker roost.

Age

If most of the adult females produce two young per year, and if adults and young are equally catchable, then the number of nonadults in our mist-netted sample in any year should approach twice the number of adult females. Only in 1979 did we have a high level of effort distributed fairly evenly through the year (Table 5-15). In 1979, all captures of adult *A. jamaicensis* totaled 2609 (Table 5-16). This translates to 1435 females (2609 × 0.55) and the prospect of 2870 young (1435 × 2). Our actual catch of juvenile and subadult *A. jamaicensis* for the year was 2891. In spite of poor distribution of catch effort during some years, the proportion of young to adult females in three of the five years was reasonably close to 2 : 1 (3 : 1 in the other two years).

To determine actual distribution of age in the population of *A. jamaicensis* on BCI and vicinity (Figure 5-8) we looked at a small subset ($n = 225$) that had been marked as juveniles or subadults and recaptured at least four times. Among these known-age bats, the mean age at last capture was 2.02 yrs in the females ($n = 110$) and 1.65 yrs in the males ($n = 115$). The average for the whole sample ($n = 225$) was 1.83 yrs. Of the 115 males in the sample, 91 were two years old or less at last capture and only 24 were more than two years old. Fifty-nine of the females were two years old or less at last capture and 51 were more than two years old. The oldest male was four and the oldest female was 6.5 years old.

Mortality

With potentially great longevity, and with most female *A. jamaicensis* producing two young each year, the species must be under severe population controls. Death rate must be high, but its causes are speculative. Mortality factors include starvation, predation, accidents, and disease, but we have little direct evidence and we have seen few dead bats.

INFANT MORTALITY.—Because of our study methods, we have no direct information on preflight mortality in wild bats. Most deaths in our NZP colony occurred in the preflight interval. On BCI we sometimes captured numbers of postlactating females along with lactating bats when few or no juveniles were caught. This could be explained in part as a function of natal and postnatal mortality.

Our capture of only ten juveniles among 152 *A. jamaicensis* on 11 and 12 April 1979 is a typical example. The condition of the adult females was 38 lactating, 22 postlactating, five pregnant, and two lacking evidence of reproductive activity. We suspect that the postlactating females were primiparous and that their lack of success in rearing young (hence the postlactating condition) paralleled the situation found among first-time mothers in our NZP colony, but this is not supported by the record. For those postlactating females whose capture record is sufficient, the majority had been neither juveniles nor subadults during the previous half year. We do not know if the inevitable loss of young by primiparous mothers recorded in the NZP colony was an artifact of their captive environment or also occurs in natural populations.

STARVATION.—Young of the spring are weaned in a season of abundant food (April–June) at the beginning of the rainy season. Young of the second birth group (summer), on the other hand, are weaned in a season of food scarcity (August–October) in the middle of the rainy season when they could easily starve before becoming efficient foragers.

According to Eisenberg and Wilson (1978), it may be more difficult for frugivorous bats to find enough food than it is for insect eaters. This would correlate with a relatively large brain in fruit eaters. Frugivores rely on a food source that is strongly pulsed in time and space. Their behavior and the nature of their diet adds risk because fruit-eating bats often congregate in large numbers at food sources where predators may be attracted and wait for prey (Howe, 1979; Morrison, 1978a). These bats also have to search out food sources for the future while still harvesting a current source (Morrison, 1978b). In contrast,

TABLE 5-14.—Sex ratio in *Artibeus jamaicensis* captured on BCI between October 1976 and October 1980.

Sex	Jul-Dec* 1976	Jul-Dec* 1977	Jan-Jun† 1978	Jul-Dec* 1978	Jan-Jun† 1979	Jul-Dec* 1979	Jan-Jun† 1980	Jul-Dec* 1980	Total	Percent
Adult									7594	
male	173	534	522	256	514	658	318	430	3405	44.8
female	213	587	579	296	771	666	533	544	4189	55.2
Subadult									4856	
male	201	421	232	200	232	409	132	364	2191	45.1
female	251	495	241	354	197	535	49	543	2665	54.9
Juvenile									4516	
male	95	285	48	101	308	506	367	625	2335	51.7
female	76	247	56	85	307	397	401	612	2181	48.3
Total	1009	2569	1678	1292	2329	3171	1800	3118	16,966	

* Wet season

† Dry season

TABLE 5-15.—Tabulation of captures of bats (all species) on BCI by year and month, 1975–1980, a total of 28,874 captures. Capture effort is measured as monthly and yearly percentages of the total catch for the six years.

Year	JAN	FEB	MAR	APR	MAY	JUN	JUL	AUG	SEP	OCT	NOV	DEC	Total	Percent per year
1975														
Total		166	385										551	
Percent		0.575	1.333											2
1976														
Total		46	669							777	1046		2538	
Percent		0.159	2.317							2.691	3.623			9
1977														
Total							656	409	352	942	255	1333	3947	
Percent							2.272	1.417	1.219	3.262	0.883	4.617		14
1978														
Total	769	755	878	462	184	274	257	143		115	995	778	5610	
Percent	2.663	2.615	3.041	1.600	0.637	0.949	0.890	0.495		0.398	3.446	2.694		19
1979														
Total	595	36	803	1085	778	438	31	1	1411	1594	1998	426	9196	
Percent	2.061	0.125	2.781	3.758	2.694	1.517	0.107	0.003	4.887	5.521	6.920	1.475		32
1980														
Total	587	352	204	660	1170			1234	1388	1437			7032	
Percent	2.033	1.219	0.707	2.286	4.052			4.274	4.807	4.977				24
1975–1980														
Total	1951	1355	2939	2207	2132	712	944	1787	3151	4865	4294	2537	28874	
Percent per month	6.757	4.693	10.179	7.644	7.383	2.466	3.269	6.189	10.913	16.849	14.872	8.786		100

TABLE 5-16.—Profile of ages of *Artibeus jamaicensis* captured on BCI between October 1976 and October 1980.

Age	Number marked	Percent of total marked	Number recaptured	Percent of total recaptures	Total captures	Percent of total captures
1976						
adult	329	37	57	48	386	38
subadult	400	45	52	44	452	45
juvenile	162	18	9	8	171	17
Total	891		118		1009	
1977						
adult	665	37	456	61	1121	44
subadult	658	36	258	34	916	36
juvenile	497	27	35	5	532	21
Total	1820		749		2569	
1978						
adult	682	40	971	77	1653	56
subadult	746	44	281	22	1027	34
juvenile	281	16	9	1	290	10
Total	1709		1261		2970	
1979						
adult	962	32	1647	66	2609	47
subadult	748	25	625	25	1373	25
juvenile	1285	43	233	9	1518	28
Total	2995		2505		5500	
1980						
adult	503	18	1322	62	1825	37
subadult	568	21	520	24	1088	22
juvenile	1705	61	300	14	2005	41
Total	2776		2142		4918	

most insectivorous bats harvest a fairly static resource while flying a stereotyped, generally solitary search pattern.

Young *A. jamaicensis* have much to learn when trying to find enough to eat. For young females this may be less traumatic, for they probably stay for a longer time in the natal harem. They may learn while foraging in the company of their mothers if they succeed in leaving the day roost with them (see Section 9, Foraging Behavior).

Young males, in contrast, may be ejected early from the harem, perhaps even before weaning has been completed (see Section 8, Roosting Behavior). Unless these young males are able to join a bachelor group quickly and learn to forage with it (we do not know if males forage in groups) they may be at greater risk when they leave their natal harem. Alternatively, juvenile males might remain near the harem roost and forage with its members even though not roosting with them.

Only occasionally did we capture frail-looking young *A. jamaicensis,* but this is not surprising. With a high metabolic rate and a low protein, high carbohydrate diet one would expect starvation to be abrupt, perhaps occurring in a single night. In our bat handling we have observed that the smaller the frugivore (in terms of age, size, and species) the less tolerant it is of stress. The small-size species of *Artibeus* and *Vampyressa*

may become lethargic and unable to fly, lose control of body temperature, chill, and die in as little as an hour under stress and food deprivation. Warming a stressed bat only delays death, but feeding it a high energy meal, such as a sugar solution, quickly revives it.

Rain must increase the risk of starvation in young *A. jamaicensis* weaned in the rainy season. A series of rainy nights (perhaps even one) with little or no opportunity for foraging or reconnoitering for the next night's food source might be lethal to inexperienced young bats. Wetting greatly increases the hazard of chilling and loss of temperature control. Stenodermatines commonly avoid flying in rain, but might be driven to do so by hunger. Whereas experienced adults might be able to forage quickly and effectively during brief intervals when the rain slackens or stops, the inexperienced bat might not be quite as efficient. Nightlong rains, particularly a series of them, could be devastating to whole populations of frugivorous bats, particularly the young, in a season of fruit scarcity. In summary, at least three factors might contribute to death by starvation: nocturnal rains, scarcity of suitable food, and early ejection of juvenile males from the maternity roost.

PREDATION.—Arboreal snakes are probably the most serious predators of *A. jamaicensis*. Roosts in holes in trunks of large

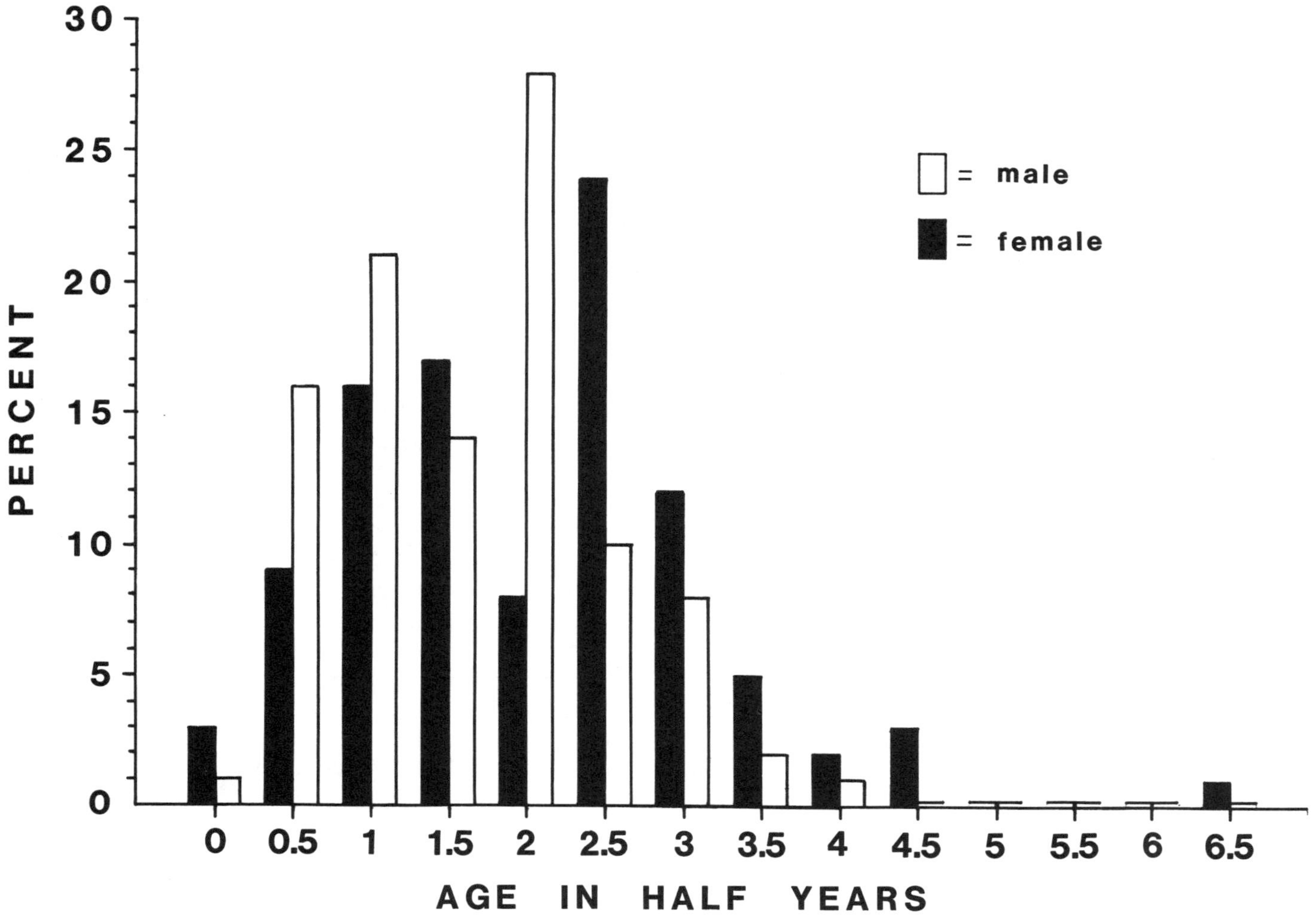

FIGURE 5-8.—Age at last capture of 110 female and 115 male *Artibeus jamaicensis* marked on BCI and adjacent mainland as juveniles or subadults and recaptured at least four times. Black bars represent females and white bars represent males.

trees offer complete protection from snakes unless, as Morrison observed (see Section 8, Roosting Behavior), vines or saplings provide access to roosts. Thomas (1974) reported on *Boa constrictor* preying on *A. jamaicensis* in a cave on Isla Providencia. Handley saw a large snake (probably *Spilotes*) at dawn on the roof of Boys' House on BCI, above the entrance of a *Molossus* and *Myotis* roost, with about 18 inches of its head and body extended beyond the eaves. This was not quite enough extension to reach the roost entrance under the eaves, but was probably adequate to catch bats coming into the roost, especially when the bats made several sorties before actually entering. Foliage roosts, on the other hand, offer much less protection from snakes, and most foliage roosters are probably vulnerable to snake predation. We have seen snakes large enough to take *A. jamaicensis* ascend palm fronds, lie almost invisible along the top side of the stem, and move without causing visible motion to the frond.

Large opossums (*Didelphis marsupialis* and *Philander opossum*) are efficient bat killers when given the opportunity. In spite of their formidable teeth, bats in mist nets seem defenseless against these predators. An opossum crushes the bat's head with one bite and then consumes the whole bat or all but the wings. However efficient the opossum may be, it probably doesn't often have access to bats. Coatimundis (*Nasua narica*) might be another predator of bats. We often have seen coatis in banana plants and in the outermost branches of fig and breadfruit trees on BCI.

The large carnivorous *Vampyrum spectrum* eats bats as well as birds and rodents and might capture *A. jamaicensis* at feeding roosts. A captive individual consumed an *A. jamaicensis* that Handley offered to it. However, *V. spectrum* seems too rare on BCI to be much of a mortality factor.

Owls are probably important predators of bats, and bats milling around a fruit tree might be particularly vulnerable to attack. We have attracted the Spectacled Owl (*Pulsatrix perspicillata*) to nets by imitating bat squeaks, and have caught these owls in nets set for bats. Handley watched a Mottled Owl (*Ciccaba virgata*) ambushing *Myotis nigricans* and *Molossus coibensis* at their roost at dawn when the bats were milling about before entering. The owl stationed itself in a tree off to the side and higher than the roost and made frequent sorties down through the group of milling bats, usually successfully. We caught Mottled Owls fairly frequently in nets and occasionally have caught the Vermiculated Screech-Owl (*Otus guatemalae*). At least seven species of owls are known to occur on BCI, and some of them are rather abundant.

The Bat Falcon (*Falco rufigularis*), mainly a crepuscular feeder and known to be an effective predator of bats (Ridgely, 1976), is rare or absent on BCI but nests nearby. The timing of the departure from and return to day roosts by *A. jamaicensis* makes it susceptible to predation by the Bat Falcon. Diurnal hawks often take bats, especially when they have been disturbed in their roost during the day. Several African hawks regularly intercept bats along their flight paths (Black et al., 1979; Fenton, Cumming, et al., 1977).

Habits characteristic of *A. jamaicensis* such as lunar phobia and picking fruit from a tree and carrying it a hundred or more meters away to be eaten at a feeding roost, are presumably defensive strategies (Howe, 1979). The frequent use of the fronds of the spiny-trunked black palm (*Astrocaryum standleyanum*) and the arching fronds of the palm *Oenocarpus panamanus* as dining roosts by *A. jamaicensis* may be a defensive behavior to protect itself from predators (see Section 8, Roosting Behavior). Response to distress calls of bats that August (1979) called "mobbing" may also be a defensive mechanism. At any rate, there are enough seemingly defensive behavioral traits in *A. jamaicensis* to suggest that it has considerable exposure to predators.

ACCIDENTS.—Many netted *A. jamaicensis* have damaged wings. Nevertheless, they appear to survive accidents that tear wing membranes or break fingers. These parts heal remarkably rapidly and flight is not notably impaired, as we have seen in our NZP colony. Common sources of injury are fighting among individuals and encounters with sharp plant spines or thorns. Bite wounds are most frequent in males, particulary young males trying to get access to females (observed in the NZP colony). Judging by netting results, bats may avoid areas in the forest where spines and thorns are particularly prevalent; although many, if not most, of the tears in flight membranes that we have observed must originate from this source. Broken bones may result from struggles to get free when wings get tangled. A broken forearm or humerus, rendering the bat flightless, is invariably fatal.

DISEASE.—*Artibeus jamaicensis* is known to have been infected with rabies and trypanosomiasis on the Pacific side of the Canal de Panamá area, but disease in bats is unknown on BCI.

Summary

On BCI *A. jamaicensis* made up about 66% of the total nightly catch of bats year-round. Numbers tended to be highest when young were first volant and lowest when the population was composed mostly of subadults and adults.

The entire sample of *A. jamaicensis* marked between October 1976 and June 1980, consisted of 8907 bats captured a total of 15,728 times. Fifty-seven per cent (5061 individuals) were caught only once, leaving 3846 captured two or more times. Subsequent captures per bat diminished until we had a single individual captured 11 times. The probability of recapturing a bat was roughly the same regardless of the number of previous captures.

Overall survivorship into the second half year by juvenile *A. jamaicensis* was 31%. Survivorship from the second to sixth half year after marking among bats marked as juveniles was lower than that of adults and subadults. Dispersal during the first year is the most likely explanation for this high rate of disappearance. More frequent recapture of males in the half year following marking suggests behavioral differences between the sexes that increase susceptibility to earlier and more frequent recapture among males.

Juveniles of either sex marked in the first half year are known to be alive in the second half year in about equal numbers (30% for males versus 32% for females). But by the third and fourth half years, the survivorship rate is higher in females, suggesting that males are more likely to emigrate, have greater mortality, and possibly learn to avoid nets because of their higher actual recapture rate during the first year after mark. Overall, females have higher survival (residency) rates than males.

Frequency of capture seems to have less to do with age than with sex. Apparently most bats are easier to capture the first time, regardless of age: 57% caught once, 25% caught only twice, 18% caught more than twice, and thereafter the capture curve levels out.

In our sample of almost 17,000 records of *A. jamaicensis* (October 1976 to October 1980, adult and subadult females outnumbered adult and subadult males 55 : 45, while juvenile females are outnumbered by juvenile males 48 : 52. The recapture histories of bats marked as juveniles also support the contention that females have higher survival (or residency) potential than males.

Among known-age bats, the mean age at last capture was 2.02 years in the females ($n = 110$) and 1.65 yrs in the males ($n = 115$). The average for the entire sample ($n = 225$) was 1.83 years. The oldest male was four and the oldest female was 6.5 years old. Subsequent to 1980 Handley caught two *A. jamaicensis* that were nine years old.

With potentially great longevity, and with most female *A. jamaicensis* producing two young each year, the species must be under severe population controls. Death rate must be high, but its causes are speculative. Mortality factors include starvation, predation, accidents, and disease, but we have little direct evidence and we have seen few dead bats.

Young of the "spring" birth group are weaned in a season of abundant food (April–June) at the beginning of the rainy season. Young of the "summer" group, on the other hand, are weaned in a season of food scarcity (August–October) in the middle of the rainy season when they could easily starve before becoming efficient foragers.

Young *A. jamaicensis* face a number of risks. With a high metabolic rate and a low protein, high carbohydrate diet one would expect starvation to be abrupt, perhaps in a single night if food sources are not adequate. Young *A. jamaicensis* and all ages of the smaller-size species of *Artibeus* and *Vampyressa* become unable to fly, lose control of body temperature, chill, and die in as little as an hour under stress and food deprivation. Rain must increase the risk of starvation in young weaned in the rainy season.

Artibeus jamaicensis relies on a food source that is strongly pulsed in time and space. It must search out resources for the future while harvesting a current source. Its behavior and diet adds risk from predators that may be attracted to the food source.

Snakes and owls are important predators of bats. Falcons, opossums, coatimundis, and the large carnivorous bat *Vampyrum spectrum* also are potential predators. There are enough seemingly defensive behavioral traits in *A. jamaicensis* to suggest that it has considerable exposure to predators.

Artibeus jamaicensis is a strong, robust bat that usually survives accidents that tear the wing membranes or break fingers. These parts heal remarkably rapidly and flight is not notably impaired. Many netted *A. jamaicensis* have damaged wings. Common sources of injury are fighting among individuals and encounters with sharp plant spines or thorns.

6. Population Estimates

Egbert G. Leigh, Jr., and Charles O. Handley, Jr.

Between 20 October 1976 and 19 May 1980, Bat Project participants marked 8907 *Artibeus jamaicensis*, which had been captured a total of 15,728 times by 20 October 1980, the last night of field work covered by this section. Can this record tell us how long individual *A. jamaicensis* live, and how many there are on Barro Colorado Island (BCI)? For purposes of the ensuing analyses, we will assume that we are sampling the entire population of *A. jamaicensis* on BCI, and that it is a closed population.

A Simple Estimate

AVERAGE LIFETIME.—Suppose *A. jamaicensis* has an exponential "life table," in which the probability of a bat living past age y is e^{-my}, where e (the base of Napierian logarithms) is 2.71828, and m is the annual death rate per capita. Then, if the capture effort, averaged over a period during which a quarter of the bats are replaced, does not vary excessively from one such period to another, we may estimate the average lifetime 1/m of these bats by the average time elapsed between first and last captures of bats captured more than once. We accomplish this as follows:

1. Let cdt be the probability that a given bat, alive between time t and time t + dt, is captured during this time interval. Here, c is the capture rate and dt denotes an "infinitesimal" time interval. Then the probability of a bat that lives L years being caught n times in its life is

$$[(cL)^n / n!] \exp (-cL)$$

2. If the probability that a bat's total lifetime lies between L and L + dL years is (mdL) exp −(mL), then the probability that a bat will be caught n times in its life is

$$P(n) = \int_o^\infty (mdL) \exp (-mL)[(cL)^n / n!] \exp (-cL)$$

$$= \int_o^\infty [(cL)^n / n!](mdL) \exp [-(m + c)L]$$

$$= (m/c)[c/(c + m)]^{n+1}$$

3. The average age at which a bat is first captured is

$$\frac{\int_o^\infty (ctdt) \exp - (m + c)t}{\int_o^\infty (cdt) \exp - (m + c)t} = 1 / (m + c)$$

4. This time interval 1/(m + c) is also the average time interval between captures, as long as capturing a bat does not alter its behavior. Thus, the average time between first and last captures is

$$\frac{[P(2) + 2P(3) + 3P(4) + 4P(5) + \ldots]}{(m + c)[P(2) + P(3) + P(4) + P(5) + \ldots]} = 1/m,$$

where 1/m is the average lifetime.

The average time elapsed between first and last capture of bats captured more than once is an exact estimate of average lifetime, if the life table is exponential. Although the bats we caught more than once live longer than our records suggest, the bats we captured more than once tended to be the bats that live longer than average in the first place, and the errors cancel.

We have calculated the average time between first and last captures by Bat Project participants of bats first marked by Morrison (1978b) and Bonaccorso (1979) between 1972 and 1974. Judging from the average interval between first and last capture by Bat Project personnel of the 33 females and eight males originally marked by either Morrison or Bonaccorso and captured more than once during the BCI Bat Project (Table 6-1), the average lifetime of such bats was 1.5 years. If we consider only females, the estimate of average lifetime is 1.6 years. These bats were two or more years old when first caught by Bat Project personnel, but if the life table is exponential, the average expectation of further life does not depend on the initial age of the bat.

The average time between first and last capture for 301 bats caught five or more times during the Bat Project is 1.6 years. This is a surprisingly low figure for bats caught so many times, because in theory the more times a bat is caught, the greater the time between first and last capture.

EXPONENTIAL LIFE TABLE.—Our basis for assuming an exponential life table is as follows. The numbers $n(x)$ of the

Egbert G. Leigh, Jr., Smithsonian Tropical Research Institute, Unit 0948, APO AA 34002-0948, or Apartado Box 2072, Balboa, Republic of Panamá.
Charles O. Handley, Jr., National Museum of Natural History, Smithsonian Institution, Washington, D.C. 20560.

TABLE 6-1.—Interval between first and last capture by the Bat Project of *Artibeus jamaicensis* first marked on BCI by Morrison and Bonaccorso.

Category	Number of years between first and last capture								
	0.25*	0.50	1.00	1.50	2.00	2.50	3.00	5.00	Total
Both sexes	6	9	2	9	4	4	6	1	41
Females only	5	6	2	8	2	3	6	1	33

* Minimum assumed elapsed time for bats caught for the first time and recaptured for the last time in the same half year.

TABLE 6-2.—Observed and expected frequency of capture of *Artibeus jamaicensis* marked on BCI by Morrison and Bonaccorso, 1972–1974, and by the Bat Project, 1976–1980.

	Number of times caught										
	≥1	≥2	≥3	≥4	≥5	≥6	≥7	≥8	≥9	≥10	≥11
MORRISON-BONACCORSO BATS											
Observed ($N = 146$)	69	39	22	11	3	1	1				
Expected $(69)(0.5274)^{x-1}$	69	36	19	10	5	3	1.5	0.8	0.4	0.2	0.1
BAT PROJECT BATS											
Observed ($N = 15728$)	8907	3846	1683	734	317	146	65	21	6	2	1
Expected $(8907)(0.4337)^{x-1}$	8907	3863	1675	727	315	137	59	26	11	5	2

bats marked by Morrison and Bonaccorso and caught at least x times during the bat project form a geometric series where $n(x) = 69(77/146)^{x-1}$ (see Table 6-2). Moreover, one of the most striking features of the Bat Project's capture records is that the number $n(x)$ of bats captured at least x times forms a nearly perfect geometric series: $n(x + 1) = n(1)A^x$, where $n(1) = 8907$ is the total number of bats caught at least once, and A is the total number of captures of bats caught at least twice, divided by the total number of captures, or $6821/15728 = 0.4337$ (see Table 6-2).

If the probability that a bat alive at time t is caught between times t and t + dt is cdt (where capture rate c is constant and dt is the length of a "short" time interval), then if all bats live the same lifetime (L), the probability, following the law of Poisson, that a bat may be caught n times in its life is

$$[(cL)^n / n!] \exp(-cL)$$

On the other hand, if the bats have an exponential life table with an average lifetime of $1/m$, the chance $P(x)$ that a bat will be captured x times in its life is

$$P(x) = [m/(c + m)] [c/(c + m)]^x,$$

a geometric series in x. This agrees with our observation if

$$A = c/(c + m).$$

If these bats have an exponential life table, their age should not affect their prospects of further life. Is an average lifetime of 1.6 years reasonable for female *A. jamaicensis*? Notice that

if the average lifetime $1/m$ of females is 1.6 years, then $m = 1/1.6 = 0.625$. Thus, survival rate per year is $\exp(-m) = 53.5\%$, and survival rate per half year is $\exp(-m/2) = 73.2\%$.

Half of the females that live long enough to reproduce can bear a young a half year after they are born (they are products of the second reproductive episode); all can bear a young every half year thereafter. If all that can bear a young do so, and if half the young are female, the expected number of young females a newborn female bat will bear in her life is

$$(1/4)(0.732) + (1/2)(0.535) [1 + 0.732 + (0.732)^2 + \ldots]$$
$$= 0.183 + (0.2675) / [1 - 0.732] \approx 1.18.$$

If, as our records show, one of every ten mature female bats in each breeding season fails to reproduce (see Section 3, Reproduction in a Captive Colony, and Section 4, Reproduction on BCI), the population will be very nearly in balance. Therefore, the average expectation of further life for these bats probably matches the average lifetime of all female *A. jamaicensis* rather closely and suggests that their life table is indeed exponential.

NUMBER OF BATS.—For the sake of analysis, we shall make two more assumptions, even though they are not completely true (see Section 5, Survival and Relative Abundance).

(1) Capturing a bat does not affect the probability of capturing it again later.
(2) Bats are sampled from a pool in which all individuals are equally liable to capture.

Then we may estimate the total number of bats available for capture during the Bat Project from the number of bats actually marked, divided by the chance a bat in the pool will be marked, where the latter is assumed equal to the proportion of marked bats that are recaptured at least once. In other words, the total number available during the project equals the number of bats marked, divided by the chance of marking a bat (equals the number of bats marked divided by the number of bats recaptured).

The number of bats available for capture at any one time is the total number available over the duration of the project, multiplied by the average bat lifetime $1/m$ (including both sexes), and divided by the total duration of the sampling period (4 years). Another way of saying this is: the total number of bats available at a given time equals the total available during the project, times the mean lifetime of a bat, divided by the duration of the project.

If *Artibeus jamaicensis* has an exponential life table, and if all bats "available for capture" are equally likely to be caught, then the total number of available bats during the four years of the sampling period is $(8907)/(0.4337) = 20535$. The number available at any one time is the total, times the lifetime of these bats (averaged for both sexes), divided by the four years' duration of the project, or $[(20535)(1.5)]/4 = 7701$.

An equivalent way to calculate population size of these bats is to find the capture rate c by setting $A = 0.4337$ equal to $c/(c + m)$, and to assume that the mortality rate m for both sexes concurrently is 0.667. We find $c = 0.5106$, which implies that, on the average, $(0.5106)/12$ or 4.166% of the bats are caught each month. As $(15,728)/48$, or 328, bats were caught per month, on the average, the total population of *A. jamaicensis* on BCI is $328/(0.04166) = 7873$.

VALIDITY OF ASSUMPTIONS.—How valid are the assumptions behind these calculations? If all available bats are equally liable to capture and if all bats have equal prospects of further lifetime, regardless of current age, then the chances of recapture of all bats marked within a given half year will be the same, regardless of age or sex.

This is not true. Adults marked in the fall of 1976 and 1977 were at least as likely to be recaptured as the juveniles or subadults marked at that time, although adults marked later on were recaptured much less often than were juveniles or subadults (Table 6-3; and Section 5, Survival and Relative Abundance, Table 5-9). Few of the adults newly marked in the fall of 1979 and spring of 1980 were caught again, while recapture rates were more nearly normal for juveniles and subadults. Many of these newly marked adults were caught from peninsulas on the mainland surrounding BCI, where we netted far less often than on the island. In general, the ratio of the proportion of newly marked juveniles and subadults, to the

TABLE 6-3.—Percentages, by sex and age class, of *Artibeus jamaicensis* marked in successive half years on BCI and subsequently recaptured; proportion of young to adult recaptures; and proportion of adults marked among total captures of adults (marks and recaptures) caught in each half year.

Age class	Fall 1976	Fall 1977	Spring 1978	Fall 1978	Spring 1979	Fall 1979	Spring 1980
				FEMALES			
Percent recaptured							
Juveniles	0.3784	0.3814	0.7273	0.3855	0.5551	0.4743	0.3723
Subadults	0.4185	0.4413	0.5404	0.4879	0.5315	0.3113	0.2105
Adults	0.5225	0.5055	0.3586	0.2524	0.3668	0.2194	0.1237
Ratio of percentage recaptured							
Juveniles and Subadults / Adults	0.7820	0.8259	1.6372	1.8427	1.4918	1.8373	2.9450
Proportion of new marks among captured bats							
Number adults marked / Total captures of adults	1.0000	0.8044	0.5380	0.3787	0.5303	0.2476	0.3811
				MALES			
Percent recaptured							
Juveniles	0.2955	0.4215	0.6154	0.3093	0.5368	0.4220	0.3109
Subadults	0.3526	0.4600	0.5778	0.3459	0.4935	0.3163	0.2154
Adults	0.3841	0.5050	0.5364	0.3271	0.2934	0.2893	0.2055
Ratio of percentage recaptured							
Juveniles and Subadults / Adults	0.8677	0.8754	1.0966	1.0151	1.7762	1.3404	1.4383
Proportion of new marks among captured bats							
Number adults marked / Total captures of adults	1.0000	0.7910	0.4314	0.4632	0.5000	0.3710	0.2219

TABLE 6-4.—Total captures of *Artibeus jamaicensis* on BCI in successive half years by sex and age class, by proportion of young to adult females, by proportion of new marks to total captures (marks and recaptures) among adult females.

Age class	Fall 1976	Fall 1977	Spring 1978	Fall 1978	Spring 1979	Fall 1979	Spring 1980	Fall 1980
				FEMALES				
Juveniles	76	247	56	85	307	397	401	612
Subadults	251	495	241	354	197	535	49	543
Adults	213	587	579	296	771	666	533	544
				MALES				
Juveniles	95	285	48	101	308	506	367	625
Subadults	201	421	232	200	232	409	132	364
Adults	173	534	522	256	514	658	318	430
				TOTALS				
All bats	1009	2569	1678	1292	2329	3171	1800	3118
Nonadult bats	623	1448	577	740	1044	1847	949	2144
				PROPORTIONS				
Juveniles and Subadults / Adult females	2.925	2.467	0.997	2.500	1.354	2.773	1.781	3.941
Adult female marks / Adult female captures	0.836	0.624	0.496	0.348	0.477	0.233	0.364	0.248

$$\text{Total fall nonadults / Total fall adult females} = \frac{6802}{2306} = 2.950$$

proportion of newly marked adults recaptured subsequently, was higher the greater the proportion of adults already marked (Table 6-3).

Our assumption that all available bats are equally liable to recapture is clearly wrong. It seems, rather, that as more bats were netted, we reached a point where most of the adult bats on BCI had been marked, so most bats available for marking on the island were juveniles and subadults. Thereafter, many of the bats marked as adults resided in localities where prospects of recapture were not great or they learned to avoid the nets.

Not only are bats in certain places more liable to recapture, younger bats are more liable to recapture than older ones. The ratio of the total number of juveniles and subadults of both sexes to the number of adult females handled (counting each instance of capture) in the fall of 1976 was 2.925; combining all the fall catches of the sampling period together, it was 2.950 (Table 6-4). Because adult females can have no more than two young a year, adult females must be nearly twice as hard to catch as juveniles of either sex.

A Refined Estimate

Population size in these bats can be better estimated after two intermediate steps.

(1)   Calculating mortality rates for both sexes.
(2)   Calculating the total numbers of marked bats of each sex alive in successive half years (Dowdeswell et al., 1940; Fisher and Ford, 1947).

The first step enables the second, because the total number of individuals M(t) alive in half-year t of the Bat Project that were marked earlier is the number $m'(1)$ marked in the first half year (fall 1976), times the proportion $p(t-1)$ surviving to half-year t, plus the number $m'(2)$ marked in the second half year (spring 1977) times the proportion $p(t-2)$ of those surviving to half-year t, and so on. In summary:

$$M(t) = m'(1)p(t-1) + m'(2)p(t-2) + \ldots + m'(t-1)p(1).$$

MORTALITY ESTIMATES.—If we assume that the numbers of *A. jamaicensis* on BCI do not change substantially from year to year, the simplest way to calculate the mortality of adults is to consider the recaptures of a sample of marked bats as long after they were first marked as possible. The bats marked by Morrison (1978b) and Bonaccorso (1979) provide a suitable sample. Between 1972 and 1974 they marked 1212 *A. jamaicensis,* of which about half were female. Of these females, 47 were caught by the Bat Project after 1 July 1977. Because 93 of the 178 adult female *A. jamaicensis* marked by the Bat Project in the fall of 1976 were recaptured after 1 July 1977, it seems reasonable to assume that 93/178 of the females marked by Morrison and Bonaccorso, and still living in the fall of 1976, were recaptured after 1 July 1977. If so, then 47(178/93), or 90, of Morrison and Bonaccorso's female bats were alive in the fall of 1976, implying a survival rate of $(90/606)^{1/3.5}$, or 58%, a year, and an expectation of further life of 1.835 years. Similarly, 16 of their marked male bats were caught after 1 July 1977. As 58 of 151 adult males marked by the Bat Project in the fall of 1976 were caught after 1 July 1977, roughly 16(151/58), or 42, of the bats marked by Morrison and Bonaccorso were alive during the fall of 1976, implying a survival rate of 46.7% a year. If the population were declining, these estimates would be too high, and vice versa.

A second way to calculate survival rate is to consider the dates of first capture of the bats recaptured in a given half year. Let $n(t, x)$ be the number of bats captured in half-year t that were first marked in half-year x, and let $m'(x)$ be the total number of bats marked in half-year x (Tables 6-5 and 6-6). Then $m'(x)p(t-x)$ is the total number of bats marked in half-year x surviving to half-year t. If a fraction c(t) of the bats alive in half-year t are captured then, $n(t, x) = c(t)m'(x)p(t-x)$. If the bats have an exponential life table, then

$$p(t-x) = \exp\,[-m(t-x)/2],$$

where exp (–m) is the average survival rate per year. Moreover,

$$n(t, x)/m'(x) = c(t)\,\exp\,[-m(t-x)/2],$$

suggesting that we calculate m as the coefficient of regression of log $[n(t, x)/m'(x)]$ on x.

This method of estimating m suffers from the disadvantage that bats marked later in the project are more likely to be from infrequently sampled sites. Thus, bats first marked in half-year 7 are less likely to be recaptured in half-year 8 than already marked bats caught in half-year 7. On the other hand, this estimate of m does not depend on the stability of the bat population as a whole.

Nonetheless, estimates (Table 6-7) of survival rates of female bats based on recaptures for half-year 7 and half-year 9, for which the correlation between the half-year x of marking and the logarithm of the proportion $n(t, x)/m'(x)$ of bats marked then that were recaptured in half-year t is relatively close, agree with each other, and with the estimate based on bats marked by Morrison and Bonaccorso and recaptured by the Bat Project. The average of the former two estimates is 57.4%, compared with 57.99% from the Morrison and Bonaccorso bats.

On the other hand, estimates of survival rates of male bats based on recaptures for half-years 7 and 9 average 37.2% a year, markedly lower than the 46.7% estimated from recaptures of male bats marked by Morrison and Bonaccorso. Do older males survive better? The 46.7% figure is based on a rather small sample of recaptures. Another piece of evidence is the recapture in the fall of 1981 of a male marked by Morrison and Bonaccorso. If 37.2% of the males survive each year, a male has one chance in 4000 of surviving the 8.5 years to the fall of 1981, so the chances are 1 in 7 that one of their 606 bats would still be living, and perhaps half that, had it been alive, we would have caught it then. That capture record suggests, but does not prove, that at least some older males survive rather better.

Finally, if the population is stable, we may calculate adult mortality rates from the proportions $n(x, t)/[m'(x) + n(x)]$ of bats marked in a given half-year t among the total number of bats—both marks, $m'(x)$, plus recaptures, $n(x)$—caught in later half years. If $m'(x) + n(x) = c(x)N(x)$, where $c(x)$ is the proportion of the N(x) bats alive in half-year x that were caught during that half year, while

$$n(x, t) = c(x)m'(t)\,\exp\,-m(x-t)/2,$$

then $n(x, t)/[m'(x) + n(x)] = [m'(t)/N(x)]\,\exp\,-m(x-t)/2$. If N is constant, the regression on x of the logarithm of this proportion is m/2. If the bat population is growing by a factor exp (r/2) per half year, then $N(x) = N(t)\,\exp\,r(x-t)/2$, and our regression gives (m + r)/2.

Mortality estimates for bats, both male and female, marked in the fall of 1976 and the spring of 1978, agree with each other, and are only slightly lower than those calculated from marking dates of bats recaptured in the fall of 1979 and the fall of 1980. Mortality estimates for bats marked in the fall of 1977 are much lower, but like those for bats marked in the spring of 1978, the fall 1977 estimates are based on strong correlations, illustrating the uncertainties in our calculations.

If the chances of capturing a bat depend on its age class, then this estimate requires that the age composition of bats captured in successive half years be the same. If the chances of capturing a bat depend on whether it is already marked, then the proportion of recaptures among the bats handled should also be constant. Finally, if chances of recapture vary from place to place, netting effort should be distributed over the island in the same way during successive half years.

Are our figures true mortalities? Although bats do occasionally move between BCI and the surrounding mainland, this exchange does not seem to be great (see Section 7, Movements). We do not think we are mistaking emigration for mortality. However, some bats do lose their necklaces. Adding together the intervals between time of first necklacing and of last recapture for each bat concerned, our project has monitored bats carrying Morrison and Bonaccorso wing bands for over fifty bat years. During this interval, five bats lost their necklaces, suggesting that the survival rate of necklaces is 90% a year. If so, the average survival rate of *A. jamaicensis* is 11% higher than our regressions suggest, perhaps between 62% and 65% a year for adult females.

Given the survival rate of juveniles relative to adults, and the breeding rate of adult females, the survival rate of adult females must be near 60% a year if the population is to be in balance. To show this, we make the following assumptions and observations. Half of the females (those that are born in the second reproductive episode) can bear a young a half year after birth; all can bear a young when they are a year old, and all can breed once each half year thereafter; half their offspring are female. Twenty-eight of the 74 juvenile females caught in the fall of 1976 were recaptured in successive half years, while 87 of the 178 adult females marked then were recaptured later (see Section 5, Survival and Relative Abundance, Table 5-9). Thus, a marked juvenile female had 0.7242 times the chance of being recaptured as an adult as did an adult female marked then. If in general, the chance of a juvenile female surviving through its

TABLE 6-5.—Numbers of female *Artibeus jamaicensis* marked on BCI in successive half years by age class, by numbers n(x, y) of these recaptured in later half-years y, and by number *n*(x) of recaptures in half-years x of these bats marked in previous half years; and based on these values, proportions n(t, x)/*m*′(x) of these bats recaptured in selected half-years t among those marked in earlier half-years x, and proportions *n*(x, t)/[*m*′(x) + *n*(x)] of bats marked in selected half years among those caught in later half years.

Females	Half-year x							
	1 Fall 1976	3 Fall 1977	4 Spring 1978	5 Fall 1978	6 Spring 1979	7 Fall 1979	8 Spring 1980	9 Fall 1980
Marked								
Juveniles	74	236	66	83	272	331	368	500
Subadults	227	358	198	289	141	257	19	286
Adults	178	366	290	103	368	155	194	135
Total [*m*′(x)]	479	960	554	475	781	743	583	921
Recaptures *n*(x, y) in half-year x of bats first caught in half-year y								
Recaptures from half-year y								
1		91	68	27	36	33	23	16
3			181	81	102	88	47	41
4				61	74	67	31	38
5					114	100	32	46
6						183	75	76
7							107	115
8								157
Total		91	249	169	326	471	315	489
Proportions								
Recaptured/Marked								
n(9, x)/*m*′(x)	16/479	41/960	38/554	46/475	76/781	115/743	157/583	
n(8, x)/*m*′(x)	23/479	47/960	31/554	32/475	75/781	107/743		
n(7, x)/*m*′(x)	33/479	88/960	67/554	100/475	183/781			
n(x, 1)/[*m*′(x) + *n*(x)]		91/1051	68/803	27/644	36/1107	33/1214	23/898	16/1410
n(x, 3)/[*m*′(x) + *n*(x)]			181/803	81/644	102/1107	88/1214	47/898	41/1410
n(x, 4)/[*m*′(x) + *n*(x)]				61/644	74/1107	62/1214	21/898	38/1410

first year is 0.7242 times the chance (p^2) of an adult female surviving a full year, and if a female has a 1/4 chance of bearing a female young when it is a year old, and a 1/2 chance of doing so each half year thereafter, then the number (R_o) of female young a newborn female *A. jamaicensis* can expect to bear during her life is

$$R_o = 0.7242 \ [(1/4)p + (1/2)p^2 + (1/2)p^3 + \dots]$$
$$= 0.1810p + 0.362p^2 / (1 - p).$$

If survival rate p^2 per year is 0.5625 (so $p = 0.75$), then $R_o = 0.95$; if $p^2 = 0.6$ (so $p = 0.7746$), then $R_o = 1.10$; and if $p^2 = 0.64$ (so $p = 0.8$), then $R_o = 1.30$.

In fact, not all females breed in each breeding season. In six reproductive episodes in our captive colony wild-caught adult females were pregnant in 66 of 72 chances (91.7%), multiparous captive-born females were pregnant in 12 of 14 chances (85.7%), and primiparous captive-born females were pregnant in 14 of 20 chances (70%) (see Section 3, Reproduction in a Captive Colony, Tables 3-4 and 3-5). On BCI in the month of April, when all females should be reproductive, we invariably caught a few nonreproductive adult females and some subadults who failed to reproduce on time (undoubtedly 9-month-old young of the previous July–August birth-group). For example, in April 1978, 9 of 102 adult females caught were nonreproductive, and we caught 22 subadult females; in April 1979, 7 of 47 newly marked adult females were nonreproductive. We can make a minimum correction for this by multiplying our values of R_o for different survival rates by 0.89, the percentage (133/149) of adult females caught in 1978 and

TABLE 6-6.—Numbers of male *Artibeus jamaicensis* marked on BCI in successive half years by age class, by numbers $n(x, y)$ of these recaptured in later half-years y, and by number $n(x)$ of recaptures in half-years x of these bats marked in previous half years; and based on these values, proportions $n(t, x)/m'(x)$ of these bats recaptured in selected half-years t among those marked in earlier half-years x, and proportions $n(x, t)/[m'(x) + n(x)]$ of bats marked in selected half years among those caught in later half years.

Males	Half-year x							
	1 Fall 1976	3 Fall 1977	4 Spring 1978	5 Fall 1978	6 Spring 1979	7 Fall 1979	8 Spring 1980	9 Fall 1980
Marked								
Juveniles	88	261	52	97	272	410	341	496
Subadults	173	300	135	159	154	196	65	198
Adults	151	299	220	107	242	197	73	101
Total [$m'(x)$]	412	860	407	363	668	803	479	795
Recaptures $n(x, y)$ in half-year x of bats first caught in half-year y								
Recaptures from half-year y								
1		79	55	14	19	9	9	3
3			235	50	65	56	18	10
4				60	57	44	22	13
5					101	61	31	15
6						164	50	46
7							126	90
8								114
Total		79	290	124	242	334	256	291
Proportions								
Recaptured/Marked								
$n(9, x)/m'(x)$	3/412	10/860	13/407	15/363	46/668	90/803	114/479	
$n(8, x)/m'(x)$	9/412	18/860	22/407	31/363	50/668	126/803		
$n(7, x)/m'(x)$	9/412	56/860	44/407	61/363	164/668			
$n(x, 1)/[m'(x) + n(x)]$		79/939	55/697	14/487	19/910	9/1137	9/735	3/1086
$n(x, 3)/[m'(x) + n(x)]$			235/697	50/487	65/910	56/1137	18/735	10/1086
$n(x, 4)/[m'(x) + n(x)]$				60/487	57/910	44/1137	22/735	13/1086

1979 that were actually breeding.

There may be other errors in our calculations. On the one hand, spring-born juveniles should survive better than fall ones; on the other, our figures for juvenile mortality do not include deaths before the juveniles are old enough to fly into our nets. All in all, we believe the survival rate of adult female *Artibeus jamaicensis* is between 60% and 64% a year.

ESTIMATE OF SURVIVING MARKED BATS IN SUCCESSIVE HALF YEARS.—Let $M_{af}(t)$ represent the number of marked adult female bats alive at the beginning of half-year t, and $M_{sf}(t)$ be the number of marked subadult female bats also alive then. In addition, p represents the survival rate of adult females per half year.

Eighty-three of the 227 subadult bats marked in the fall of 1976 were recaptured in succeeding half years, while 87 of 178 adults marked then were recaptured later (Section 5, Survival and Relative Abundance, Table 5-9), suggesting that a subadult female was 0.8010 times as likely as an adult to survive her next half year. If this is true in any half year, then a juvenile female is $(0.7242)/(0.8010) = 0.9041$ times as likely as an adult female to survive her next half year. If we make the convention that a fraction $\sqrt{p}$ of the adults, a fraction $0.8010 \sqrt{p}$ of the subadults, and a fraction $0.9041 \sqrt{p}$ of the juveniles marked in a given half year survive to the beginning of the next half year, and that juveniles become subadults, and subadults become adults, at that time, then we may assume the number $M_{sf}(t)$ of marked subadult females at the beginning of a half year to be $0.9041 \sqrt{p}\, m_{jf}(t - 1)$, where $m_{jf}(t - 1)$ is the number of juvenile females marked in half-year $t - 1$.

The marked adult females alive at the beginning of half-year

TABLE 6-7.—Estimates of mortality rate (m) and survival rate e^{-m} per year of *Artibeus jamaicensis* marked on BCI and recaptured in several half years.

	Coefficient of regression			
	$(1/2)\,m$	$1/m$	e^{-m}	r^2
	FEMALES			
Regression on x of				
$\log n(9, x)/m'(x)$	0.2911	1.718	0.559	0.9492
$\log n(8, x)/m'(x)$	0.1814	2.756	0.696	0.8154
$\log n(7, x)/m'(x)$	0.2647	1.889	0.589	0.9338
$\log [m'(x) + n(x)]/n(x, 1)$	0.3178	1.573	0.530	0.9274
$\log [m'(x) + n(x)]/n(x, 3)$	0.3746	1.335	0.437	0.9781
$\log [m'(x) + n(x)]/n(x, 4)$	0.3175	1.575	0.530	0.9869
	MALES			
Regression on x of				
$\log n(9, x)/m'(x)$	0.5019	0.996	0.367	0.9746
$\log n(8, x)/m'(x)$	0.3417	1.463	0.505	0.8590
$\log n(7, x)/m'(x)$	0.4876	1.025	0.377	0.9944
$\log [m'(x) + n(x)]/n(x, 1)$	0.5452	0.917	0.336	0.9150
$\log [m'(x) + n(x)]/n(x, 3)$	0.6416	0.779	0.277	0.9669
$\log [m'(x) + n(x)]/n(x, 4)$	0.5401	0.926	0.340	0.9701

t include

(1) $\sqrt{p}\ m_{af}(t-1)$ adult females marked the preceding half year (assumed marked in the middle thereof, so that they have had half a half year in which to die, so a fraction $\sqrt{p}$ have survived to the beginning of half-year t),

(2) $0.8010\ \sqrt{p}\ m_{sf}(t-1)$ subadult females marked in half-year $t-1$,

(3) a proportion p of the $M_{af}(t-1)$, marked adults alive at the beginning of half-year $t-1$, and

(4) $0.8010\ pM_{sf}(t-1)$ adults, which already were marked and subadult at the beginning of half-year $t-1$.

To summarize:

$$M_{sf}(t) = 0.9041\ \sqrt{p}\ m_{jf}(t-1)$$

$$M_{af}(t) = p[0.8010\ M_{sf}(t-1) + M_{af}(t-1)]$$
$$+ \sqrt{p}[0.8010\ m_{sf}(t-1) + m_{af}(t-1)].$$

If $p = 0.7477$, so that the survival rate of adult females, with their necklaces, is $p^2 = 55.9\%$ a year, then the maximum number of marked adult females was 1534 on 1 July 1980.

If we conclude from Table 5-9 (Section 5, Survival and Relative Abundance) that juvenile males marked in half-year t have probability $0.7693\ q^{3/2}$ of surviving to the beginning of half-year $t + 2$, while subadult males alive at the beginning of half-year $t + 1$ have probability $0.9180\ q$ of living another half year, where q is the survival rate of adult males per half year, then the number of marked subadult males alive at the beginning of half-year t is

$$M_{sm}(t) = 0.8380\ \sqrt{q}\ m_{jm}(t-1)$$

and the number of marked adult males alive at the beginning of half-year t is

$$M_{am}(t) = q[0.9180\ M_{sm}(t-1) + M_{am}(t-1)]$$
$$+ \sqrt{q}\ [0.9180\ m_{sm}(t-1) + m_{am}(t-1)].$$

If $q = 0.6058$, so that survival rate of marked males with their necklaces is $q^2 = 36.7\%$ a year, then the maximum number of marked adult males was 829 on 1 January 1980.

How do our estimates compare with calculations for other species? Most available data are for temperate-zone species. Rice (1957) proposed an annual survival rate of 46% for *Myotis austroriparius* in Florida, the rate necessary to maintain a balanced population given his calculated birth rate for the population. Davis et al. (1962) assumed a constant mortality rate for *Tadarida brasiliensis* after the initiation of flight and estimated an adult survival rate of 70%–80% with a maximum age of 15 years. Stebbings (1966) calculated a survival rate of 75% and a life expectancy of four years for two species of *Plecotus* in England. Similarly, Dwyer (1966) suggested 75% for *Miniopterus schreibersii* in Australia.

Mean annual survival rates for 20 temperate zone species ranged mainly from 40% to 80%, but extremes of 4% and 98% were also reported (Tuttle and Stevenson, 1982). All of these were based on banding studies of individuals of unknown ages. Studies based on known-age cohorts are available mainly for summer roosts of temperate-zone animals, and estimates of survivorship are similarly variable. A summary of the results and some discussion of the problems inherent in such studies were presented by Tuttle and Stevenson (1982).

Fleming (1988) used three separate sets of data to analyze

survivorship in *Carollia perspicillata* in Costa Rica. Based on recapture data, he concluded that relatively few individuals remain in the population after five years, but his was an open population, with no way of separating dispersal from mortality. Young females disappeared at a higher rate than young males, a difference he attributed to dispersal rather than mortality. He found no differences in survivorship of young born at different times of the year (different birth groups). Adult males and females had similar survivorship rates (around 60%).

Annual survivorship estimates based on census data for a single cave population of *C. perspicillata* ranged from 40% to 82% (Fleming, 1988). Again, the data are confounded with the problem of dispersal, and his estimates based on survival across two years yielded estimates of 57% for males and 43% for females. A survivorship curve based on toothwear data yielded values within the ranges of the estimates based on the other methods (Fleming, 1988).

ESTIMATING POPULATION SIZE.—The number of adult bats actually marked provides minimum estimates of the number of adult bats in the population from which we were netting. Tables 6-8 and 6-9 suggest that there were at least 1500 adult females and 800 adult males in the population between 1 January 1980 and 1 January 1981.

Another minimum estimate of the number of adult female bats is the number required to produce the numbers of juvenile and subadult bats marked in a given year. In 1980, we marked 1173 female and 1100 male juvenile and subadult bats (Table 6-8). If a female produces 1.78 young a year (corresponding to the 89% pregnancy rate of adult females netted in April 1978 and April 1979), it would take about 1300 adult females to produce this many young.

Another seemingly more accurate way to estimate the number of adult females is to divide the number of adult females marked before half-year t and presumed to be alive in the middle of that half year, $0.8802\,M_{af}(t)$, by the proportion of adult females caught that half year that were marked in previous half years. Averaging such estimates for those half years when fewer than half the adult females caught were unmarked, we were netting from a population of about 1800 adult females. Similarly, we were netting from a pool of about 850 adult males.

If we assume that, averaging from year's end to year's end, there are roughly as many nonadult bats as adult females, then our minimum estimates suggest we were netting from a pool of

TABLE 6-8.—Number of *Artibeus jamaicensis* marked on BCI alive in successive half years.

| | Half-year t | | | | | | | | |
| | 1 | 3 | 4 | 5 | 6 | 7 | 8 | 9 | 10 |
	Fall 1976	Fall 1977	Spring 1978	Fall 1978	Spring 1979	Fall 1979	Spring 1980	Fall 1980	
FEMALES									
New marks									
Juveniles $m_{jf}(t)$	74	236	66	83	272	331	368	500	
Subadults $m_{sf}(t)$	227	358	198	289	141	257	19	286	
Adults $m_{af}(t)$	178	366	290	103	368	155	194	135	
Recaptures*, $n_f(t)$		91	249	169	326	471	315	489	
At beginning of half year†,‡									
Total marked adults $M_{af}(t)$		282	793	1126	1200	1394	1532	1534	1480
Total marked subadults $M_{sf}(t)$			188	53	66	216	263	293	398
MALES									
New marks									
Juveniles $m_{jm}(t)$	88	261	52	97	272	410	341	496	
Subadults $m_{sm}(t)$	173	300	135	159	154	196	65	198	
Adults $m_{am}(t)$	151	299	220	107	242	197	73	101	
Recaptures*, $n_m(t)$		79	290	124	242	334	256	291	
At beginning of half year♣,♦									
Total marked adults $M_{am}(t)$		178	555	698	639	721	829	754	800
Total marked subadults $M_{sm}(t)$			170	34	63	177	267	222	326

* Each bat recaptured in a given half year is counted once, regardless of how often it was caught in that half year.

† Total number $M_{af}(t)$ of marked adult females at beginning of half-year t is $[M_{sf}(t-1)\,0.801 + M_{af}(t-1)]\,0.7747 + [m_{sf}(t-1)\,0.801 + m_{af}(t-1)]\,0.8802$.

‡ Total number $M_{sf}(t)$ of marked subadult females at beginning of half-year t is $(0.9041)(0.8802)\,m_{jf}(t-1)$, where $(0.8802)^2 = 0.7747$.

♣ Total number $M_{am}(t)$ of marked adult males at beginning of half-year t is $[0.918\,M_{sm}(t-1) + M_{am}(t-1)]\,0.6058 + [0.918\,m_{sm}(t-1) + m_{am}(t-1)]\,(0.6058)$.

♦ Total number $M_{sm}(t)$ of marked subadult males at beginning of half-year t is $(0.8380)(0.7783)\,m_{jm}(t-1)$, where $(0.7783)^2 = 0.6058$.

TABLE 6-9.—Number of *Artibeus jamaicensis* marked on BCI: number of non-adult bats marked, number of adults already marked at beginning of half year; proportion of these marked bats recaptured, proportion of adult bats caught that had been recaptured in previous half years, and estimated number of adults in successive half years.

	Half-year t						
	3	4	5	6	7	8	9
	FEMALES						
Number marked that half year							
Total marked, $m_f(t)$	960	554	475	781	743	591	921
Nonadults marked, $m_f(t) - m_{af}(t)$	594	264	372	413	588	387	786
Adults already marked, $M_{af}(t)$	282	793	1126	1200	1394	1532	1534
Proportion of marked bats recaptured that half year, $P_f(t) = n_f(t)/[M_{af}(t) + M_{sf}(t)]$	0.323	0.254	0.143	0.258	0.293	0.175	0.268
Proportion of adults previously recaptured, $Q_f(t) = n_f(t)/[n_f(t) + m_{af}(t)]$	0.199	0.462	0.621	0.470	0.752	0.619	0.784
Ratio of number marked to efficiency of recapture, $m_f(t)/1000P_f(t)$	2.97	2.14	3.32	3.03	2.19	3.38	3.44
Estimated population size, $\sqrt{p}\,M_{af}(t)/Q_f(t)$	1247	1511	1596	2247	1632	2178	1722
	MALES						
Number marked that half year							
Total marked, $m_m(t)$	860	407	363	668	803	479	795
Nonadults marked, $m_m(t) - m_{am}(t)$	561	187	256	426	606	406	694
Adults already marked, $M_{am}(t)$	178	555	698	639	721	829	754
Proportion of marked bats recaptured that half year, $P_m(t) = n_m(t)/[M_{am}(t) + M_{sm}(t)]$	0.444	0.400	0.172	0.345	0.372	0.234	0.298
Proportion of adults previously recaptured, $Q_m(t) = n_m(t)/[n_m(t) + m_{am}(t)]$	0.209	0.569	0.537	0.500	0.629	0.778	0.742
Ratio of number marked to efficiency of recapture, $m_m(t)/1000P_m(t)$	1.92	1.02	2.11	1.94	2.16	2.05	2.67
Estimated population size, $\sqrt{q}\,M_{am}(t)/Q_m(t)$	663	759	1012	995	892	829	791

3800 bats (1500 adult females, 800 adult males, and 1500 juveniles and subadults), whereas the latter method suggests 4450 (1800 adult females, 850 adult males, and 1800 juveniles and subadults).

These estimates are riddled with assumptions, of which the most conspicuous are (1) marking does not affect an adult's prospects of subsequent capture, and (2) all adults are equally liable to capture. The first assumption is not very plausible, although it is admittedly difficult to distinguish between the greater difficulty of catching older bats, which is obvious from our data, and the greater difficulty of catching marked bats. If capture effort were spread evenly enough to assure that all live bats of a given age were equally liable to capture, one might expect the ratio of the number of bats marked, or at least of the number of nonadult bats marked, to the proportion of marked bats recaptured, to be constant, or nearly so. It is not (Table 6-9). Despite these problems, these estimates are close enough to the minimum estimates that further refinement of our counts of adult bats seems unnecessary.

RELATIONSHIP OF POPULATION SIZE AND DISTRIBUTION.—The Bat Project apparently was netting from a pool of about 4000 *A. jamaicensis*. How big an area did this population represent? To find out, we calculated the number of bats alive on 1 July 1979 that were marked at various points on the mainland, and asked how many of these were recaptured in the fall of 1979 and where they were caught. Tables 6-10 and 6-11 show that, although there is some exchange of bats between BCI and nearby mainland sites, bats from these areas do not form a common pool.

The average maximum distance between recaptures of *A. jamaicensis* captured three or more times is about 1.5 km (see Section 7, Movements). As BCI is roughly a rectangle 3 km by 5 km, and as half the perimeter of the island is surrounded by nearby mainland approximately a kilometer distant, we shall assume that our 4000 bats represent a land area of 3.5 km by 5.5 km, or roughly 20 km^2, implying 200 bats per square kilometer. We accordingly conclude that BCI supports an average of 3000 *A. jamaicensis*.

TABLE 6-10.—Exchange of *Artibeus jamaicensis* between BCI and the mainland.

Locality of first marking	Number* of marked bats still alive 1 July 1979	Number of marked bats recaptured fall 1979	
		BCI	Bohio Peninsula
Bohio Peninsula	136	2	2
Mona Grita Point	17	1	0
Old Frijoles Road	107	1	0
Barro Colorado Island	1855	675	2

* Number calculated assuming a survival rate of 50% a year. If there are two marked females for every marked male, as is true for the population at large, then the annual survival rate is $^2/_3$ (0.56) + $^1/_3$ (0.37) = 0.50.

TABLE 6-11.—Proportion of recaptures to marks among *Artibeus jamaicensis* netted on the mainland near BCI.

	Bohio			Mona Grita	Frijoles
	Jan 1978	Dec 1978	Apr 1979	Jan 1979	Apr 1979
Number marked	114	67	105	33	151
Number recaptured					
from Standley 21	8	1	12	0	0
from elsewhere on BCI	9	1	13	3	4
first marked at Bohio					
in January 1978	0	7	4	0	0
in December 1978	0	0	9	0	0

Summary

During the five years of field work covered by this volume, we marked 8,907 individual *Artibeus jamaicensis* and captured them a total of 15,728 times. Data on multiple captures form a nearly perfect geometric series, suggesting an exponential life table. Assuming an exponential life table, the average lifespan of these bats is 1.6 years, based on the average time between first and last capture of individuals captured more than once. If average lifespan is 1.6 years, and if adult females bear a young in each breeding season, the average female will produce 1.18 young in her lifetime. Our records indicate that one of every ten females fails to reproduce in each reproductive season, a figure that yields a population that is nearly in balance.

Survival rates based on bats marked between 1972 and 1974 and recaptured during 1976–1980 were 58% for females and 46.7% for males, if the population was stable during this period. Similar rates calculated for bats recaptured during a given half year, independent of the stability of the population as a whole, are similar for females (57.4%), but different for males (37.2%). One possible explanation for this difference is that older males survive better. Our data on necklace loss, coupled with reproductive data suggest that the actual survival rates are somewhat higher, perhaps on the order of 60%–64% for females.

Population estimates based on mark-recapture methodology and on capture rate are surprisingly close, at 7701 and 7873, but both are flawed by assumptions of equal probability of capture that clearly are not valid. More refined estimates of surviving marked bats in successive half years suggest that there were a maximum of 1,534 marked females and 829 marked males in 1980. In that year we marked 2,273 juvenile and subadult bats, suggesting a minimum of 1,300 adult females to produce that many young. Mark-recapture methodology for successive half years yields estimates of 1,800 adult females and 850 adult males. If we assume there are roughly as many nonadults as adult females during a year's time, then the minimum estimates suggest 3,800 bats and the higher ones about 4,500.

Assuming a pool of 4,000 bats taken from a land area of about 20 square kilometers that includes BCI and parts of the surrounding mainland, we estimate there are about 200 *A. jamaicensis* per square kilometer, or a population of about 3,000 on the island itself.

7. Movements

Charles O. Handley, Jr., Alfred L. Gardner,
and Don E. Wilson

With our abundant capture and recapture data we hoped to trace the movements of *Artibeus jamaicensis* on Barro Colorado Island (BCI) and to estimate the extent of these movements around each capture locality. However, conventional mark and recapture analyses are not as productive for bats as they are for some other organisms, partly because bats are comparatively less frequently recaptured. The recapture rate for our 8907 *A. jamaicensis* averaged 1.8 per bat. Less than 50% (3846) were recaptured at all, and less than 20% (1683) were captured three or more times (see Section 5, Survival and Relative Abundance, Table 5-2). Therefore, only a few individuals were useful in describing distributions and patterns of movements.

A mark at one locality and recapture elsewhere does not reveal the focal "home base" or roost locality. Roosts are difficult to locate except by radio tracking, which we did not routinely do. Except for maternity roosts in caves, tree holes, and man-made structures, the "home base" is likely to be temporary anyway. *Artibeus jamaicensis* frequently traveled long distances (up to about 6 km) between capture sites on BCI and mainland. The distance from the center of BCI to any point on its perimeter is only about 3 km, and the maximum distance between points on the perimeter is about 5 km. Therefore, the potential home range of an *A. jamaicensis* on BCI includes the entire island and perhaps parts of the adjacent mainland as well.

It proved fruitless to distinguish between mark and recapture records and between mark and recapture localities on an individual basis when describing distributions. Therefore, we pooled all records for analysis rather than treating only the records of individual bats. This greatly increased the sample size for each capture locality and reduced some kinds of biases such as that caused by unusually long movements. We converted these records to percentages of the total for a particular locality so as to examine movements for that site. We

then tested and manipulated our data to explore effective means of describing home ranges and movement patterns.

Measures of Movements and Capture Effort

We tabulated captures by species and by half years on a matrix of locality groups, totalled the number of *A. jamaicensis* captures recorded at each locality group, and summarized the entire marking interval (October 1976 through October 1980; Figure 7-1, Table 7-1). We then analyzed the records of all individuals recorded from a locality that were caught again in the same half year at that and other localities. These we called "other-captures" (tabulated by the column number of the locality of subsequent capture along the horizontal axis of Table 7-1). For example, other-captures at Lutz (642) of the 5309 Lutz records for *A. jamaicensis,* are under column 1 and other-captures of Lutz bats at Shannon-AMNH (107) are in line 1, column 3.

We found that the proportion of other-captures to the total number of captures from any given locality was extremely variable (from 1% to 43%; Figure 7-2 and Table 7-2). Contributing to this variation were factors such as the proximity of netting sites to other netting locations, and the frequency of netting (netting effort) at each locality.

We also found that the proportion of other-captures to total captures was not always dependent on netting effort. For instance, although both sites had similar netting effort, Shannon-AMNH (730 records of *A. jamaicensis*) and Bohio (746 records) had 42% and 17% other-captures (Table 7-2), respectively. Barbour-Hood (640 bats) and Standley End (649), also with similar netting effort, had 36% and 22% other-captures, respectively. In contrast, at Chapman with 29% and Lutz with 30% other-captures, total *A. jamaicensis* numbered 388 and 5309, respectively.

The localities that had from 36% to 43% other-captures are centrally located on BCI (Figure 7-2, Table 7-2). Localities with 17% to 35% other-captures (excepting Barbour Stream with 24% and Lutz with 30%) are all marginal locations near the lakeshore. The lowest other-capture percentages are 17% at Drayton End and the isolated mainland localities (Bohio, 17%;

Charles O. Handley, Jr., and Don E. Wilson, National Museum of
Natural History, Smithsonian Institution, Washington, D.C. 20560.
Alfred L. Gardner, NERC, U.S. Fish and Wildlife Service, National
Museum of Natural History, Washington, D.C. 20560.

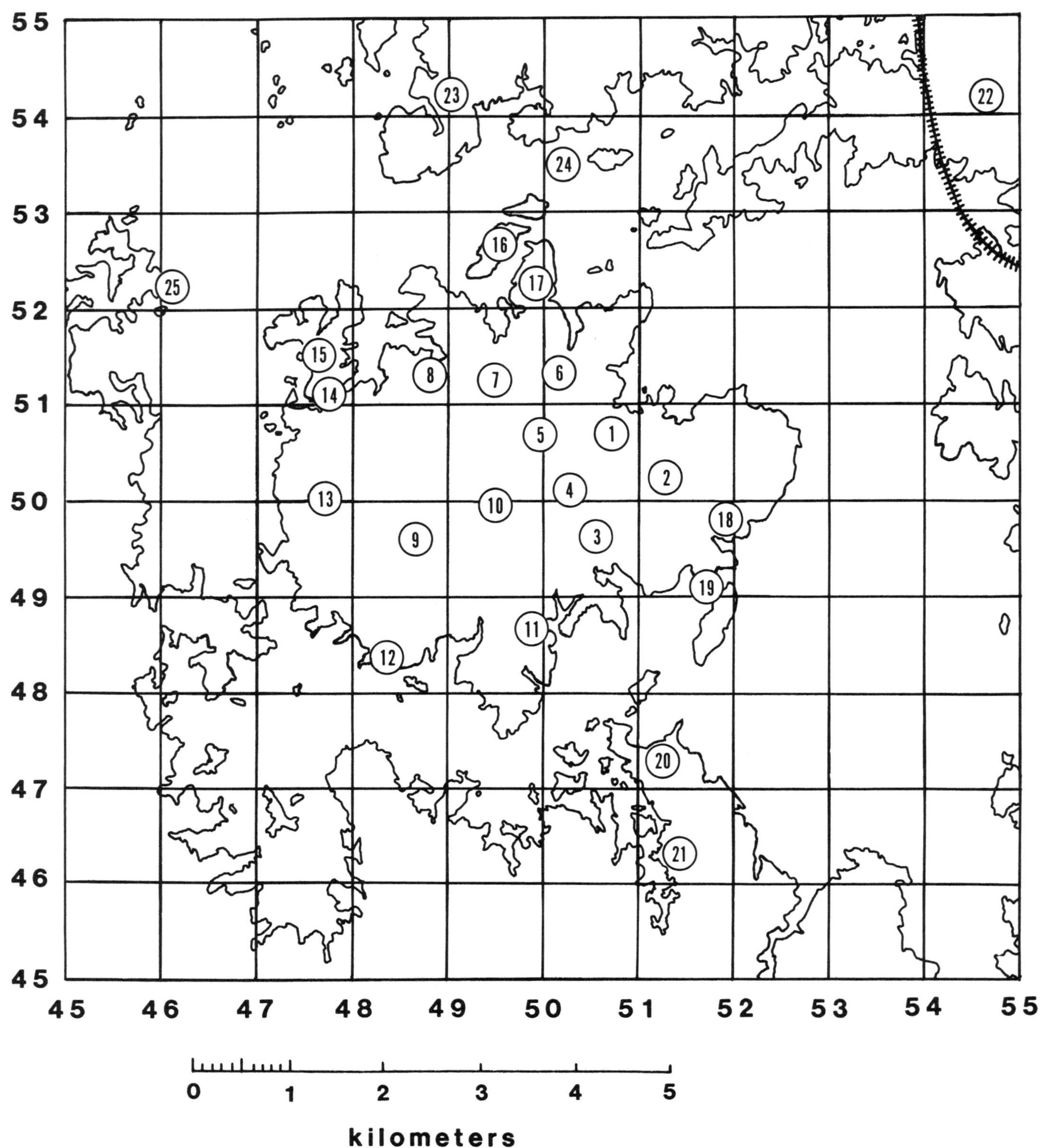

FIGURE 7-1.—Distribution of locality groups (aggregations of localities) where bats were captured on BCI and adjacent mainland (from Table 7-1). The locality groups are as follows: (1) Lutz; (2) Barbour-Hood; (3) Shannon-AMNH; (4) Shannon-Balboa; (5) Lake-Wheeler; (6) Barbour Stream; (7) Miller Ridge; (8) Fuertes; (9) Conrad; (10) Plateau; (11) Drayton End; (12) Armour End; (13) Zetek 21; (14) Standley Ridge; (15) Standley End; (16) Orchid Island; (17) Gross Point; (18) Chapman; (19) Harvard; (20) Mona Grita; (21) Gigante; (22) Frijoles Rd.; (23) Bohio; (24) Buena Vista; (25) Peña Blanca.

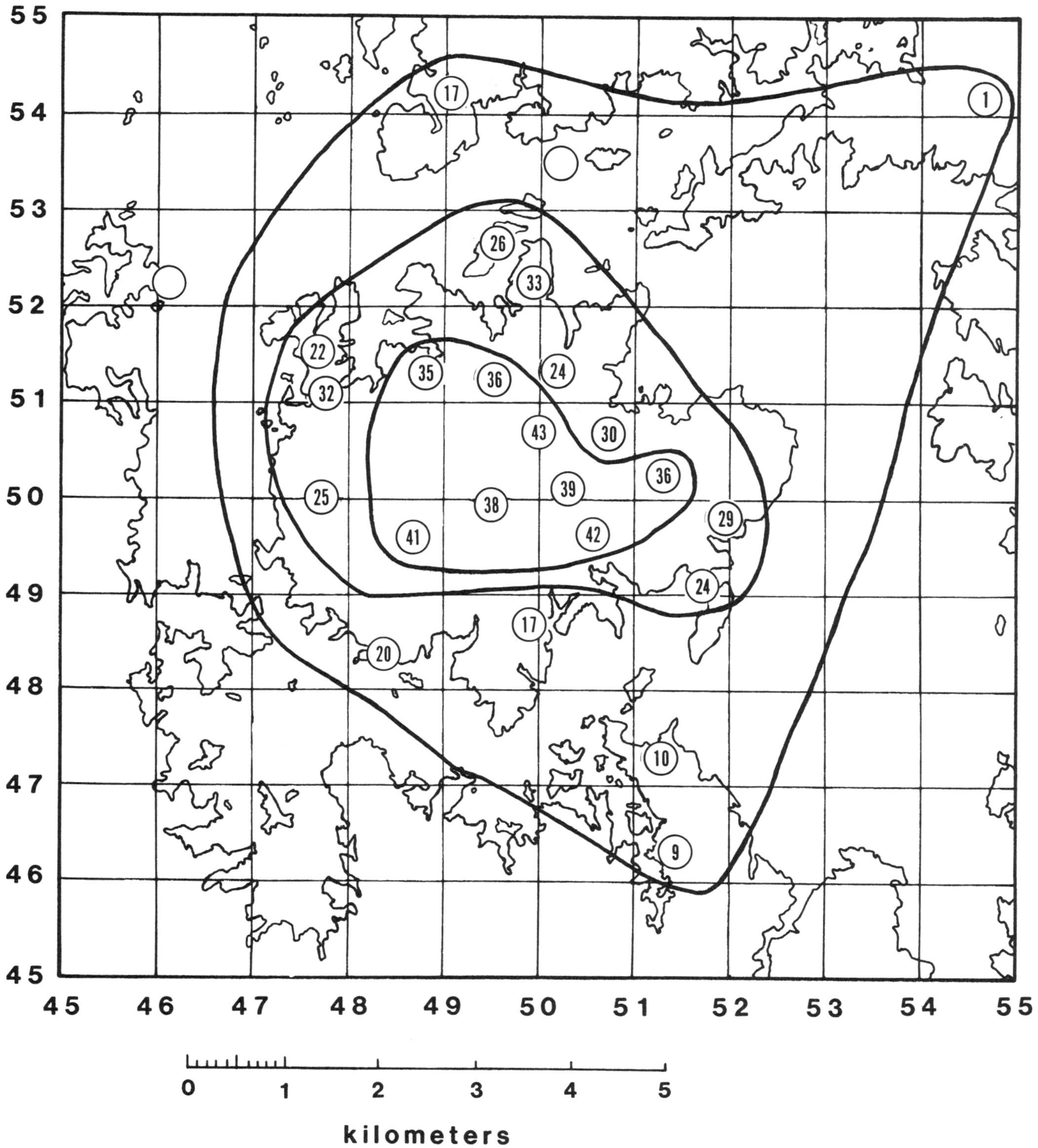

FIGURE 7-2.—Percentage of other-captures in the total records (total individuals plus other-captures of them) in capture suites of *Artibeus jamaicensis* at locality groups on BCI and the adjacent mainland, 1976–1980. Data are from Table 7-2. Isolines arbitrarily enclose locality groups with similar percentages of other-captures. Marginal numbers identify kilometer squares (see Appendix, Methods of Capturing and Marking Tropical Bats).

TABLE 7-1.—Total number of captures of *Artibeus jamaicensis* and other-captures (see text for explanation) of those individuals, recorded at 25 locality groups on BCI and the adjacent mainland (Fig. 7-1) in the eight half-year marking intervals between October 1976 and October 1980.

Locality group	Total captures	Other-captures by locality group																									Total
		1	2	3	4	5	6	7	8	9	10	11	12	13	14	15	16	17	18	19	20	21	22	23	24	25	
1. Lutz	5309	642	106	107	19	20	27	230	71	48	143	6	8	10	33	13	10	11	33	10	1	2	0	10	4	3	1567
2. Barbour-Hood	640	119	17	19	0	1	2	19	7	1	13	1	2	1	2	2	1	0	22	4	0	0	0	0	0	0	233
3. Shannon-AMNH	730	124	19	27	7	4	1	25	7	6	51	5	3	1	4	2	3	1	10	4	0	0	0	0	0	1	305
4. Shannon-Balboa	165	24	0	7	2	0	0	6	1	0	13	1	3	0	2	0	4	1	1	0	0	0	0	0	0	0	65
5. Lake-Wheeler	156	21	1	4	0	3	2	17	4	1	7	0	0	0	3	3	0	1	0	0	0	0	0	0	0	0	67
6. Barbour Stream	234	31	2	1	0	2	6	4	2	1	4	0	0	0	1	0	1	0	1	0	0	0	0	0	0	1	57
7. Miller Ridge	2499	229	18	25	6	16	3	169	140	60	99	1	14	6	51	18	8	11	2	3	0	0	0	18	1	1	899
8. Fuertes	1189	68	6	7	1	4	2	139	61	26	38	2	14	1	20	11	1	2	4	0	0	0	0	14	0	0	421
9. Conrad	578	58	1	6	0	1	1	61	28	11	39	0	7	2	16	2	0	0	3	0	0	0	0	3	0	0	239
10. Plateau	1377	150	13	49	13	7	3	99	37	38	58	2	16	1	18	5	5	4	4	0	0	0	0	1	0	1	524
11. Drayton-End	138	6	1	5	1	0	0	1	2	0	2	2	1	0	0	0	0	0	1	0	0	1	0	0	0	0	23
12. Armour-End	456	9	2	3	3	0	0	14	14	7	16	1	10	2	1	0	2	1	1	0	1	0	0	5	0	0	92
13. Zetek 21	146	10	1	1	0	0	0	6	1	2	1	0	2	4	3	4	2	0	0	0	0	0	0	0	0	0	37
14. Standley Ridge	786	33	2	4	2	3	1	53	19	16	17	0	1	3	51	26	1	0	1	0	0	0	0	14	0	1	248
15. Standley End	649	14	2	2	0	3	0	18	11	2	5	0	0	3	27	39	1	0	1	0	0	0	0	2	5	1	142
16. Orchid Island	249	10	1	3	4	0	1	8	1	0	4	0	2	2	1	1	5	15	1	0	0	0	0	2	5	0	66
17. Gross Point	145	11	0	1	1	1	0	12	2	0	4	0	1	0	0	0	15	0	0	0	0	0	0	0	0	0	48
18. Chapman	388	34	22	11	1	0	1	2	5	3	4	1	1	0	1	1	1	0	14	6	0	0	0	2	0	0	110
19. Harvard	125	10	2	4	0	0	0	3	0	0	0	0	0	0	0	0	0	0	6	4	0	0	0	0	1	0	30
20. Mona Grita Pt.	40	1	0	0	0	0	0	0	0	0	0	0	1	0	0	0	0	0	0	0	2	0	0	0	0	0	4
21. Gigante	140	2	0	0	0	0	0	0	0	0	0	1	0	0	0	0	0	0	0	0	0	10	0	0	0	0	13
22. Frijoles Road	156	0	0	0	0	0	0	0	0	0	0	0	0	0	0	0	0	0	0	0	0	0	2	0	0	0	2
23. Bohio	746	11	0	0	0	0	0	18	14	3	1	0	5	0	15	11	2	0	2	0	0	0	0	29	6	9	126
24. Buena Vista	143	4	0	0	0	0	0	1	0	0	0	0	0	0	0	1	4	0	0	1	0	0	0	5	23	0	39
25. Peña Blanca	138	3	0	1	0	0	1	1	0	0	1	0	0	0	1	1	0	0	0	0	0	0	0	7	0	15	31
Total	17,322*	1624	216	287	60	65	51	906	427	225	520	23	91	36	250	140	66	47	107	32	4	13	2	121	42	33	5388

* 15,736 band numbers.

TABLE 7-2.—Tabulation of total records, total other-captures, and percentage of other-captures (see text for explanation) of the total number of *Artibeus jamaicensis* in each locality group on BCI and adjacent mainland, 1975–1980. See Figure 7-1 for locations of locality groups.

Localities	(A) Total individuals	(B) Total other-captures	(B/A) Percent other-captures
1. Lutz	5309	1567	30
2. Barbour–Hood	640	233	36
3. Shannon–AMNH	730	305	42
4. Shannon–Balboa	165	65	39
5. Lake–Wheeler	156	67	43
6. Barbour Stream	234	57	24
7. Miller Ridge	2499	899	36
8. Fuertes	1189	421	35
9. Conrad	578	239	41
10. Plateau	1377	524	38
11. Drayton End	138	23	17
12. Armour End	456	92	20
13. Zetek 21	146	37	25
14. Standley Ridge	786	248	32
15. Standley End	649	142	22
16. Orchid Island	249	66	26
17. Gross Point	145	48	33
18. Chapman	388	111	29
19. Harvard	125	30	24
20. Mona Grita Point	40	4	10
21. Gigante	140	13	9
22. Frijoles Road	156	2	1
23. Bohio	746	126	17
24. Buena Vista	143	39	27
25. Peña Blanca	138	33	24

Mona Grita Point, 10%; Gigante, 9%; and Frijoles Road, 1%). These values show that more of the population was marked and a larger portion of the area frequented was sampled at the central localities. Bats moving in any direction from the central localities were likely to encounter other capture stations. Bats using the more isolated lakeshore localities probably foraged extensively on the mainland; hence, we sampled smaller fractions of their populations and the areas that they frequented. The effects of sampling small fractions of populations and their foraging areas are best seen on the mainland where stations were few and (except for Bohio) capture effort was low. We often netted at Bohio and it ranked sixth in numbers of *A. jamaicensis* caught. However, Bohio was nineteenth in the number of other-captures (17%, Table 7-2) and 83% of its *A. jamaicensis* were caught only once.

The lowest other-capture rates (Table 7-2) on BCI were at Drayton End (17%) and Armour End (20%). Among the most isolated BCI stations, and located in high, old forest containing few fruit trees, each of these two sites is over a kilometer from the next nearest station. Bats at the southern perimeter of BCI may forage on the adjacent mainland where the forest is younger and contains many fig trees. Among the few recaptures at Gigante and Mona Grita Point on the mainland were Armour End and Drayton End bats.

At the outset of the Bat Project we gridded BCI and planned to randomly net that grid. This plan was based on the perception that good (representative) catches night after night would result if we moved at random through our grid, using well-sited and well-set nets at choice netting stations. Our presumption was naive; on some nights we caught almost nothing, whereas on other nights we were overwhelmed with bats.

It became obvious that the nightly distribution of fruit bats was not random, but coincided with the presence of ripe fruit, which tends to be unevenly distributed in time and space. Nets set at certain fruiting trees caught bats, nets placed elsewhere often did not, and a locality might have an abundance of bats on one date and few on another. When there was only one preferred tree with ripe fruit on BCI, clearly that was the place to be for bats. There are many factors that influence the distribution and movements of fruit bats on BCI: distribution and abundance of preferred foods; rain, wind, moon phase, and cloud cover; the size, topography, and other limitations of insularity. These factors tended to overwhelm our efforts to set up meaningful measures of capture effort involving time and netting conditions.

TABLE 7-3.—Percentages of other-captures of *Artibeus jamaicensis* for selected locality groups on BCI weighted by the number of total captures of *A. jamaicensis* at each of the locality groups where the other-captures were recorded. Adjusted percentages were derived by dividing the number of other-captures (data from Table 7-1) by the number of total captures for that locality group and reducing that value to a percentage of its column total. The name of each locality group analyzed is followed by the total number of other-captures in parentheses.

Locality group	Total captures	Other-captures				
		Shannon–AMNH (305) %	Chapman (110) %	Lutz (1567) %	Miller Ridge (899) %	Standley End (248) %
1. Lutz	5309	6.35	3.51	7.09	6.05	1.25
2. Barbour–Hood	640	8.06	18.86	9.71	2.74	1.49
3. Shannon–AMNH	730	10.04	8.28	8.60	3.35	1.29
4. Shannon–Balboa	165	11.51	3.34	6.75	3.55	
5. Lake–Wheeler	156	6.95		7.52	10.02	9.20
6. Barbour Stream	234	1.17	2.36	6.77	1.25	
7. Miller Ridge	2499	2.71	0.44	5.39	6.60	3.45
8. Fuertes	1189	1.60	2.30	3.50	11.50	4.46
9. Conrad	578	2.82	2.85	4.87	14.04	1.68
10. Plateau	1377	10.26	1.59	6.09	7.02	1.72
11. Drayton–End	138	9.82	3.95	2.55	0.70	
12. Armour–End	456	1.79	1.12	1.03	3.00	
13. Zetek 21	146	1.87		4.02	4.01	9.82
14. Standley Ridge	786	1.38	0.71	2.46	6.34	16.48
15. Standley End	649	0.84	0.82	1.17	2.70	28.80
16. Orchid Island	249	3.28	2.19	2.36	3.13	1.92
17. Gross Point	145	1.87		4.45	7.41	
18. Chapman	388	7.00	19.79	4.99	0.51	1.25
19. Harvard	125	8.68	26.32	4.69	2.34	
20. Mona Grita Pt.	40			1.47		
21. Gigante	140			0.84		
22. Frijoles Road	156					
23. Bohio	746		1.48	0.79	2.35	7.04
24. Buena Vista	143			1.64	0.68	6.71
25. Peña Blanca	138	1.98		1.27	0.70	3.45
Total	17,322*					

* 15,736 band numbers.

Also, we had to adjust (prorate) capture rates because the capture effort differed among localities. We tried "recaptures per net-night" and "recaptures per net-hour," but these gave anomalous results, probably because netting effort commonly was inversely related to capture success. In other words, the poorer our catch, the more nets we maintained and the longer we worked them.

After realizing that capture effort measured in units of time was unreliable, we discovered that the numbers of *A. jamaicensis* caught per locality provided us the simplest and most useful measure of capture effort. *A. jamaicensis* regularly made up about two-thirds of our catch and was usually the commonest bat anywhere we netted. Our accumulated 17,322 mark and recapture records of this species varied between localities from a maximum of 5309 bats in the Lutz Watershed to a minimum of 40 at Mona Grita Point (Table 7-3). Using capture data from *A. jamaicensis* (Table 7-1), we derived weighted percentages of other-captures by dividing the number

of other-captures at a locality group by the total number of *A. jamaicensis* recorded at that same locality group. Each value was then converted to a percentage of its column total to produce the *A. jamaicensis*-weighted percentages plotted in Figures 7-3 through 7-8.

The efficacy of these methods for illustrating dispersion of other-capture records of *A. jamaicensis* around a sample center can be seen by comparing resulting percentages with the distance between each capture locality and the center (Table 7-4). When graphed (Figure 7-9), the data points approximate a line declining with increasing distance to zero indicating that the frequency of other-captures was proportional to distance from the sample center.

Another way we examined the correspondence of frequency of other-captures with distance from the sample center was by plotting the adjusted values on a map containing concentric circles with radii of 1, 2, and 3 km from the sample center (Figure 7-4). The Shannon-AMNH data fit rather well

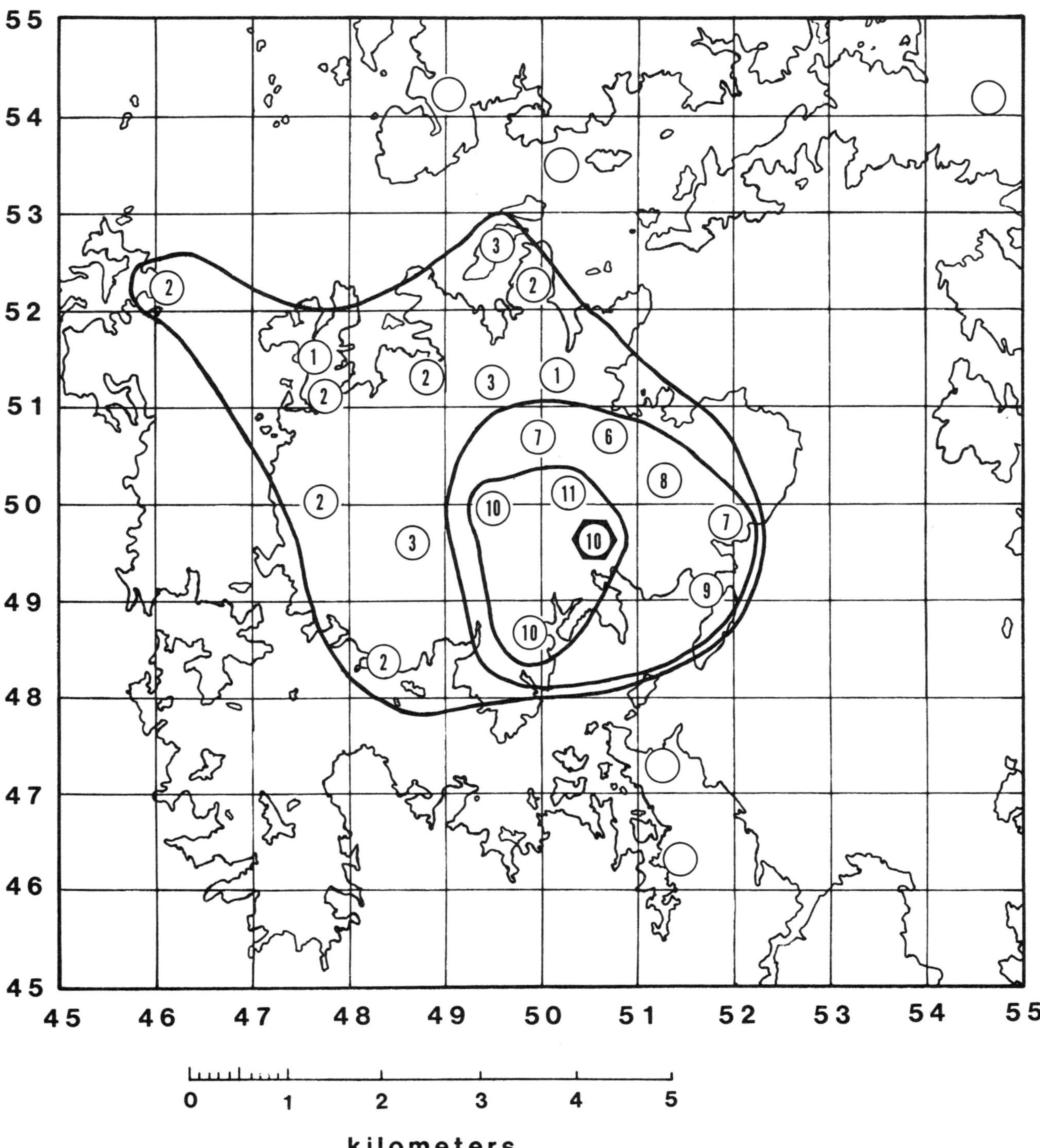

FIGURE 7-3.—Other-captures of individual *Artibeus jamaicensis* recorded at least once in the Shannon-AMNH locality group on BCI, 1976–1980 (expressed as a percentage of total other-captures of Shannon-AMNH bats and weighted by *A. jamaicensis* capture means; data are from Table 7-3; Shannon-AMNH is outlined with a hexagon). Isolines arbitrarily enclose locality groups with similar percentages of other-captures. Innermost line encloses core area. Marginal numbers identify kilometer squares (see Appendix, Methods of Capturing and Marking Tropical Bats).

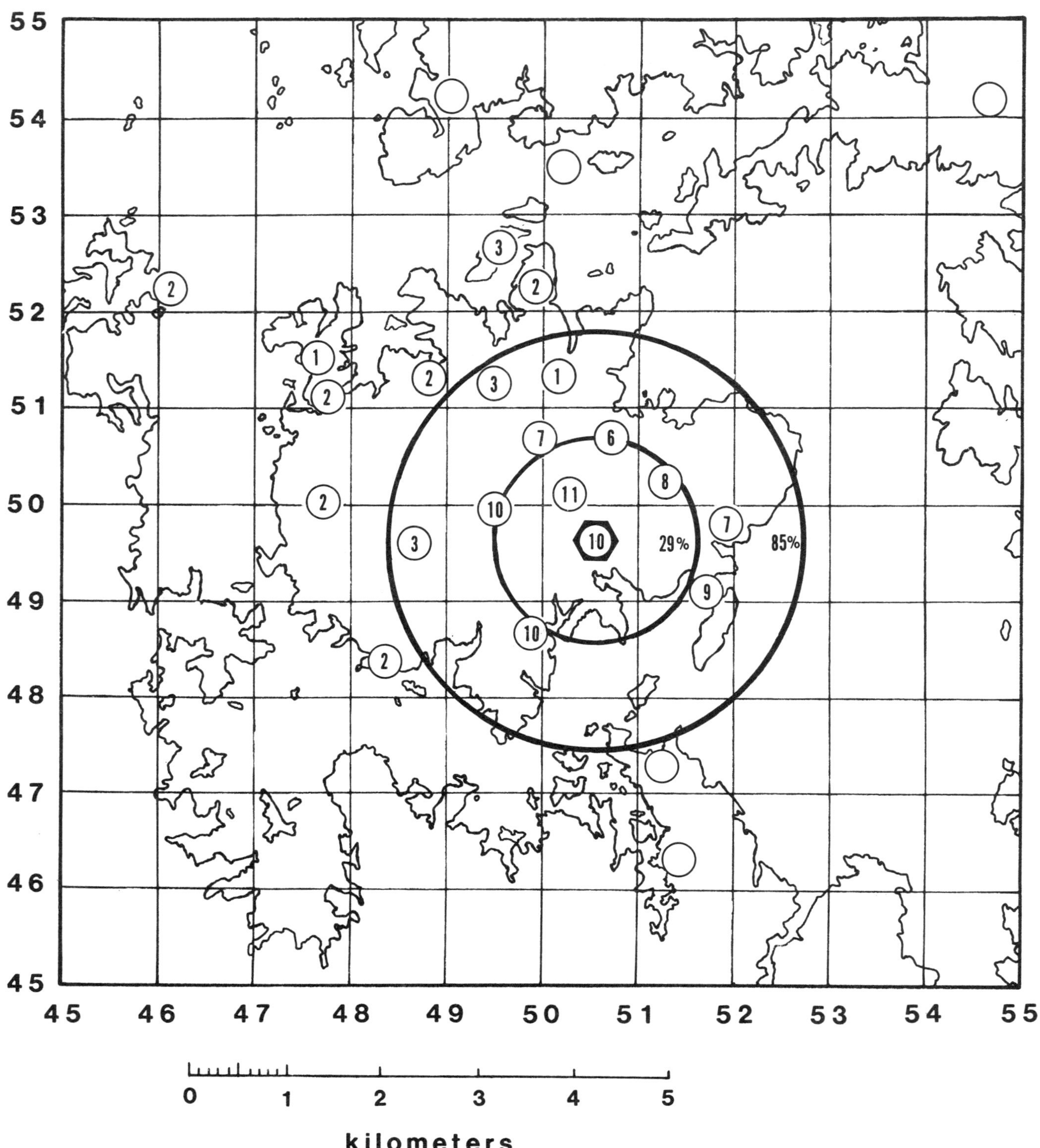

FIGURE 7-4.—Other-captures of individual *Artibeus jamaicensis* recorded at least once in the Shannon-AMNH locality group on BCI, 1976–1980 (expressed as a percentage of total other-captures of Shannon-AMNH bats and weighted by *A. jamaicensis* capture means). Circles at one- and two-kilometer intervals encompass 29% and 85%, respectively, of all other-captures. Shannon-AMNH is outlined with a hexagon.

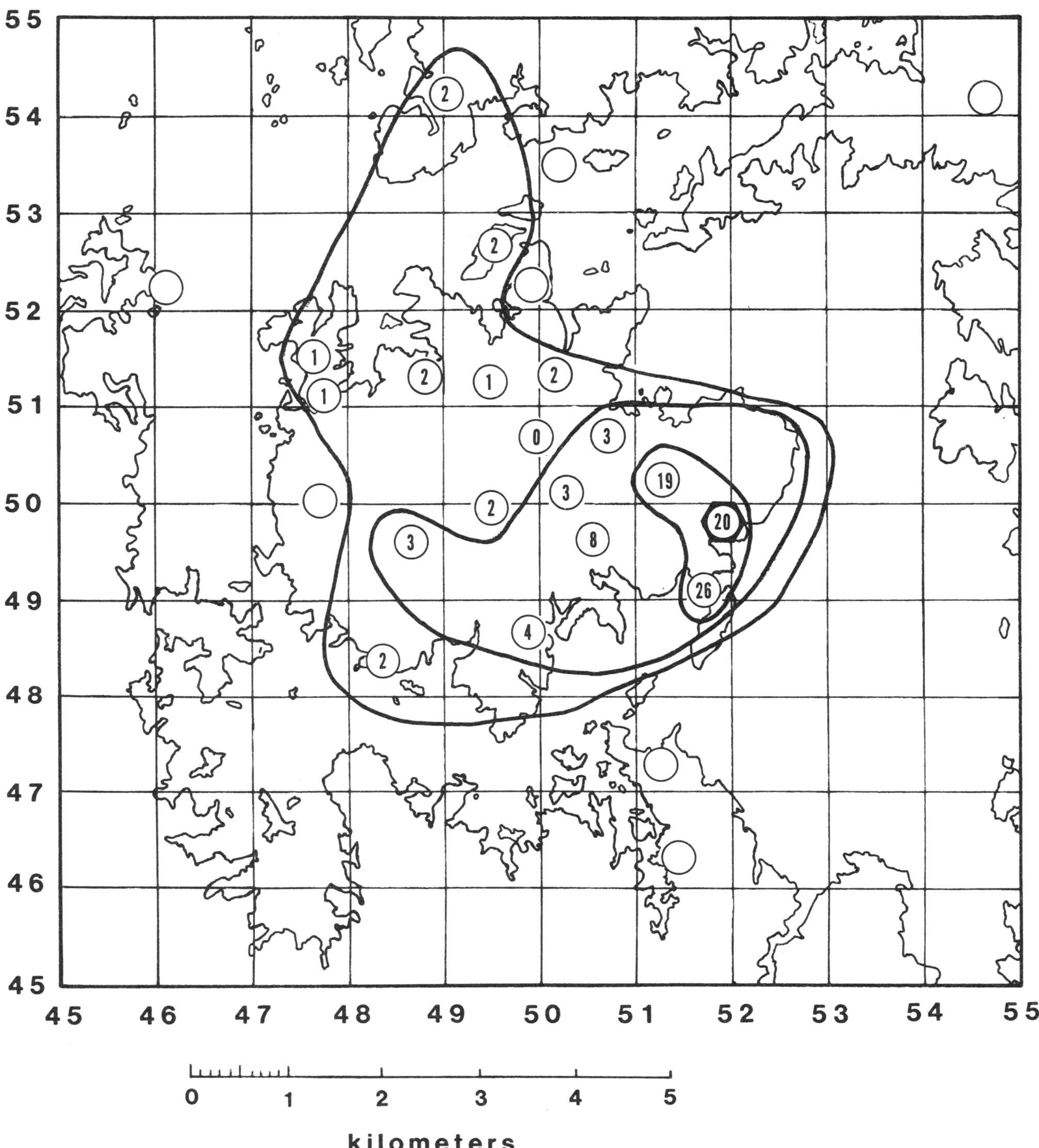

FIGURE 7-5.—Other-captures of individual *Artibeus jamaicensis* recorded at least once in the Chapman locality group on BCI, 1976–1980 (expressed as a percentage of total other-captures of Chapman bats, and weighted by *A. jamaicensis* capture means). Chapman is outlined with a hexagon. Isolines arbitrarily enclose locality groups with similar percentages of other-captures. Innermost line encloses core area. Marginal numbers identify kilometer squares (see Appendix, Methods of Capturing and Marking Tropical Bats).

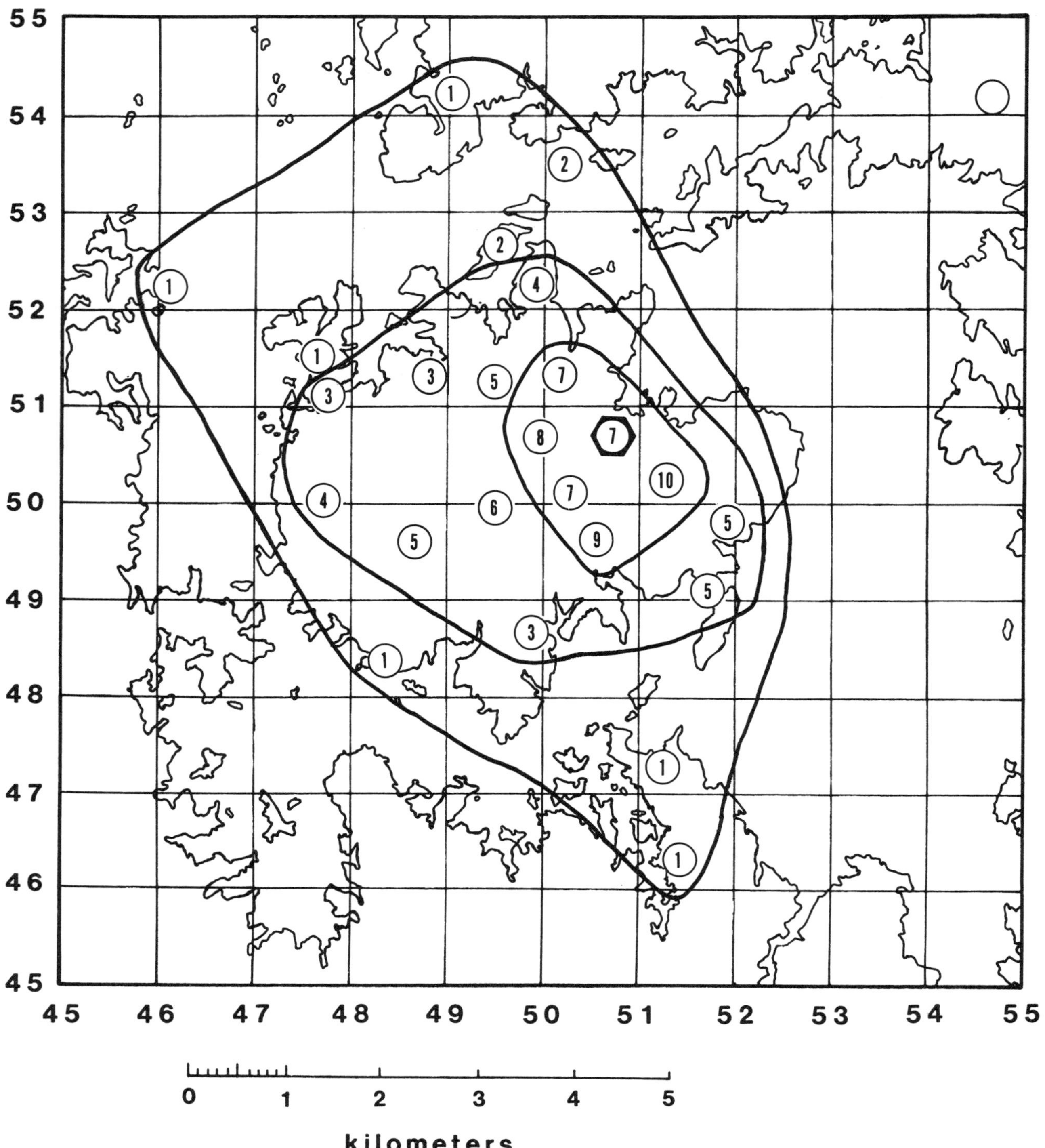

FIGURE 7-6.—Other-captures of individual *Artibeus jamaicensis* recorded at least once in the Lutz locality group on BCI, 1976–1980 (expressed as a percentage of total other-captures of Lutz bats, and weighted by *A. jamaicensis* capture means). Lutz is outlined with a hexagon. Isolines arbitrarily enclose locality groups with similar percentages of other-captures. Innermost line encloses core area. Marginal numbers identify kilometer squares (see Appendix, Methods of Capturing and Marking Tropical Bats).

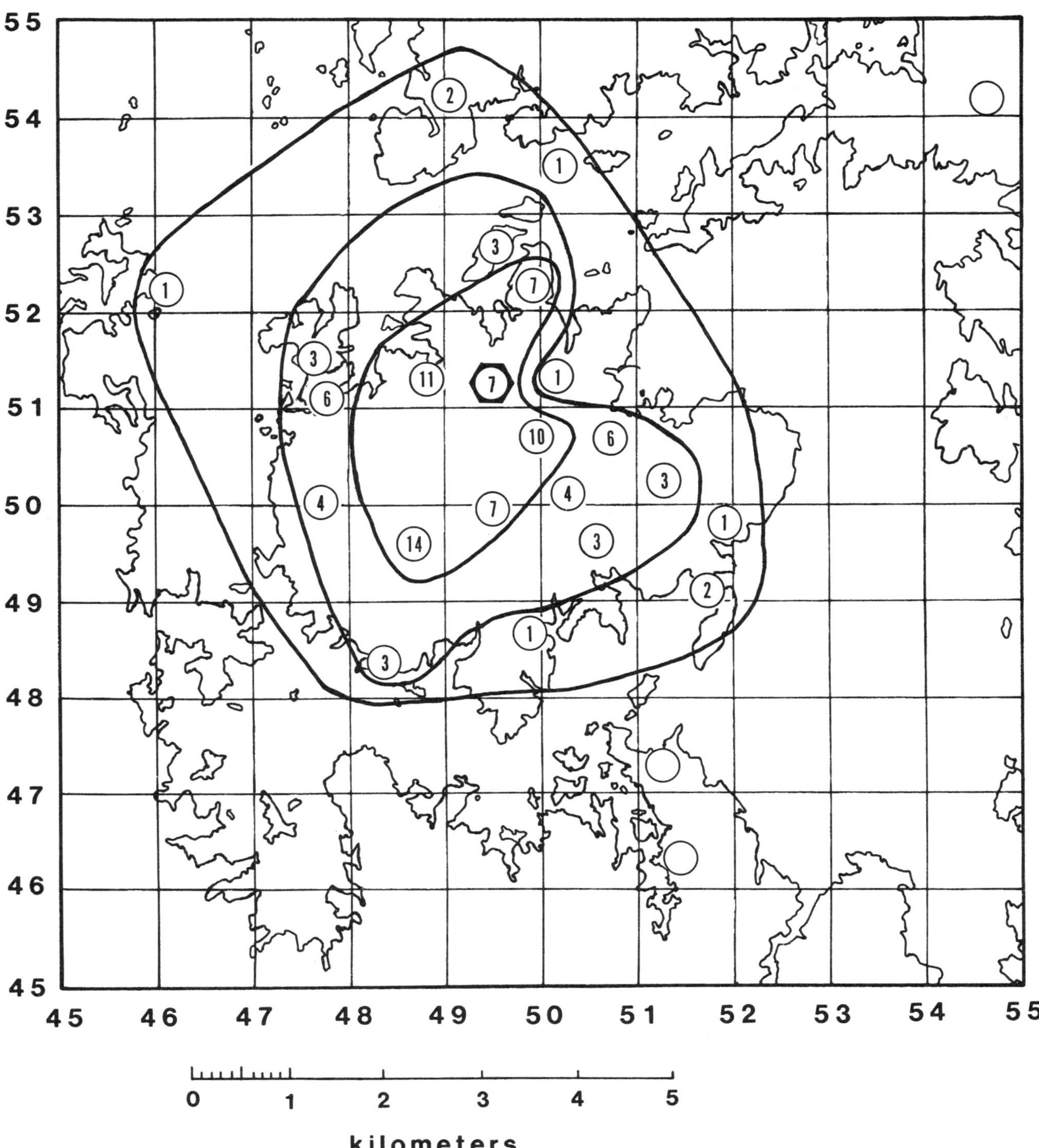

FIGURE 7-7.—Other-captures of individual *Artibeus jamaicensis* recorded at least once in the Miller Ridge locality group on BCI, 1976–1980 (expressed as a percentage of total other-captures of Miller Ridge bats, and weighted by *A jamaicensis* capture means). Miller Ridge is outlined with a hexagon. Isolines arbitrarily enclose locality groups with similar percentages of other-captures. Innermost line encloses core area. Marginal numbers identify kilometer squares (see Appendix, Methods of Capturing and Marking Tropical Bats).

  SMITHSONIAN CONTRIBUTIONS TO ZOOLOGY

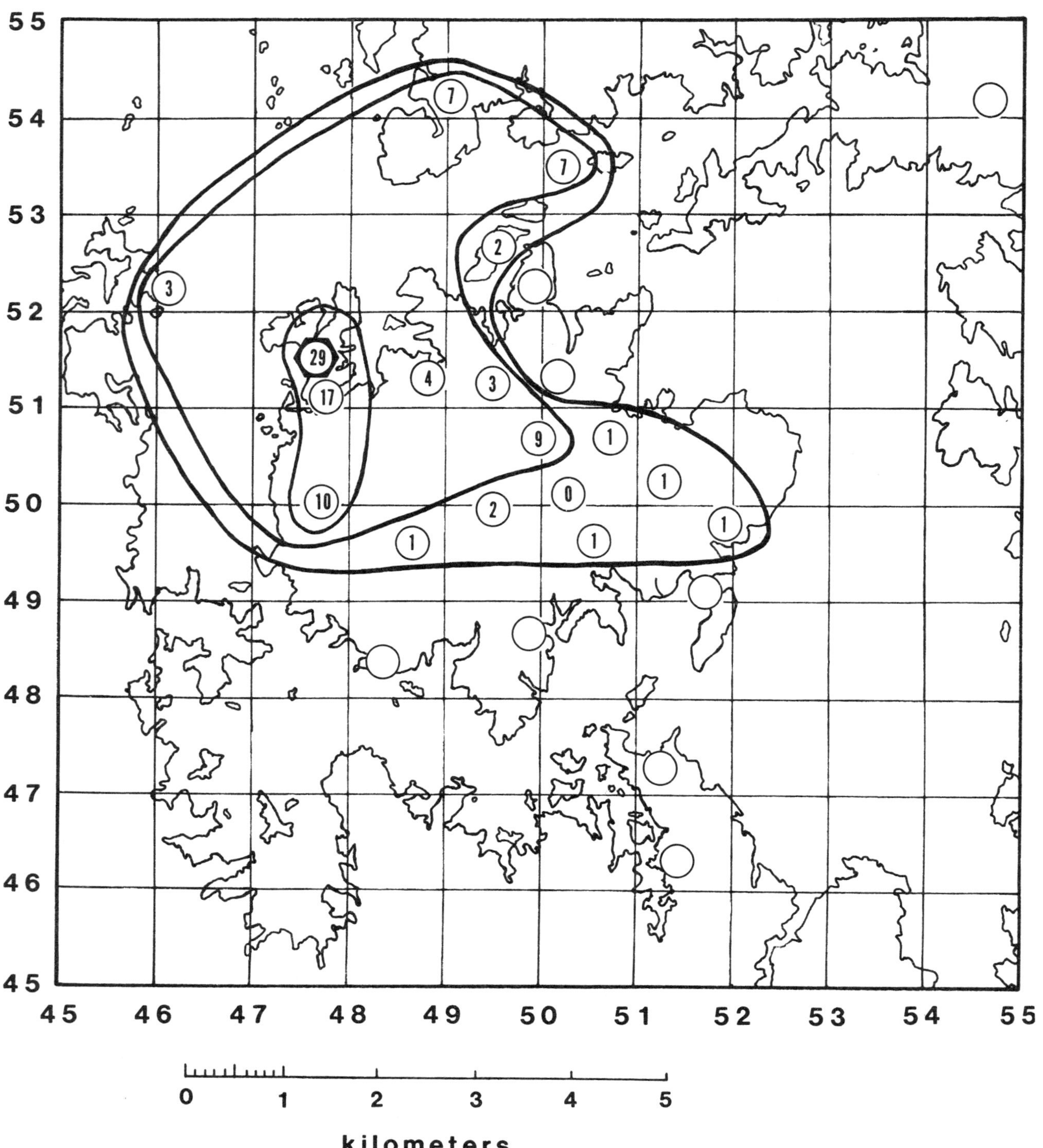

FIGURE 7-8.—Other-captures of individual *Artibeus jamaicensis* recorded at least once in the Standley End locality group on BCI, 1976–1980 (expressed as a percentage of total other-captures of Standley End bats, and weighted by *A. jamaicensis* capture means). Standley End is outlined with a hexagon. Isolines arbitrarily enclose locality groups with similar percentages of other-captures. Innermost line encloses core area. Marginal numbers identify kilometer squares (see Appendix, Methods of Capturing and Marking Tropical Bats).

TABLE 7-4.—Measures of dispersal of *Artibeus jamaicensis* caught at least once at Shannon–AMNH. See Figure 7-1 for locality group locations.

| Locality group | Distance* | Total (unweighted) | | Percent (A. jamaicensis weighted) | Percent (netting-night weighted) | Percent (net-hour weighted) |
		N†	%‡			
3. Shannon–AMNH	0	27	9	10	11	10
4. Shannon–Balboa	0.6	7	2	12	16	16
2. Barbour–Hood	1.0	19	6	8	9	12
11. Drayton–End	1.1	5	2	10	5	5
10. Plateau	1.1	51	17	10	11	8
1. Lutz	1.2	124	41	6	6	7
19. Harvard	1.3	4	1	9	3	4
5. Lake–Wheeler	1.3	4	1	7	3	3
18. Chapman	1.4	10	3	7	6	8
6. Barbour Stream	1.8	1	0.5	1	1	1
9. Conrad	2.0	6	2	3	4	3
7. Miller Ridge	2.1	25	8	3	4	4
20. Mona Grita Pt.	2.3	0	0	0	0	0
8. Fuertes	2.4	7	2	2	3	3
12. Armour–End	2.5	3	1	2	3	2
17. Gross Point	2.8	1	0.5	2	2	1
13. Zetek 21	2.9	1	0.5	2	2	2
14. Standley Ridge	3.2	4	1	2	2	1
16. Orchid Island	3.3	3	1	3	7	9
21. Gigante	3.4	0	0	0	0	0
15. Standley–End	3.5	2	1	1	1	1
24. Buena Vista	3.9	0	0	0	0	0
23. Bohio	4.8	0	0	0	0	0
25. Peña Blanca	5.2	1	0.5	2	1	0
22. Frijoles Road	6.2	0	0	0	0	0

* Distance in kilometers from Shannon–AMNH.

† Number of other-captures at each locality group of those individuals caught at least once at Shannon–AMNH.

‡ Other-captures of individuals recorded at least once at Shannon–AMNH (expressed as a percentage of total other-captures of Shannon–AMNH bats).

(compare Figures 7-3 and 7-4), with 29% of the localities within 1 km, 85% inside 2 km, and 93% within 3 km of the sample center.

Core Areas

Maps showing other-captures of bats taken at Chapman (an eastern locality; Figure 7-5), Lutz (an east-central locality; Figure 7-6), Miller Ridge (a west-central locality; Figure 7-7) and Standley End (a western locality; Figure 7-8) reveal little difference among the outlines of maximum extent of movements. Given sufficient time and a large enough sample of records, we suspect that bats of any locality on BCI could be caught at every other locality on the island as well as on the mainland. However, the decline of other-captures with distance in each data set (Figures 7-4 and 7-9) shows that the population of *A. jamaicensis* on BCI is not an amorphous, unstructured mass of mobile bats that fly randomly throughout the island. However mobile, individuals were more likely to be found in a particular part of the island than elsewhere.

The mapped distributions of other-captures (Figures 7-5 through 7-8) each contain a central locality whose records were analyzed for dispersion. Each map has three concentric isolines around the central locality, enclosing all localities with other-captures of bats caught at the central locality. The focal localities of the four core areas (Figures 7-5 through 7-8; combined in Figure 7-10) are spaced at intervals of about 1.5 to 2 km from east to west across BCI. Although the eastern (Chapman) and western (Standley End) core areas are smaller, the core areas overlap each other to a similar extent.

The smaller sizes of the Chapman and Standley End core areas most likely resulted from undersampling the mainland parts of their ranges. The area bound by the second isoline (88% of other-captures) of Standley End (Figure 7-8) includes several points on the mainland, but much less of BCI than is covered by the frequently used areas of Miller Ridge (Figure 7-7) and Lutz (Figure 7-6) populations. The Chapman (Figure 7-5) area of frequent use is almost as limited on BCI as the

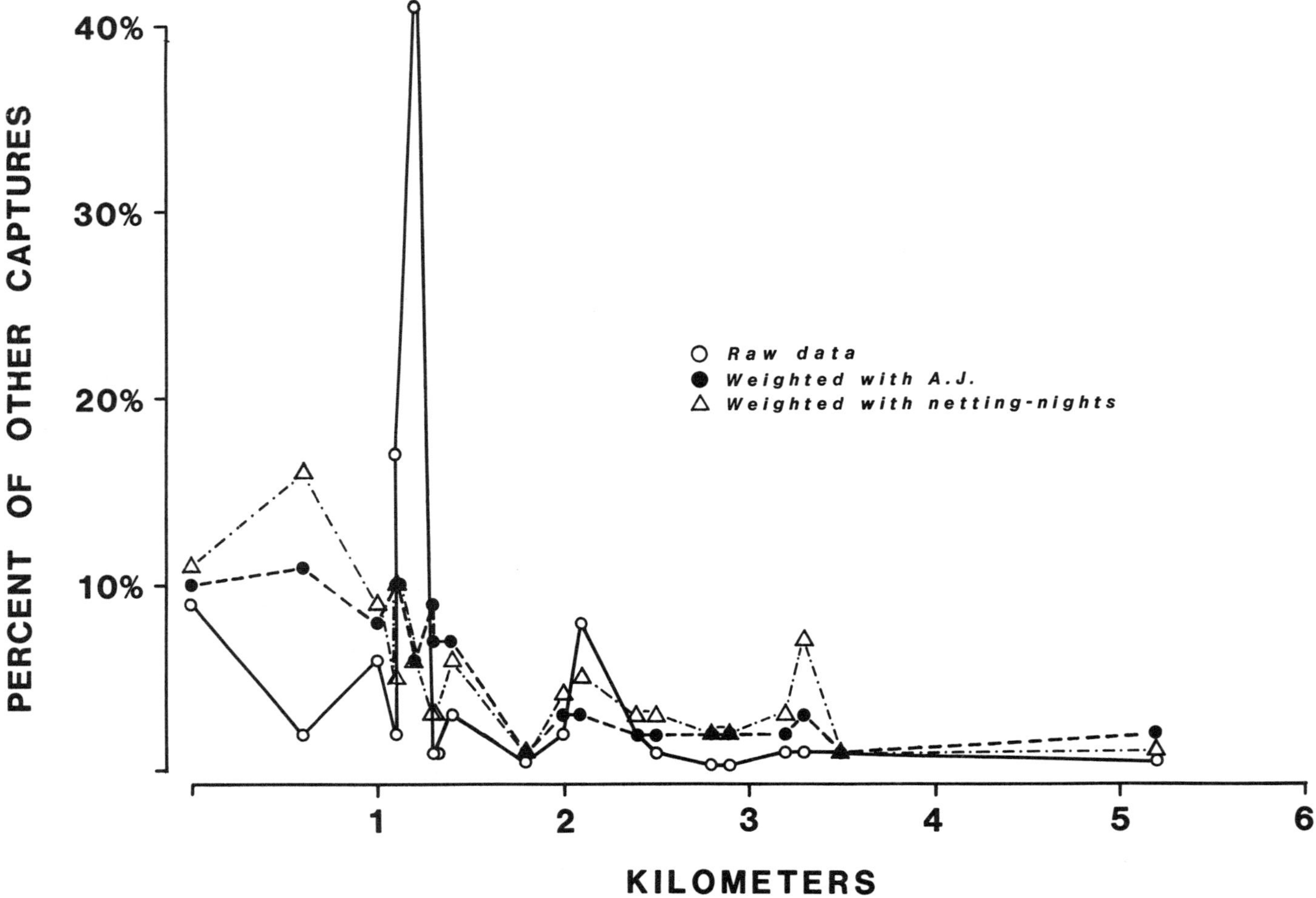

FIGURE 7-9.—Comparison of three methods of illustrating distribution of percentages of other-captures of individual *Artibeus jamaicensis* recorded at least once in the Shannon-AMNH locality group on BCI, and captured again there and elsewhere, 1976–1980. The graph correlates frequency of other-captures with distance from Shannon-AMNH at zero. Data points are from Table 7-4.

corresponding area of Standley End. We do not have records from the northeastern corner of BCI and the mainland to the east where we never netted.

The core areas of most of our capture localities are plotted in Figures 7-11 through 7-13. The convoluted shapes of some of the areas such as those of Orchid Island (Figure 7-11), Shannon-Balboa (Figure 7-12) and Armour End (Figure 7-13) are the products of inadequate sample sizes from some of the localities within the core other than the central locality. When the plotted core areas (Figures 7-1 through 7-13) are superimposed, even disregarding Orchid Island, Shannon-Balboa, and Armour End because of incomplete sampling, there is little of BCI left outside of a core area. Ten cores overlap the Lutz core, 12 overlap the Plateau core, and 13 overlap the Miller Ridge core.

Although the central localities of Lutz, Barbour-Hood, and Shannon-AMNH are close together (each about a kilometer from the other) and their core areas overlap (Figure 7-11), each seems to represent a slightly different group of *A. jamaicensis*. In contrast, the core areas of Fuertes, Conrad, and Miller Ridge are neatly nested as though they all represented the same group of bats (Figure 7-13).

Perhaps the most discrete units in the population of *A. jamaicensis* on BCI are the harem groups permanently based in tree holes. The bats netted at a particular locality would be expected to include those that have day roosts there, those that come to forage, and those that pass by enroute between day roosts and feeding areas elsewhere. Discreteness of core areas may distinguish among these bats.

We compared the percentages of other-captures at each

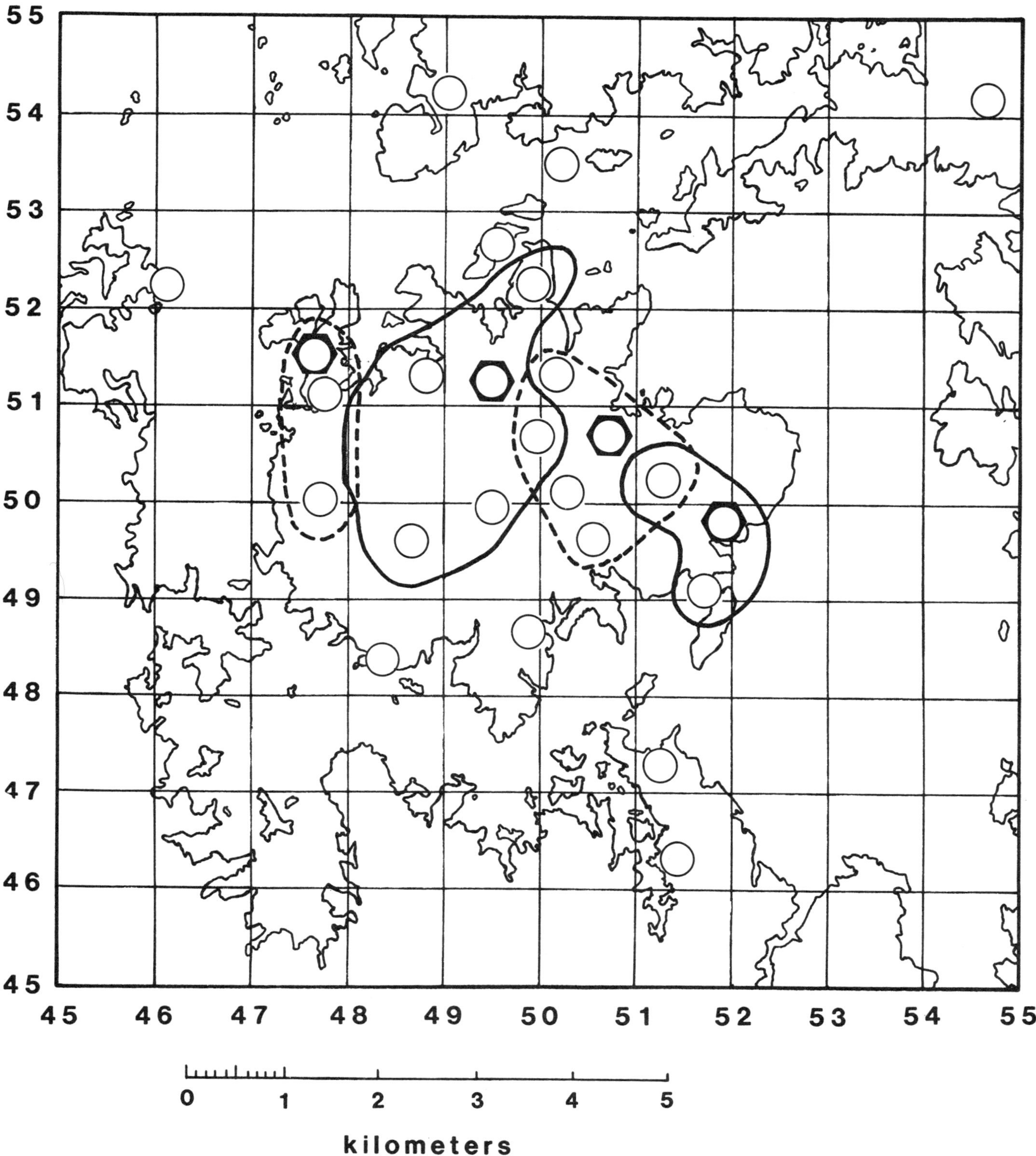

FIGURE 7-10.—Overlap of core areas shown in Figures 7-5 through 7-8.

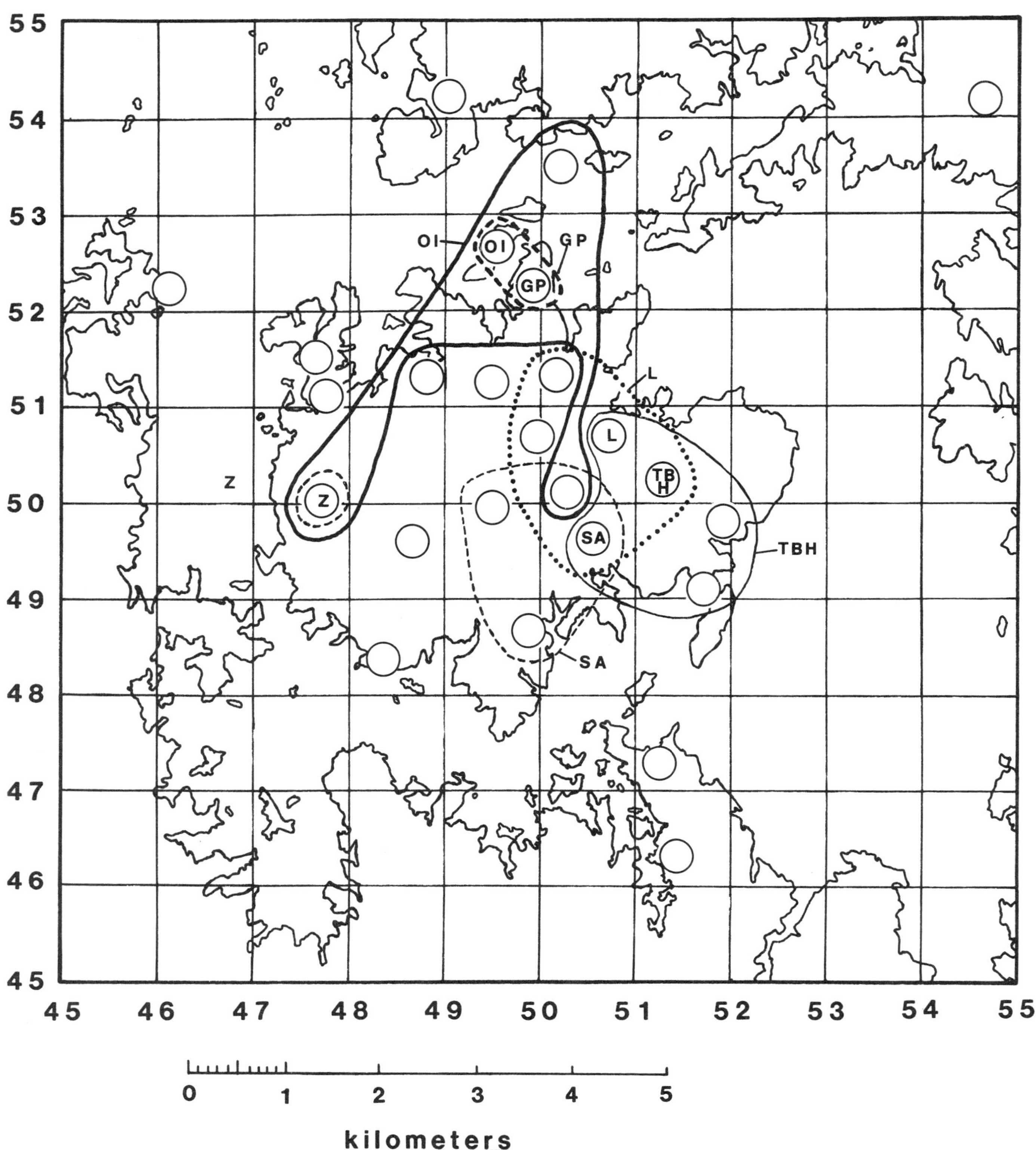

FIGURE 7-11.—Overlap of core areas on the East-central and peripheral portions of BCI. GP = Gross Point; L = Lutz; OI = Orchid Island; SA = Shannon-AMNH; TBH = Thomas Barbour-Hood; Z = Zetek.

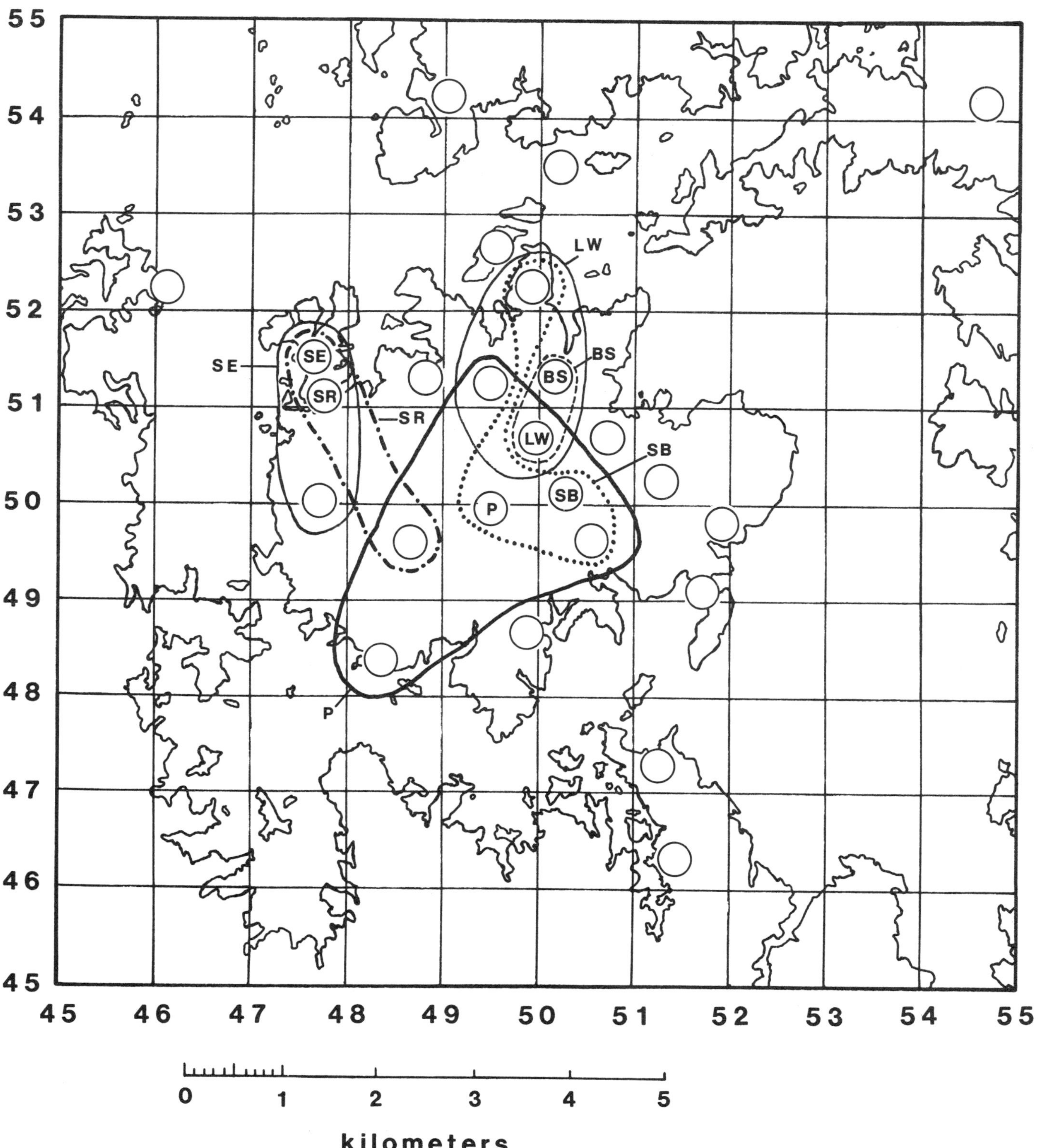

FIGURE 7-12.—Overlap of core areas in central and peripheral portions of BCI. BS = Barbour Stream; LW = Lake-Wheeler; P = Plateau; SB = Shannon-Balboa; SE = Standley End; SR = Standley Ridge.

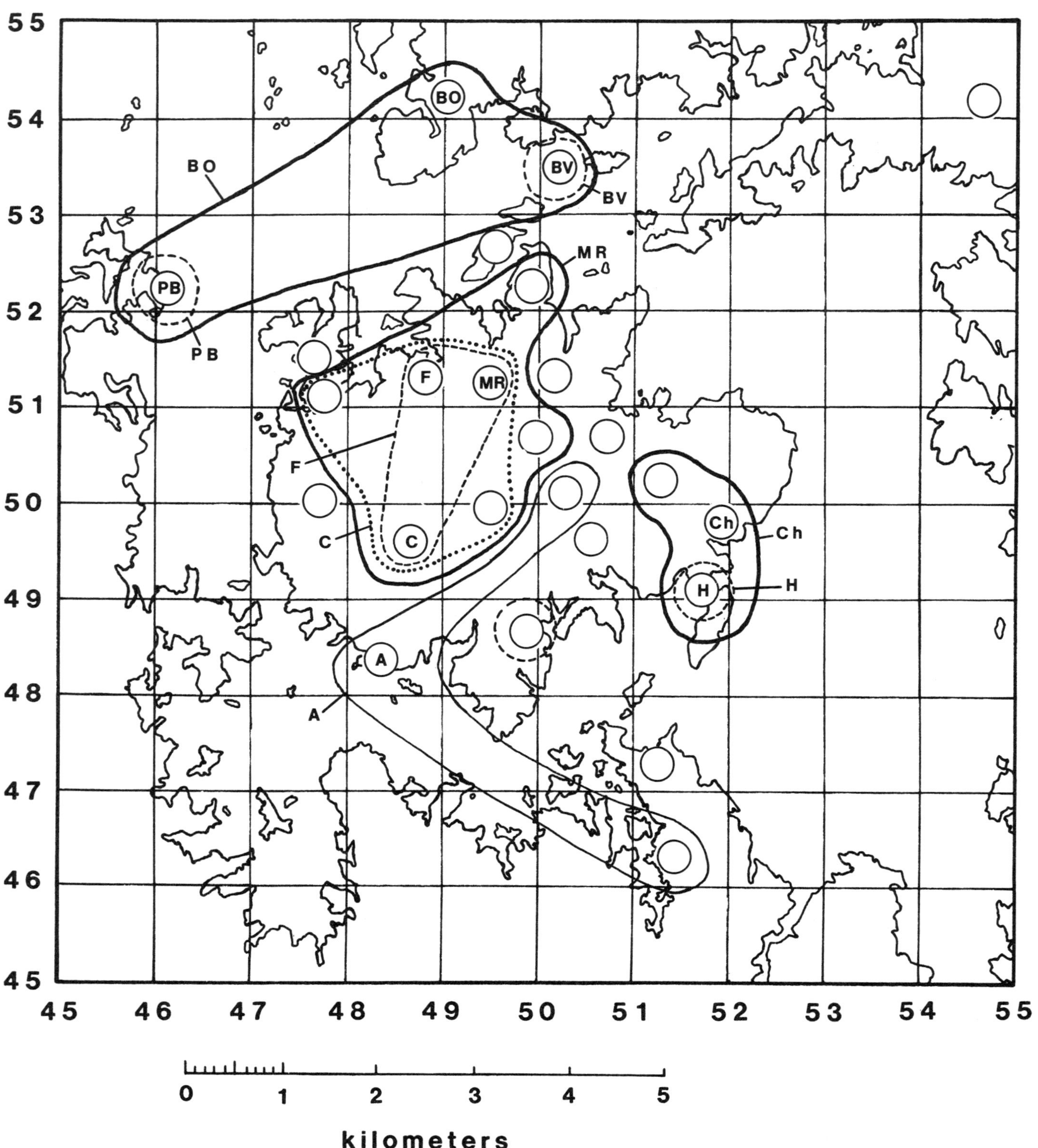

FIGURE 7-13.—Overlap of core areas in West-central and peripheral portions of BCI. A = Armour End; BO = Bohio; BV = Buena Vista; CH = Chapman; C = Conrad; D = Drayton End; F = Fuertes; MR = Miller Ridge; PB = Peña Blanca.

TABLE 7-5.—Core areas on BCI and vicinity in which the central locality had a lower percentage of other-captures (see text for explanation) of *Artibeus jamaicensis* (captured at least once there) than some other localities in the core. Each entry includes the central locality (CAPITALIZED), percentage of other-captures* there, rank among all localities in terms of number of individual *A. jamaicensis* caught, and percentages of other-captures* at other localities within the core. * = *A. jamaicensis*-weighted.

Locality	Total captures	Percent other-captures	Rank
LUTZ	5309	7	1
Barbour-Hood		10	
Shannon-AMNH		9	
Lake-Wheeler		8	
Barbour Stream		7	
Shannon-Balboa		7	
MILLER RIDGE	2499	7	2
Fuertes		11	
Lake-Wheeler		10	
Conrad		14	
Gross Point		7	
Standley Ridge		6	
Plateau		7	
PLATEAU	1377	6	3
Shannon-Balboa		14	
Shannon-AMNH		12	
Conrad		11	
Lake-Wheeler		8	
Miller Ridge		7	
Armour-End		6	
FUERTES	1189	13	4
Miller Ridge		14	
Conrad		11	
BOHIO	746	16	6
Peña Blanca		27	
Buena Vista		18	
SHANNON-AMNH	730	10	7
Shannon-Balboa		11	
Plateau		10	
Drayton-End		10	
BARBOUR-HOOD	640	11	9
Chapman		24	
Harvard		14	
Shannon-AMNH		11	
Barbour Stream		10	
CONRAD	578	10	10
Plateau		15	
Miller Ridge		13	
Fuertes		12	
Standley Ridge		11	
ARMOUR-END	456	13	11
Mona Grita Point		15	
Shannon-Balboa		12	
CHAPMAN	388	20	12
Harvard		26	
Barbour-Hood		19	
ORCHID ISLAND	249	9	13
Gross Point		44	
Buena Vista		15	
Shannon-Balboa		10	
Zetek 21		9	
SHANNON-BALBOA	165	14	15
Orchid Island		19	
GROSS POINT	145	†	19
Orchid Island		70	

† No other-captures recorded.

TABLE 7-6.—Core areas on BCI and nearby mainland in which the central locality had a higher percentage of other-captures (see text for explantion) of *Artibeus jamaicensis* (captured at least once there) than other localities in the core. Each entry includes the central locality (CAPITALIZED), percentage of other-captures* there, rank among all localities in terms of number of individual *A. jamaicensis* caught, and percentages of other-captures* at other localities within the core. * = *A. jamaicensis*-weighted.

Locality	Total captures	Percent other-captures	Rank
STANDLEY RIDGE	786	23	5
Standley-End		14	
Conrad		10	
STANDLEY-END	649	29	8
Standley Ridge		16	
Zetek 21		10	
BARBOUR STREAM	234	36	14
Lake-Wheeler		18	
LAKE-WHEELER	156	26	16
Barbour Stream		12	
Miller Ridge		10	
Gross Point		10	
ZETEK 21	146	42	18
BUENA VISTA	143	82	20
DRAYTON-END	138	31	22
PEÑA BLANCA	138	84	23
HARVARD	125	49	24
Chapman		24	

locality within each core (Tables 7-5 and 7-6). Percentages of other-captures at the central localities ranged from 6% to 84%. The lower percentages at Lutz, Miller Ridge, and Plateau must reflect the numbers of captures, which were highest at these three localities. If other-captures at a central locality are predominantly local residents, their representation out of all bats caught at that locality will be inversely related to the total number caught. The greater the number of individual bats involved, the greater the likelihood that other-captures will be widely dispersed, thus also reducing the proportion of other-captures at a central locality.

The high percentages of other-captures at Buena Vista and Peña Blanca simply demonstrate fidelity to those sites, which are day roosts in channel markers. Whenever we sampled these roosts, most of the bats caught were ones taken there before. The low percentages of these bats netted away from their day roosts suggest that they foraged where we did not net, probably on the mainland. A comparison of the Bohio core area (Figure 7-13) and the Standley End area of frequent use (Figure 7-8) shows that while these are places where Buena Vista and Peña Blanca bats foraged, they are probably not their main feeding areas. There are extensive areas of the mainland near these roosts where we have never netted (Figure 7-13).

There is a dichotomy of core areas into those whose central localities have both lower percentages and fewer other-captures than other localities in the core (Table 7-5) and those whose central localities have higher percentages and more other-captures than do neighboring localities (Table 7-6). This division may be correlated with the number of captures.

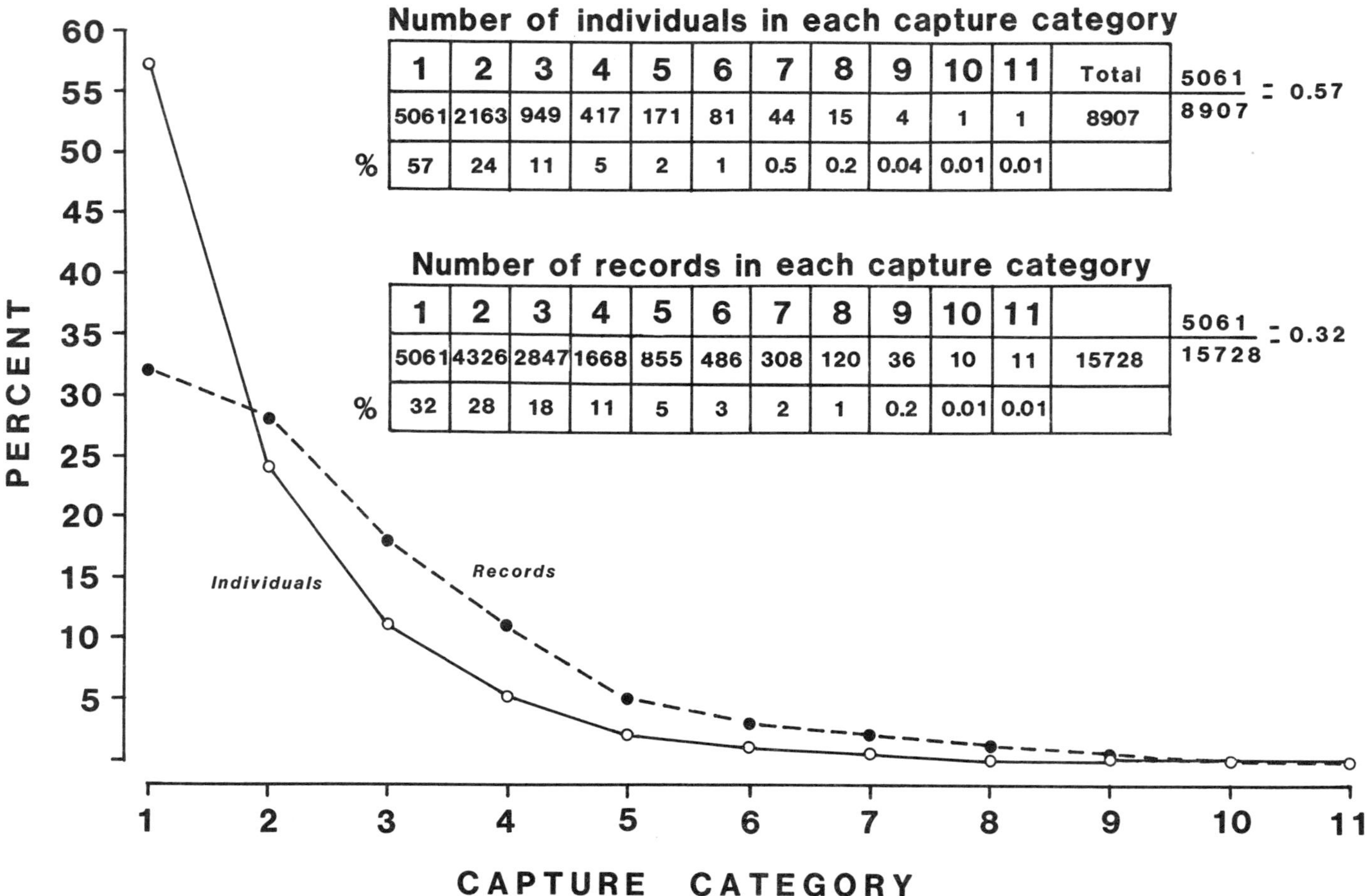

	1	2	3	4	5	6	7	8	9	10	11	Total
	5061	2163	949	417	171	81	44	15	4	1	1	8907
%	57	24	11	5	2	1	0.5	0.2	0.04	0.01	0.01	

$$\frac{5061}{8907} = 0.57$$

	1	2	3	4	5	6	7	8	9	10	11	
	5061	4326	2847	1668	855	486	308	120	36	10	11	15728
%	32	28	18	11	5	3	2	1	0.2	0.01	0.01	

$$\frac{5061}{15728} = 0.32$$

FIGURE 7-14.—Frequency of capture of *Artibeus jamaicensis* on BCI. Number of individuals caught 1, 2, 3, 4, and more times, and number of records in suites of individual bats in each capture category.

Standley Ridge and Standley End (Table 7-6) are exceptions in that comparatively high numbers of *A. jamaicensis* were caught at each (they ranked fifth and eighth, respectively), yet their numbers of other-captures are also relatively high (see discussion of Fidelity, below, in this section).

The distribution of fig trees probably exerts an influence on the relative numbers of other-captures at any locality. The first seven localities in Table 7-5 (Lutz through Barbour-Hood) have an abundance of fig trees. Combined, Lutz and Miller Ridge have over 200 fig trees, mostly *Ficus insipida,* the favored fruit of *A. jamaicensis.* Chapman, Standley Ridge, and Standley End have a fair number of fig trees, but not many *F. insipida.* The other localities in Tables 7-5 and 7-6 have few fig trees other than strangler figs (mainly *F. obtusifolia*) whose fruits are not preferred by *A. jamaicensis.*

If the number of *A. jamaicensis* harems is limited by the availability of suitable tree holes, and if trees with suitable holes are evenly distributed over BCI then there must be many harem roosts that are not near patches of fig trees. At these places we should catch the bats of resident harems and few others, thus accounting for high percentages of other-captures characterizing those sites. The bats of day roosts not located near fig trees must routinely travel to other parts of the island to feed. However, bats with day roosts near fig trees also commute to other localities to forage whenever the trees near their roosts lack ripe fruit. Nevertheless, these bats need to travel less than bats residing in tree holes in fig-poor places. Turner (1975) showed that *Desmodus rotundus* routinely changed roost sites to remain close to a spacially shifting resource. We assume that either roosting sites are limited on BCI or that *A. jamaicensis* have high fidelity to individual day roosts because we found no evidence of routine shifts even among bachelor males.

Bachelor males roost in foliage and not in tree holes. Although they might be expected to concentrate in fig-rich areas and to shift concentrations to follow fruit availability, they apparently do not, judging by the unusually high ratio of males consistently found at Standley End.

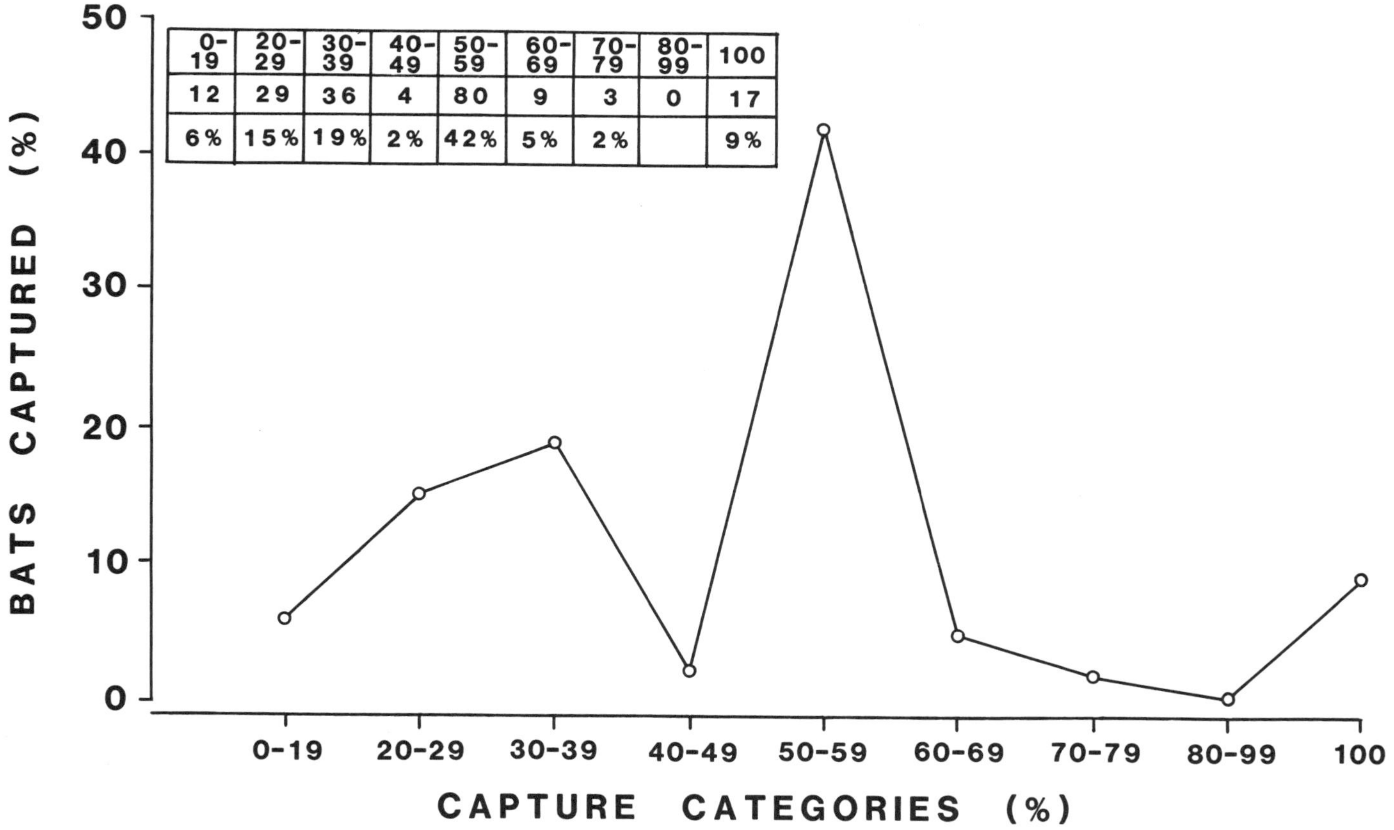

FIGURE 7-15.—Fidelity of *Artibeus jamaicensis* to the Chapman locality group on BCI. Frequency of capture of individual bats with records of multiple captures that were caught at least once at Chapman are graphed as percentages of records in various fractional capture categories ($^1/_6$ or less of records = 0%–19%, $^1/_5$–$^1/_4$ = 20%–29%, $^1/_3$ = 30%–39%, $^2/_5$ = 40%–49%, $^1/_2$ = 50%–59%, $^3/_5$–$^2/_3$ = 60%–69%, $^3/_4$ = 70%–79%, $^4/_5$–$^5/_6$ = 80%–99%, all = 100%). Row A in the table lists fractional capture categories (as percentages), row B lists number of individual bats whose records at Chapman fall into each category (n = 190), and in row C the numbers are converted to percentages. See also Table 7-7.

Fidelity

We examined fidelity to a particular locality group by analyzing suites of multiple capture records of individual bats. For each locality group we compiled records of all *A. jamaicensis* having multiple captures that were taken at least once in that particular locality group between October 1976 and October 1980. For each bat, we tabulated total captures, the number of captures in that particular locality group, and percentage of its total records in that locality group. From that tabulation we grouped those having multiple captures according to the percentage of their occurrence at that locality group among total records (e.g., Chapman, Figure 7-4).

The distribution of records among fractional categories is influenced by the number of records per individual suite. In all, 3846 *A. jamaicensis* were recaptured (Figure 7-14). Most (57%) were recaptured only once (i.e., had two captures), 24% were recaptured twice (three captures), 11% recaptured three times (four captures), and 8% recaptured four or more times. Thus, peaks would be expected (Figure 7-15) at 50% and 100% for two-record suites; 33%, 66%, and 100% for three records; and 25%, 50%, 75%, and 100% for four records, and so forth. Because most recaptured bats were caught again only once, 50% and 100% peaks per locality should occur most often.

These data indicate fidelity to a particular locality group. High fidelity means a high frequency of records in the 60%–100% interval, which represents $^3/_5$, $^2/_3$, $^3/_4$, $^4/_5$, or more captures in the same area. The hiqher the frequency of same-site captures, the higher the fidelity. Low fidelity is revealed by a higher frequency of records in the 0%–40% interval ($^2/_5$, $^1/_3$, $^1/_4$, $^1/_5$, and $^1/_6$ of captures or less) for that locality group. A simplified summary (Table 7-7) of locality-group fidelity, like that in Figure 7-15, shows three basic patterns among our data.

Pattern I (e.g., Armour End; Figure 7-16, Table 7-7) covers most of the locality groups. There is a strong similarity among

TABLE 7-7.—Fidelity of *Artibeus jamaicensis* to locality groups on BCI and nearby mainland as indicated by frequency of capture in each locality group of individual bats with records of multiple captures. Shown as percentages of records in various fractional capture categories (expressed as percentages), first for all *A. jamaicensis* caught at least once in a particular locality group and then for the subset of bats caught two or more times in that locality group. Locality groups are sorted into three patterns depending on degree of fidelity (see text). Capture effort is shown in terms of netting nights in each locality group.

Locality group	Number of bats (A)	Number of netting nights	One or more captures (Fractions of records)				Two or more captures (Fractions of records)				Percent of total bats (B/A)
			0–2/5 (0–40%)	0–1/2 (0–50%)	1/2 (50%)	3/5–all (60–100%)	0–2/5 (0–40%)	1/2 (50%)	3/5–all (60–100%)	N (B)	
PATTERN I											
Armour End	175	11	39	92	53	8	0	0	100	15	8.5
Chapman	190	17	42	84	42	16	11	11	78	35	18.4
Drayton End	59	10	37	86	49	14	0	11	89	9	15.3
Fuertes	639	26	34	87	53	13	9	21	70	117	18.3
Orchid Island	142	4	49	97	48	3	0	33	66	6	4.2
Conrad	310	14	46	94	48	6	17	29	54	35	11.3
Shannon–Balboa	117	4	63	98	35	2	25	25	50	4	3.4
Shannon–AMNH	377	24	50	91	41	9	21	11	68	47	12.5
Lake–Wheeler	102	12	57	96	39	4	29	14	57	7	6.8
Harvard	51	11	27	88	61	12	11	22	66	9	17.6
Gross Point	109	5	51	99	48	1	0	0	100	1	0.9
Barbour Stream	141	14	43	86	43	14	9	5	86	22	15.6
Barbour–Hood	347	20	49	93	44	7	17	24	59	41	11.8
Plateau	702	46	42	88	46	12	18	17	65	126	17.9
Miller Ridge	1321	52	37	82	45	18	14	18	68	338	25.6
Standley Ridge	407	16	42	85	43	15	8	21	71	86	21.1
Zetek 21	75	6	56	92	36	8	0	0	100	6	8.0
PATTERN II											
Bohio	233	14	33	70	37	30	4	4	92	77	33.0
Standley End	275	25	29	69	40	31	4	13	83	102	37.1
Lutz	2075	197	28	68	40	32	7	9	83	807	38.9
PATTERN III											
Buena Vista	50		12	48	36	52	0	7	93	28	56.0
Peña Blanca	32		12	34	22	66	4	4	92	23	71.9

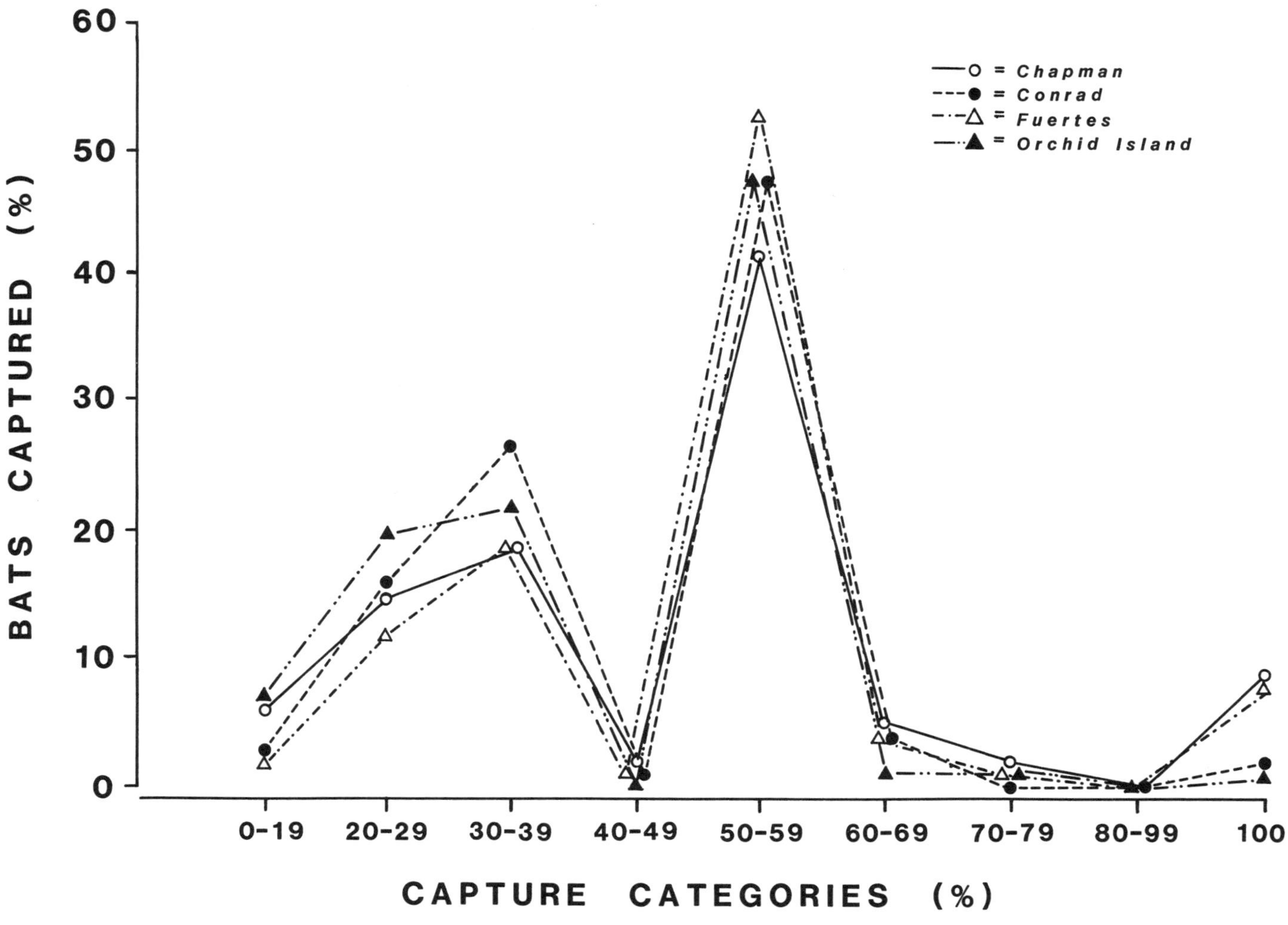

FIGURE 7-16.—Fidelity of *Artibeus jamaicensis* to Pattern I locality groups on and near BCI: Chapman, Conrad, Fuertes, and Orchid Island. See Figure 7-15 for further explanation.

the capture-category tabulations for locality groups with this pattern in spite of their diverse locations and the disparity in numbers in each suite of multiple capture records (from 51 to 1321). These locality groups share a low degree of fidelity (<19% in the 60%–100% column) and a high frequency of single captures (82%–99% in the 0%–40% and 0%–50% columns).

Pattern II (Lutz, Bohio, and Standley End; Figure 7-17, Table 7-7) comprises three seemingly unrelated locality groups whose records of fidelity are remarkably similar. All three show higher fidelity, compared with the Pattern I locality groups (30%–32% in the 60%–100% column), and fewer single records (68%–70% in the 0%–40% and 0%–50% columns). Their records may be similar for different reasons. We sampled little of the probable foraging range of the Bohio bats, so they had little risk of capture except at Bohio, thereby appearing to have high fidelity to that location. In contrast, the bats of Standley End may actually be sedentary because they were recaptured mostly at Standley End or nearby (see Figure 7-8 and discussion of Core Areas earlier in this section). Or, if

we sampled only part of their foraging areas, the situation at Standley End may be similar to that at Bohio.

We are confident that we have sampled the full extent of the foraging ranges of Lutz bats. High fidelity to Lutz may reflect a larger local resident population, or may result from catching the same individuals repeatedly because of the attraction of the fig grove in Lutz Ravine. These fig trees may adequately support local populations of *A. jamaicensis* and periodically attract bats from afar when ripe figs are scarce elsewhere. If this is true then some of the bats showing high fidelity are actually opportunistic transient foragers rather than Lutz residents.

Pattern III (Buena Vista and Peña Blanca; Figure 7-18, Table 7-7) shows the highest fidelity. The obvious explanation for this pattern is that we sampled only a fraction of the foraging ranges of these bats, and because we captured them mostly at their day roosts. We netted bats only at night at all other locality groups and, evidently, we did not net where the bats of Buena Vista and Peña Blanca most often foraged.

The degree of fidelity within patterns appears to be related to netting effort. On this basis we sorted Pattern I locality groups

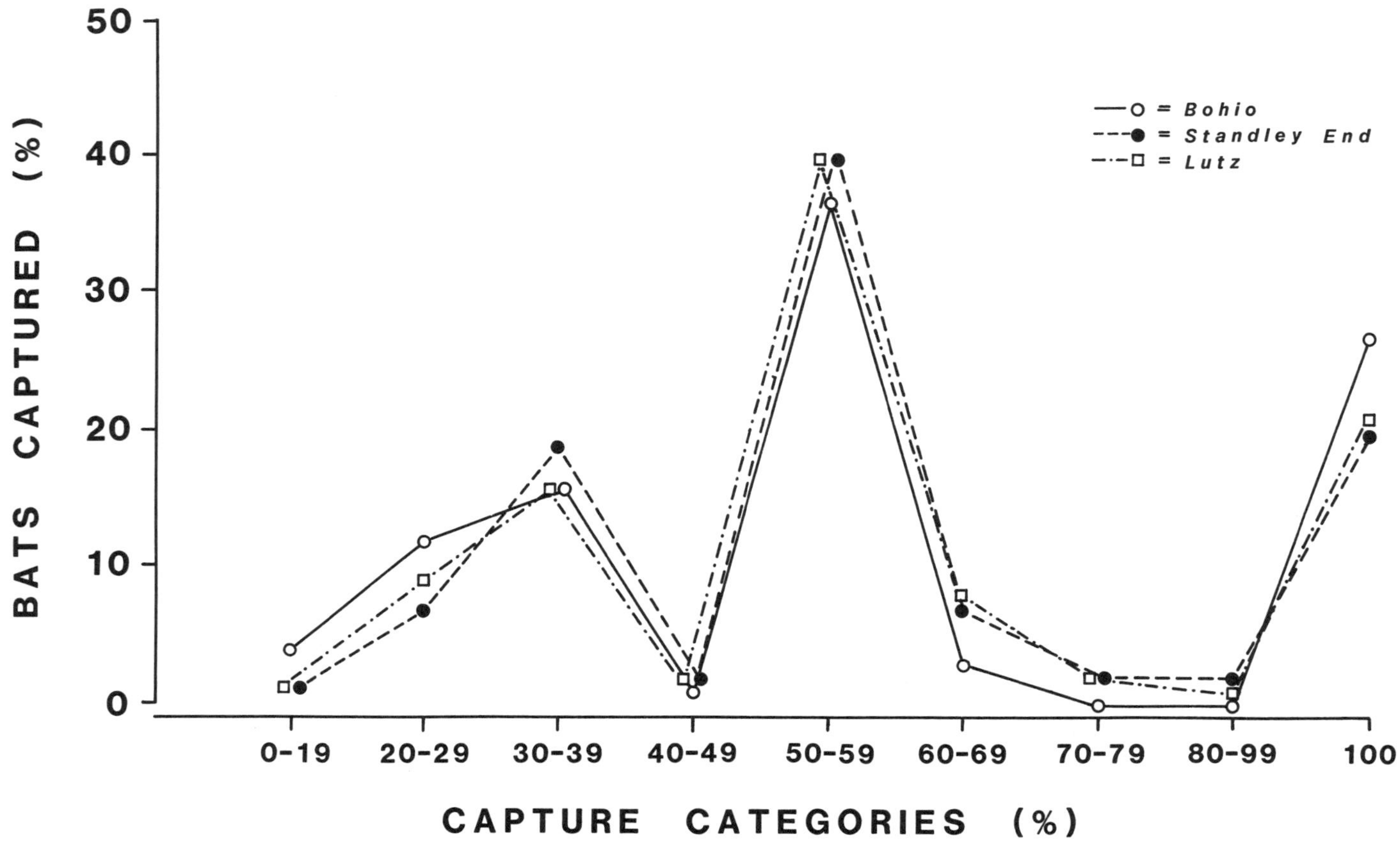

FIGURE 7-17.—Fidelity of *Artibeus jamaicensis* to Pattern II locality groups on and near BCI: Bohio, Lutz, and Standley End. See Figure 7-15 for further explanation.

into two subpatterns (Table 7-8). One with 19 or more netting-nights per locality (subpattern Ia); the other with fewer (subpattern Ib). With few exceptions (low fidelity at Conrad and Lake-Wheeler; unusually high fidelity at Drayton End), fidelity and netting effort within patterns are directly and positively correlated, suggesting that the more often a locality group is netted the more likely the bats using that locality group will be recaptured.

Nevertheless, although netting effort differed greatly between subpatterns Ia and Ib, fidelity was similar. Therefore, variation in fidelity between patterns is independent of netting effort. Subpattern Ia localities were netted from 20 to 52 nights and fidelity ranged from 7% to 18% (Table 7-8). Subunit Ib localities were netted only 6 to 17 nights apiece, yet the fidelity range was similar (8%–16%).

Fidelity was surprisingly similar (30%–32%) among pattern II localities even though netting effort ranged from 14 nights at Bohio to 197 at Lutz (Table 7-7). However, when comparing fidelity on the basis of similar netting effort between patterns I and II the results are markedly different. For example, 14 netting nights demonstrated 6% and 14% fidelity in Pattern I, and 30% fidelity in Pattern II. Locality groups with 24 and 26 netting nights in Pattern I showed 9% and 13% fidelity;

whereas fidelity was more than double (31%) after similar effort at Standley End (25 netting nights) in Pattern II.

If all bats having multiple captures that were recaptured only once in a particular locality group are excluded, we should eliminate most that were probably based elsewhere. Following that logic, we considered *A. jamaicensis* captured two or more times in a locality group more likely to be local residents. The proportion of bats with two or more captures in a locality group to the total multiple-capture cohort from that locality group varied from 0.9% to 71.9%; although they still sorted into the same patterns I, II, and III (Table 7-7). Few individuals caught two or more times confirms low fidelity at locality groups such as Conrad (11%) and Chapman (18%). The two-plus capture fraction at Conrad may contain few local bats, judging from the lower proportion (54%) of those bats in the 60%–100% column. However, the high percentage (78%) in the 60%–100% fractional capture category at Chapman indicates that the two-plus fraction, although small, may represent mostly local bats. The high proportion (78%) of bats at Peña Blanca that were caught there at least twice, and the high percentage (92%) of those in the 60%–100% bracket, suggest that this exercise is a valid means of estimating fidelity to a locality because high fidelity is to be expected at a harem day roost.

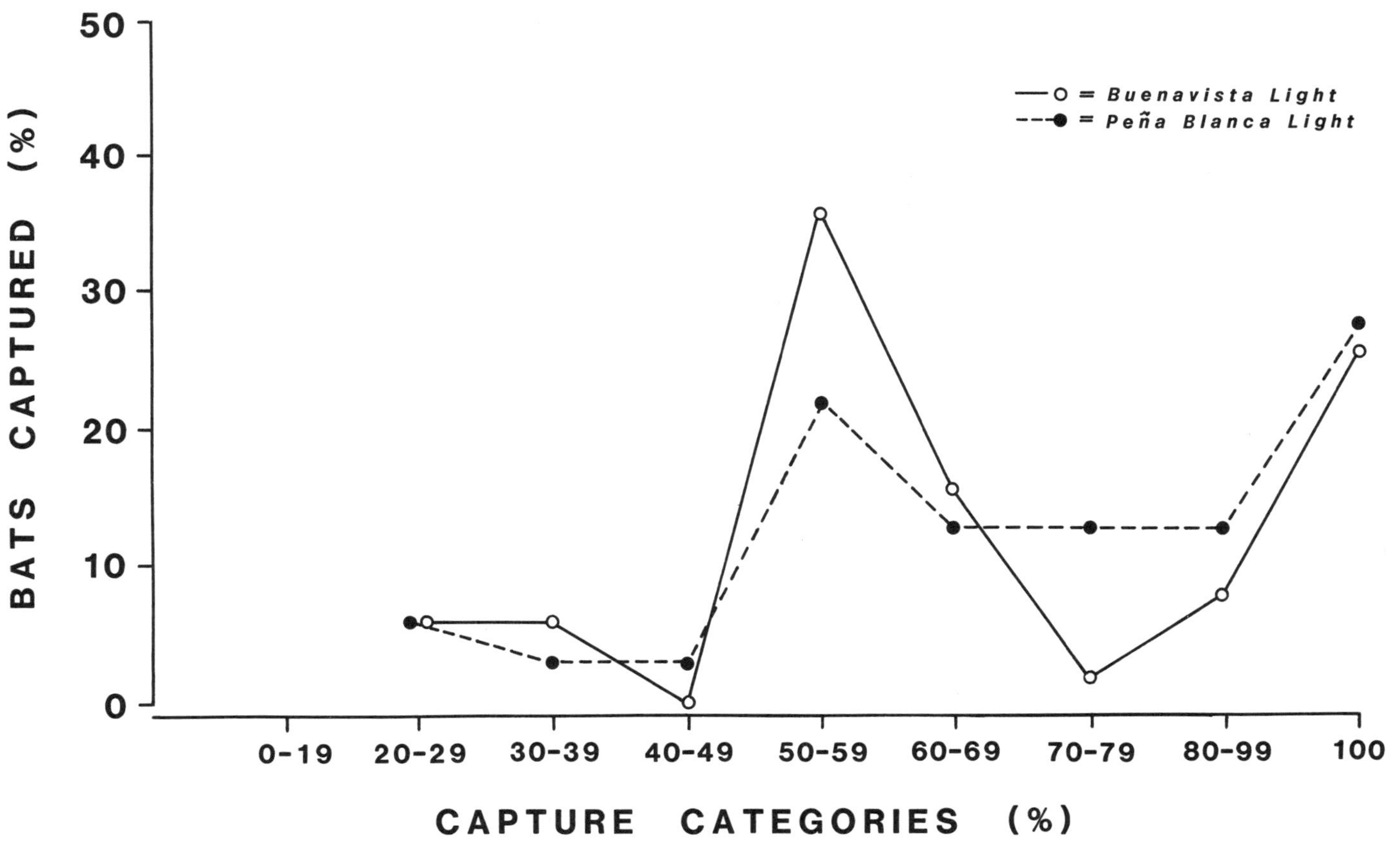

FIGURE 7-18.—Fidelity of *Artibeus jamaicensis* to Pattern III locality groups near BCI: Buena Vista and Peña Blanca. See Figure 7-15 for further explanation.

We found that 102 (37%) of the 275 *A. jamaicensis* caught at Standley End had been caught there at least twice (Table 7-7). Of these, 83% fall into the 60%–100% recapture category for the site. This is good evidence that a high proportion of the Standley End bats were local residents, a conclusion further supported when we examine their records in greater detail. Only two (7%) of the two-plus females were caught as many as two times elsewhere (at Bohio and Standley Ridge). Also, only six (8%) of the males caught two or more times at Standley End were caught as many as two times at other sites, all nearby (Fuertes, 1; Miller Ridge, 8; and Standley Ridge, 2).

The sex ratio of the Standley End *A. jamaicensis* (102) is skewed toward males (73 : 29; Table 7-9). At Lutz, another Pattern II locality with similar fidelity statistics, the ratio is 53 : 47. During three of the netting episodes at Standley End when two-plus bats were caught, males outnumbered females 30 to 1. Of males caught at least twice, 31% were also caught three or more times while only 14% of the females were caught that often. In contrast, only 21 (29%) males were caught over a relatively long period (two to four years), whereas 14 (48%) females were recaptured at Standley End during the same period. Close to three-fourths (71%) of the males were found there for a year or less (Table 7-9). Age distribution among bats

captured two or more times was similar in males and females (Table 7-10); however, a larger proportion of the young males than young females stayed in the area to be caught later as adults.

We believe that the males at Standley End were predominantly bachelors, and that most or all of the resident females were members of harems. Of the 20 females caught as adults during seasons of reproduction, 18 were pregnant, lactating, or postlactating at one or more of the captures. The greatest number of females caught as adults at Standley End in any half year was 11. Presuming that we caught most of the resident females, they probably represent from one to three harems. In the four harems of *A. jamaicensis* that Morrison (1979) studied on BCl, he found from four to 11 adult females ($\bar{X} = 6.5$).

We still do not know why fidelity was so high in the three Pattern II locality groups (Lutz, Bohio, and Standley End). Incomplete coverage of foraging ranges may be a factor at Bohio and Standley End, but not at Lutz, which is surrounded by well-netted localities that have typical low Pattern I fidelity percentages. Other peripheral locality groups such as Chapman, Standley Ridge, and Armour End show low fidelity, suggesting that peripheral location is not the reason.

TABLE 7-8.—Fidelity of *Artibeus jamaicensis* to a particular locality on BCI, ranked by number of netting nights. See Table 7-7 and text for explanation of categories.

Locality	Rank	Netting nights	60–100%
PATTERN Ia (more than 19 netting nights)			
Miller Ridge	1	52	18
Plateau	2	46	12
Fuertes	3	26	13
Shannon–AMNH	4	24	9
Barbour–Hood	5	20	7
PATTERN Ib (less than 19 netting nights)			
Chapman	1	17	16
Standley Ridge	2	16	15
Barbour Stream	3	14	14
Conrad	4	14	6
Lake–Wheeler	5	12	4
Harvard	6	11	12
Armour End	7	11	8
Drayton End	8	10	14
Zetek 21	9	6	8
Gross Point	10	5	1
Shannon–Balboa	11	4	3
Orchid Island	12	4	3
PATTERN II			
Lutz	1	197	32
Standley End	2	25	31
Bohio	3	14	30

We suggest that the high capture rate at Lutz is correlated with the unusual local abundance of fig trees. That also could be true of Bohio because our netting there generally coincided with the availability of ripe figs; however, fig trees were scarce at Standley End. The abundance of fig trees at Miller Ridge rivals that of Lutz, yet fidelity was low, as was typical of other Pattern I locality groups. Geographically, the Pattern II locality groups are dissimilar: Bohio is on the mainland, Standley End is peripheral on BCI, and Lutz is more central. In terms of size, Lutz encompasses the largest area, and because of its proximity to the Laboratory Clearing, it received the greatest sampling effort. Lacking a better explanation, one could argue that the similarities in capture frequency among the locality groups in Pattern II are merely coincidental.

Movements of Individual Bats

Thus far we have discussed movements of bats associated with a particular locality group based on pooled data. We also examined the movements of individuals captured at two or more localities away from the central locality under consideration. To illustrate this we have mapped the movements of five female *A. jamaicensis* from Peña Blanca (Figure 7-19). Each female was caught at two or more localities away from the light tower, and polygons outlining the locations where each was captured approximate their known home ranges. We then overlaid this map (Figure 7-19) with a 0.5 km grid and counted the number of times individual polygons touched each square. The resulting map (Figure 7-20) shows the frequency (from one to five) of occurrence of individual bats in each of the squares. We also mapped the same kind of information from 11 bats of the day roost in the Buena Vista light tower (Figure 7-21).

This method proved effective for illustrating movements

TABLE 7-10.—Age distribution of *Artibeus jamaicensis* captured two or more times at Standley–End, BCI.

Sex	Caught only when young	Caught first as young later as adult	Caught only as adult
Female	14%	38%	48%
Male	6%	49%	45%

TABLE 7-9.—Sexual variation in age, capture span, and number of captures of *Artibeus jamaicensis* caught two or more times at Standley–End, BCI.

Sex	N	Age during capture span					Capture span (years)				Number of captures at Standley–End				Total captures				
		J-S	J-A	SAD	S-A	AD	1	2	3	4	2	3	4	5+	2	3	4	5	6+
FEMALE																			
number	29		3	4	8	14	15	7	4	3	25	3		1	11	9	8	1	
percent			10	14	28	48	52	24	14	10	86	10		4	38	31	28	3	
									48%			14%							
MALE																			
number	73	1	11	4	24	33	52	13	7	1	50	14	8	1	31	20	10	8	4
percent		1	15	6	33	45	71	18	10	1	68	19	11	1	42	27	14	11	5
									29%			31%							

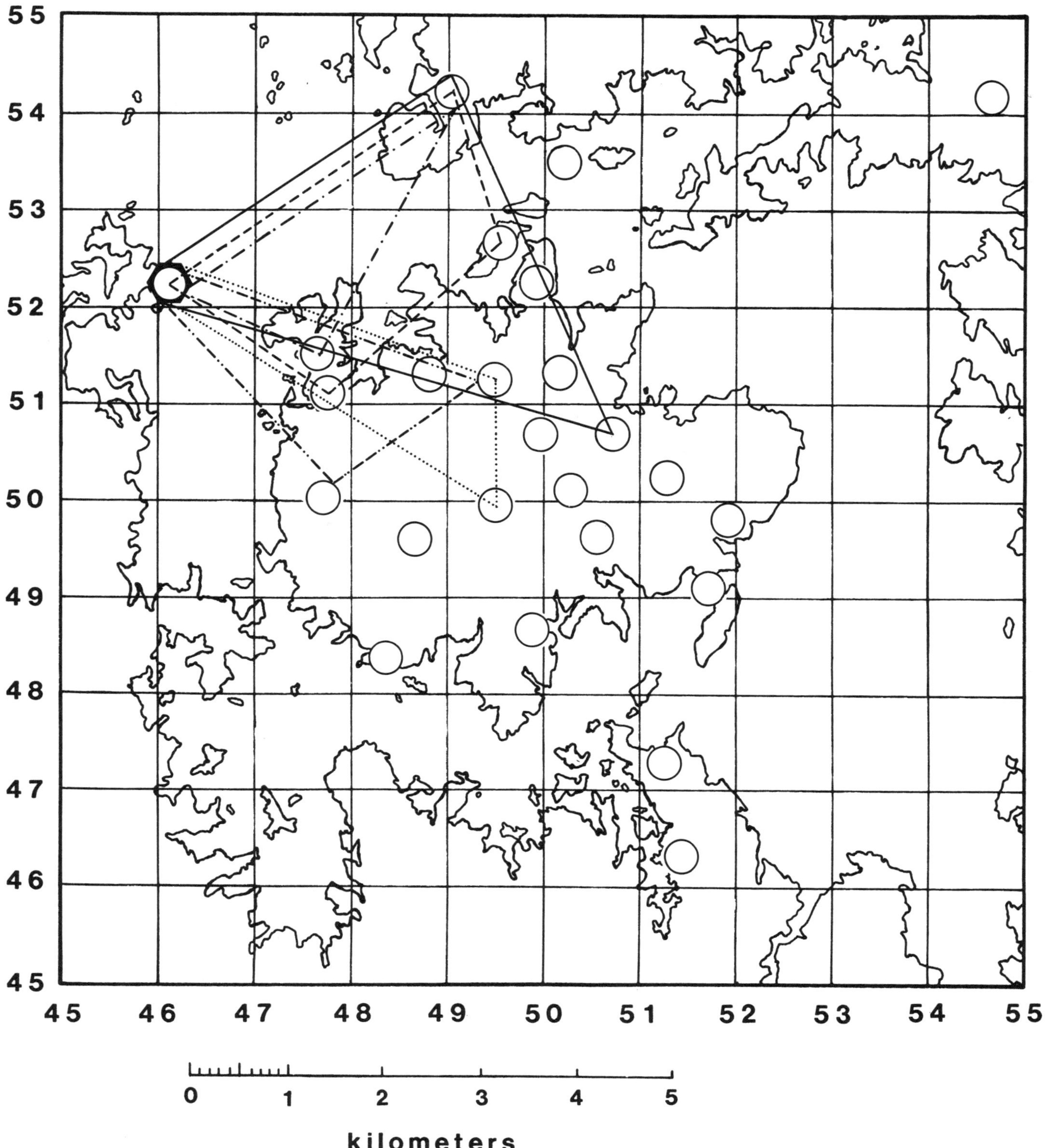

FIGURE 7-19.—Movements of five female *Artibeus jamaicensis* from the day roost in the Peña Blanca light tower to BCI and other nearby areas.

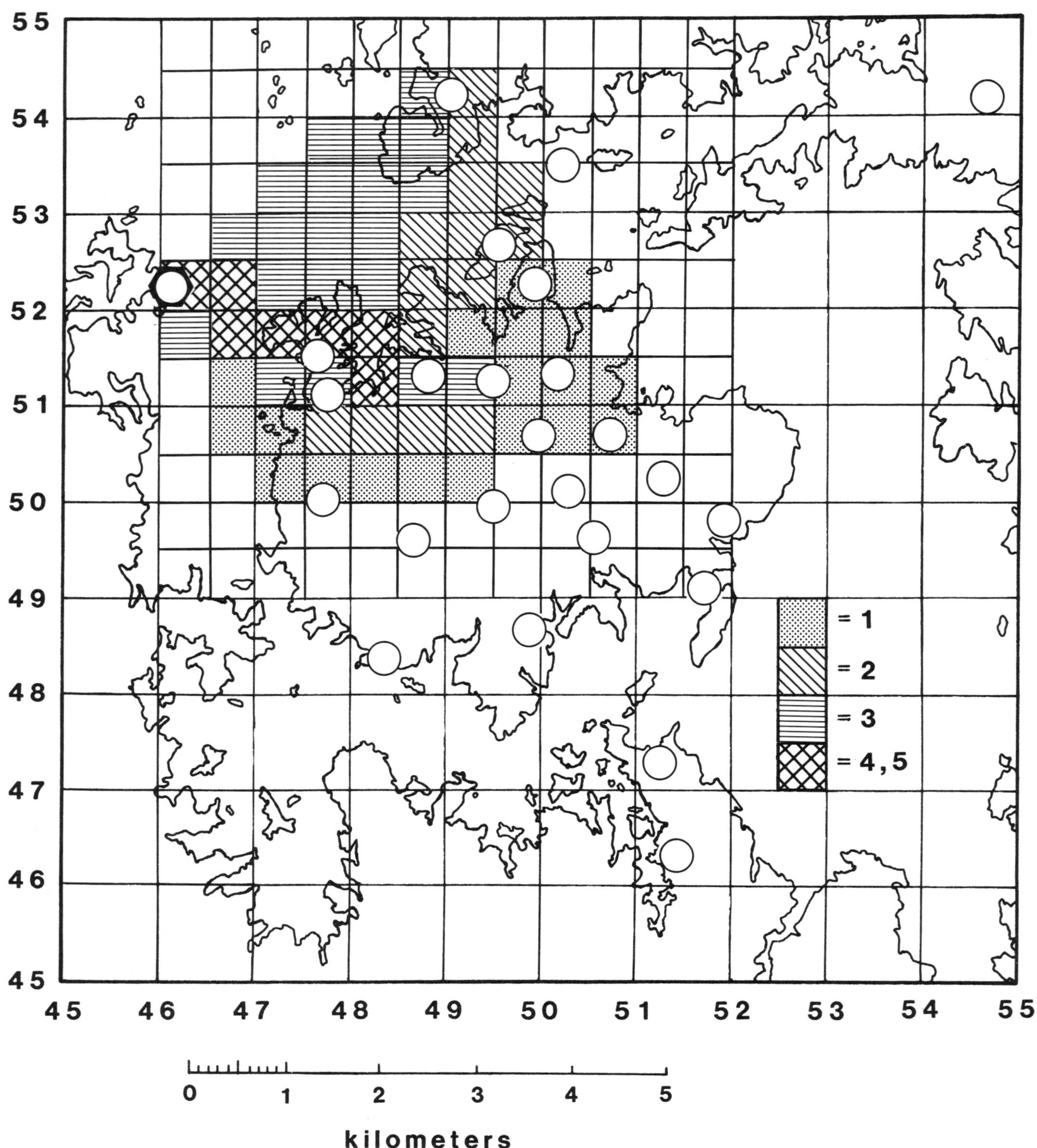

FIGURE 7-20.—Frequency of occurrence of five *Artibeus jamaicensis* from the day roost in the Peña Blanca light tower in nearby 0.5 km squares.

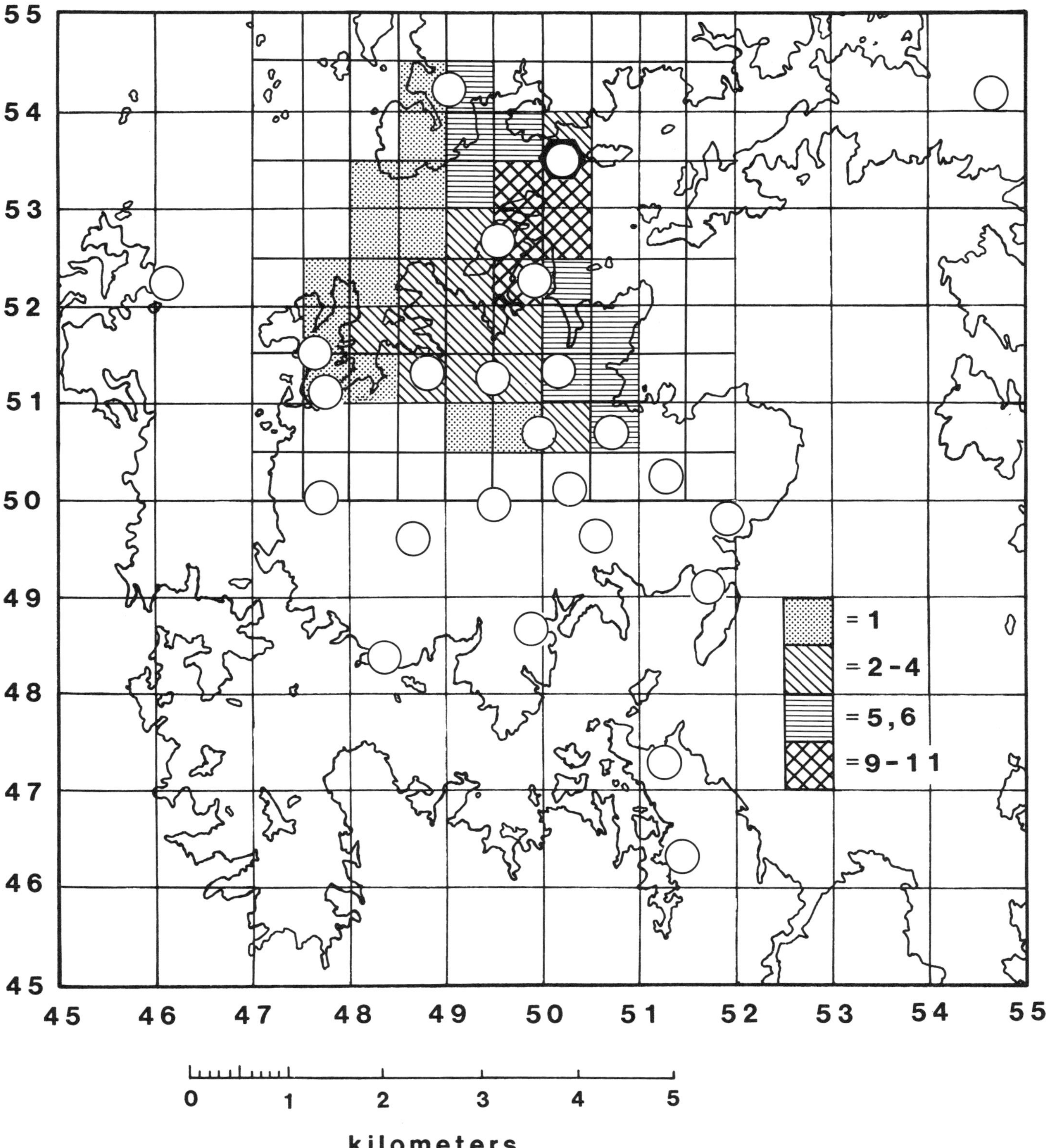

FIGURE 7-21.—Frequency of occurrence of 11 *Artibeus jamaicensis* from the day roost in the Buena Vista light tower in nearby 0.5 km squares.

even though it only approximates the area these bats use. If the average longevity of an *A. jamaicensis* on BCI is about two years (see Section 5, Survival and Relative Abundance, Figure 5-8), the average bat can be expected to fly on about 700 nights during its life. Nevertheless, we have recorded an average of only 1.8 captures per individual and a maximum of 11 captures—a small glimpse into the lives of these bats.

Mobility

There are few references in the literature on the distances that *A. jamaicensis* moves. Based on actual distances determined by radio tracking between day roosts and foraging sites, these bats are known to travel (one way) up to a kilometer (0.6 ± 0.4 km, *n* = 17) on BCI to as far as 10 km (8 ± 2 km, *n* = 4) in Jalisco, México, in a single night (Morrison 1978b). The bats in the day roosts in the canal-marker light towers also provided distances from day roost to netting sites. Of course we have no way of knowing whether a bat flew directly from the light tower to the place it was caught, or came from some other location. Nevertheless, based on the assumption that the bats moved from day roosts in the canal markers to netting sites, the average distances (Table 7-11) traveled by females from the Buena Vista roost were: Adults, 1.64 km (*n* = 32); subadults, 2.68 km (*n* = 3); juveniles, 3.36 km (*n* = 3). The average distances traveled by Buena Vista adult females is comparable to foraging distances covered between day roosts and feeding areas (1.6 km) by Costa Rican *Carollia perspicillata* as reported by Heithaus and Fleming (1978). However, the Buena Vista subadults and juveniles traveled considerably greater distances. Also greater were average distances traveled by female *A. jamaicensis* from the Peña Blanca marker: Adults, 3.21 km (*n* = 13); subadults, 3.16 km (*n* = 3). A comparison between Heithaus and Fleming's (1978) data for *C. perspicillata* and ours for *A. jamaicensis* probably is not appropriate because their data was based on radio telemetry and gathered over a short period of time (up to 19 nights) during the wet season. Because they used radio telemetry, distances between roosts and foraging sites were more easily defined and correspond more closely to Morrison's (1978b) commuting distances. Fleming and Heithaus (1986) showed that seasonal changes in food availability also influenced distances traveled by *C. perspicillata*. We do not know what our *A. jamaicensis* were doing or where they went between leaving their day roosts and being captured.

We summarized the average distance between capture sites by sex, age, and locality group based on 5542 records of recaptures of *A. jamaicensis* at sites other than where the bats were marked (Table 7-11). The distances traveled between these sites, often recorded months apart, are not exactly equivalent to Morrison's (1978b) commuting distances, which were based on bats radio tracked during a single night. The mean distances between captures (Table 7-11) do not seem to

be correlated with age or sex. Although a superficial examination of the data suggests that patterns exist, closer scrutiny failed to reveal any overall consistency.

Movements (Table 7-11) by adults outnumbered those of subadults about 2 : 1, records of subadult females outnumbered those of juvenile females by about 3 : 1, and records of subadult males outnumbered those of juvenile males by about 2 : 1. Although these ratios appear to reflect the relative duration of these age categories in the population, ratios based on the average number of months when an individual *A. jamaicensis* can be captured as either adult, subadult, or juvenile are about 15 : 8 : 1. Proportionately, however, movements by juveniles are much more frequent than expected, but those of adults are as expected. This disparity may reflect greater ease of recapturing juveniles. Although among records of movements of adults and subadults, females outnumbered males, and in movements of juveniles, males outnumbered females, this is the relationship among all captures, so it is unlikely to be significant as far as movements are concerned.

Even though the mean distances between captures seem to show no correlation with age or sex (Table 7-11) they do show a correlation with the average distance separating a capture station from all other stations. This relationship is shown in Figures 7-22 through 7-27 in which the mean distance between captures has been plotted in each locality group. With the data organized geographically, the distances traveled by females in the core area are a little greater than corresponding distances for males among both adults and juveniles. The distances flown are greatest in subadults among males; but greatest in adults and juveniles among females. Correlations between age and sex are not evident outside the core area. Three distance records of movements between Bohio and Armour End (5.93 km) are the longest recorded between any captures. These two locality groups are also the most distantly separated of any of the sites where we frequently netted.

Nightly Movements

Radio-tracking showed that *A. jamaicensis* commonly visits more than one tree in an evening (see Section 9, Foraging Behavior). We have evidence of this when, for example, we net a bat carrying a *Ficus insipida* fruit, but defecating seeds of *F. trigonata*. This bat must have fed at a *F. trigonata* on the same evening because food passage is much too rapid (we have recorded food passage times of from 7 to 45 minutes; see Section 2, Physiology) to claim that the seeds are from a fig eaten on the previous night.

We tried to demonstrate movements between fruiting trees on the same night by operating netting stations simultaneously at several localities. Although we handled 1170 *A. jamaicensis* on 13 of those nights (Table 7-12), we captured only one again the same night at a different tree. This was a postlactating female captured at 1900 h on 7 November 1979 near a *F.*

TABLE 7-11.—Mean distance (in kilometers) between actual points of capture of individual *Artibeus jamaicensis* on BCI and the adjacent mainland summarized by locality group, sex, and age. Number in each sample in parentheses.

Locality	Female			Male			Ratio (female:male)		
	JUV	SAD	AD	JUV	SAD	AD	JUV	SAD	AD
Lutz	1.13 (79)	1.27 (364)	1.04 (574)	0.93 (92)	0.93 (302)	0.91 (528)	1.13:0.93	1.27:0.93	1.04:0.91
Barbour–Hood	1.25 (8)	1.14 (35)	1.28 (84)	0.86 (8)	1.13 (23)	1.22 (35)	1.25:0.86	1.14:1.13	1.28:1.22
Chapman	0.95 (3)	1.32 (14)	1.27 (38)	1.30 (4)	0.79 (3)	1.55 (23)	0.95:1.30	1.32:0.79	1.27:1.55
Harvard		1.90 (5)	1.31 (15)		1.80 (4)	1.61 (9)		1.90:1.80	1.31:1.61
Shannon–AMNH	1.28 (28)	1.34 (69)	1.46 (77)	1.28 (29)	1.50 (48)	1.23 (66)	1.28:1.28	1.34:1.50	1.46:1.23
Drayton End			1.90 (11)		2.39 (4)	1.89 (15)			1.90:1.89
Armour End	2.81 (6)	2.22 (17)	2.18 (25)	2.97 (11)	2.09 (18)	2.41 (24)	2.81:2.97	2.22:2.09	2.18:2.41
Zetek 21		2.20 (11)	1.81 (26)		2.11 (7)	1.77 (23)		2.20:2.11	1.81:1.77
Standley Ridge	1.75 (5)	1.29 (18)	1.55 (57)	2.56 (14)	1.78 (44)	1.46 (126)	1.75:2.56	1.29:1.78	1.55:1.46
Standley End		2.59 (11)	1.73 (63)	1.50 (4)	1.77 (22)	1.84 (82)		2.59:1.77	1.73:1.84
Fuertes	1.25 (45)	1.23 (57)	1.05 (138)	1.42 (41)	1.31 (32)	0.92 (62)	1.25:1.42	1.23:1.31	1.05:0.92
Miller	1.34 (32)	1.21 (78)	1.03 (248)	1.23 (37)	1.15 (50)	1.01 (196)	1.34:1.23	1.21:1.15	1.03:0.61
Gross Point		2.01 (3)	1.21 (17)		1.47 (3)	1.15 (8)		2.01:1.47	1.21:1.15
Barbour Stream		0.77 (5)	1.01 (71)	0.83 (3)	1.15 (9)	1.42 (20)		0.77:1.15	1.01:1.42
Lake–Wheeler		0.80 (3)	1.05 (31)		1.58 (3)	0.92 (8)		0.80:1.58	1.05:0.92
Shannon–Balboa	0.95 (3)	1.07 (13)	1.22 (18)	0.72 (7)	0.77 (3)	1.07 (18)	0.95:0.72	1.07:0.77	1.22:1.07
Plateau	1.42 (32)	1.39 (73)	1.21 (209)	1.32 (51)	1.34 (88)	1.31 (154)	1.42:1.32	1.39:1.34	1.21:1.31
Conrad	2.21 (20)	1.72 (21)	1.57 (56)	2.24 (17)	1.86 (18)	1.71 (46)	2.21:2.24	1.72:1.86	1.57:1.71
Orchid Island	2.83 (5)	2.07 (23)	1.87 (60)	2.35 (5)	2.38 (13)	1.85 (26)	2.83:2.35	2.07:2.38	1.87:1.85
Bohio		3.82 (6)	1.75 (62)	4.18 (3)	2.95 (7)	2.05 (20)		3.82:2.95	1.75:2.05
Buena Vista	3.36 (3)	2.68 (3)	1.64 (32)						
Peña Blanca		3.16 (3)	3.21 (13)						
Total *n*	269	832	1925	326	701	1489			

insipida at Armour-Zetek Junction and netted again 1.2 km away at 2100 h on the same evening near another fruiting *F. insipida* at Miller 9.

Our failure to catch more bats moving between fruiting trees on the same night may have stemmed from net avoidance (adverse conditioning), from an unfortunate choice of inappropriate combinations of fruiting trees, or from low probability because of the large number of bats involved (the "drop in the bucket" principle).

When we netted the same place for two or more nights because of the continued availability of ripe figs, and our captures of *A. jamaicensis* averaged 25 or more per night ($\bar{X}$ =

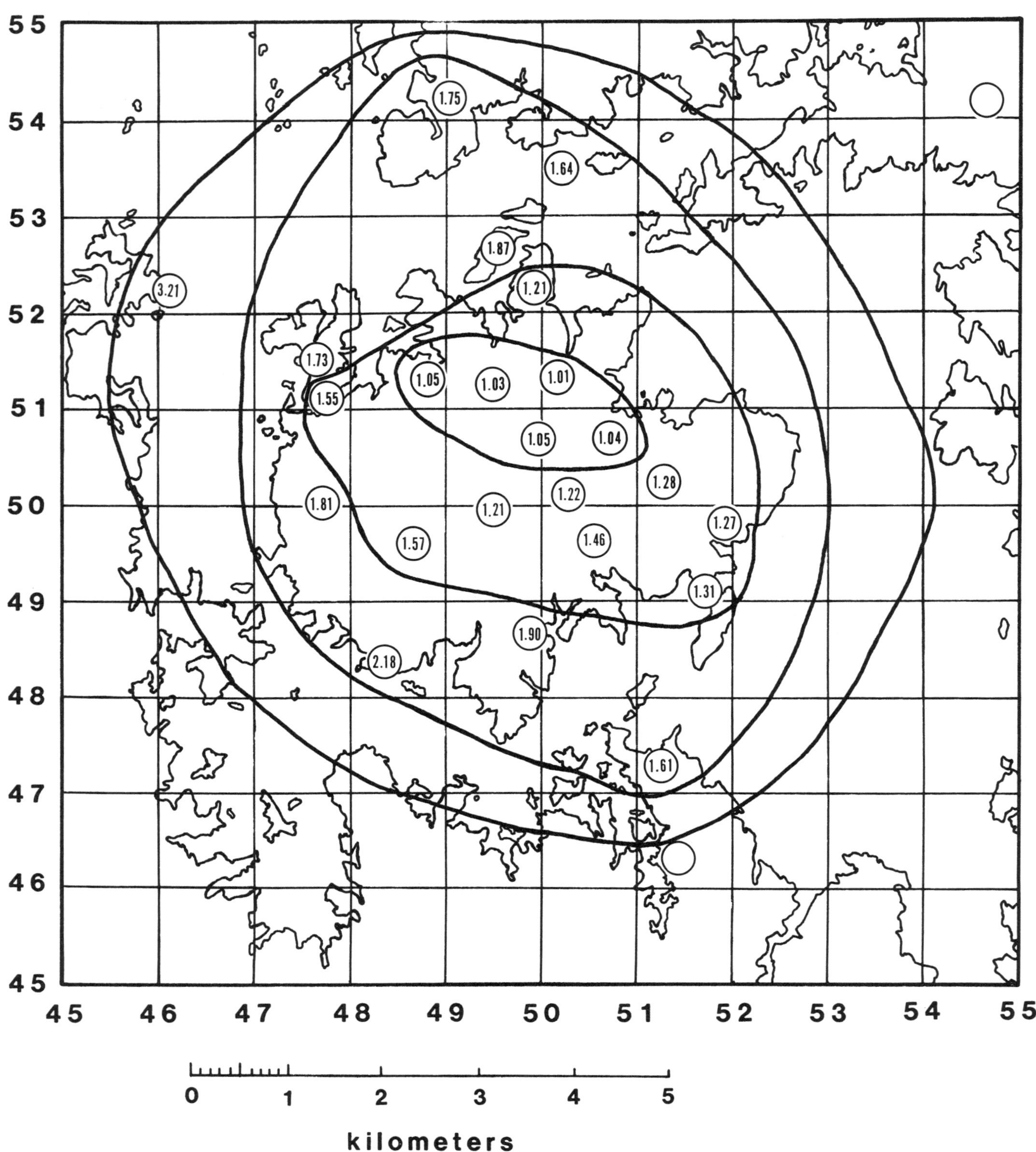

FIGURE 7-22.—Mean distance in kilometers between captures of adult female *Artibeus jamaicensis* on and near BCI. Isolines enclose locality groups with similar mean distances and reflect distances between enclosed capture stations and all other capture stations. See text for further explanation.

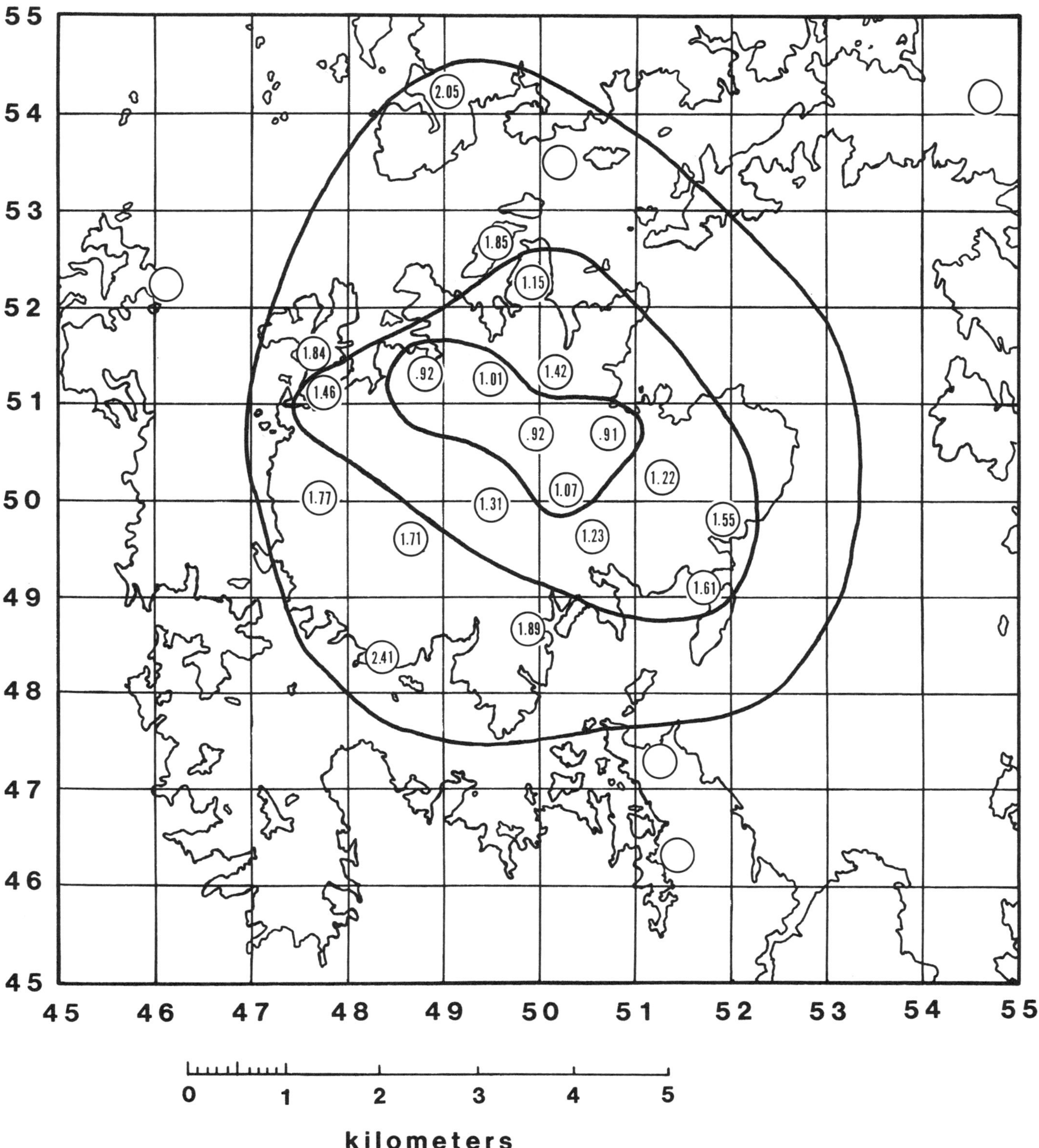

FIGURE 7-23.—Mean distance in kilometers between captures of adult male *Artibeus jamaicensis* on and near BCI. Isolines enclose locality groups with similar mean distances and reflect distances between enclosed capture stations and all other capture stations. See text for further explanation.

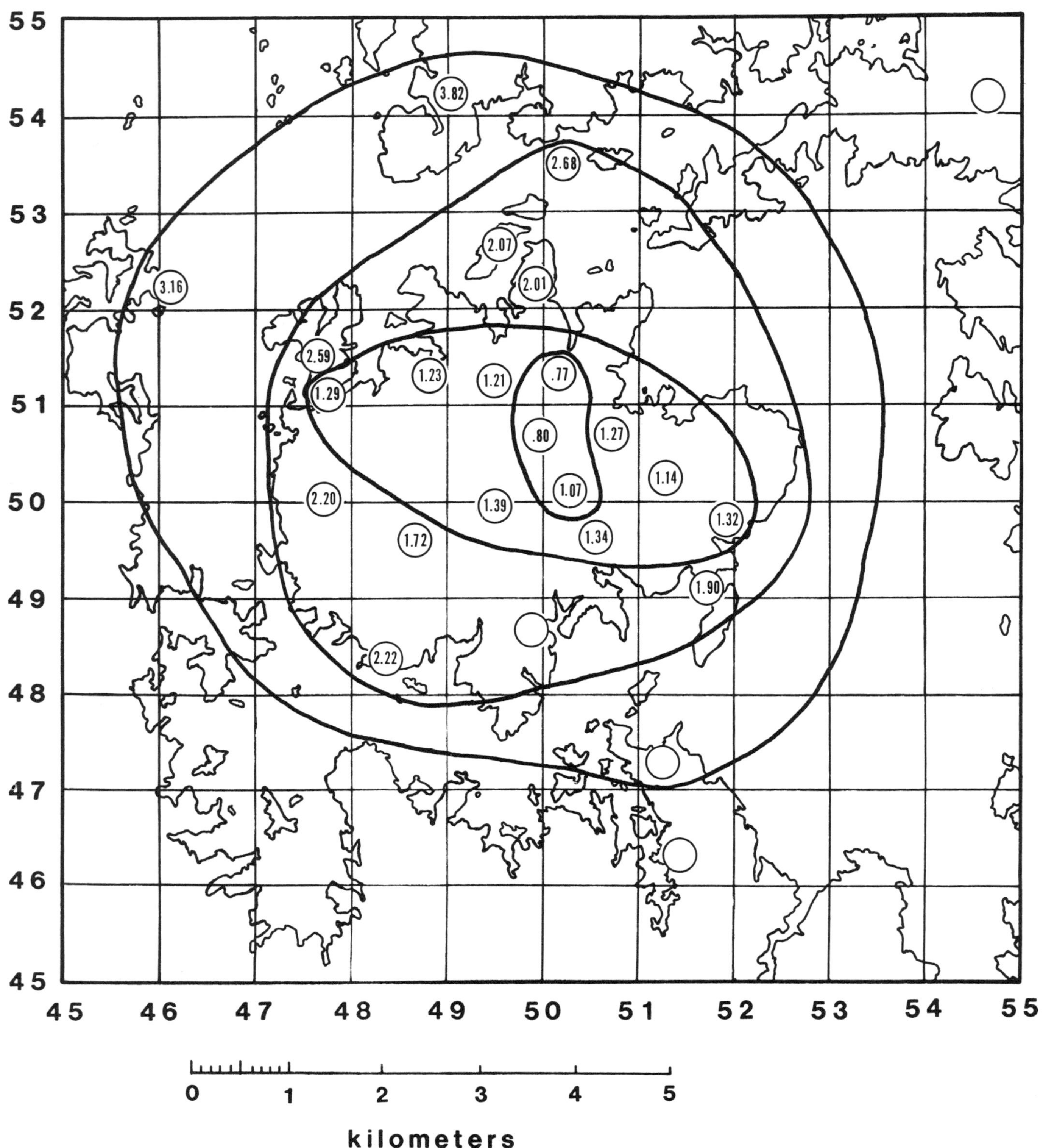

FIGURE 7-24.—Mean distance in kilometers between captures of subadult female *Artibeus jamaicensis* on and near BCI. Isolines enclose locality groups with similar mean distances and reflect distances between enclosed capture stations and all other capture stations. See text for further explanation.

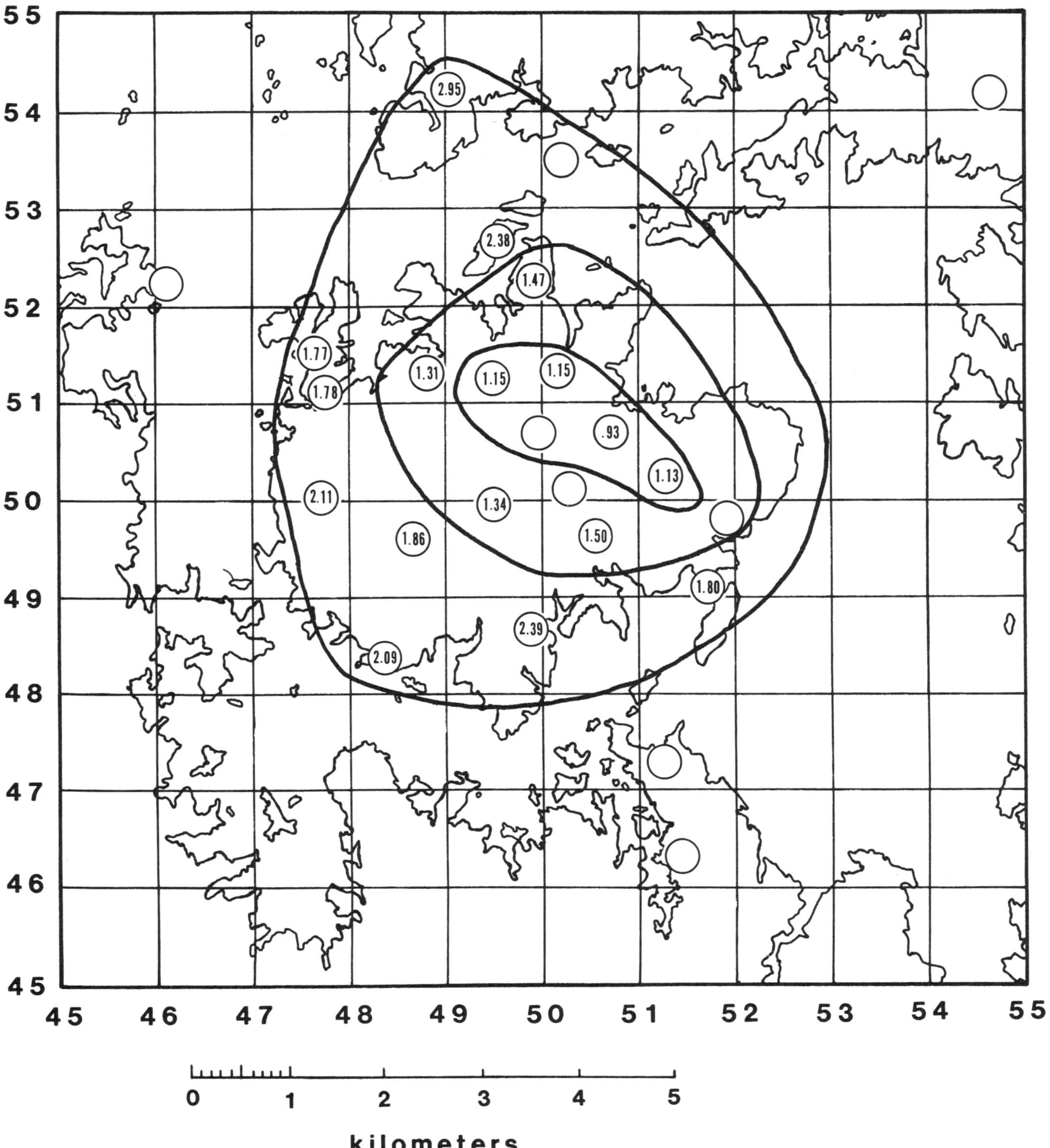

FIGURE 7-25.—Mean distance in kilometers between captures of subadult male *Artibeus jamaicensis* on and near BCI. Isolines enclose locality groups with similar mean distances and reflect distances between enclosed capture stations and all other capture stations. See text for further explanation.

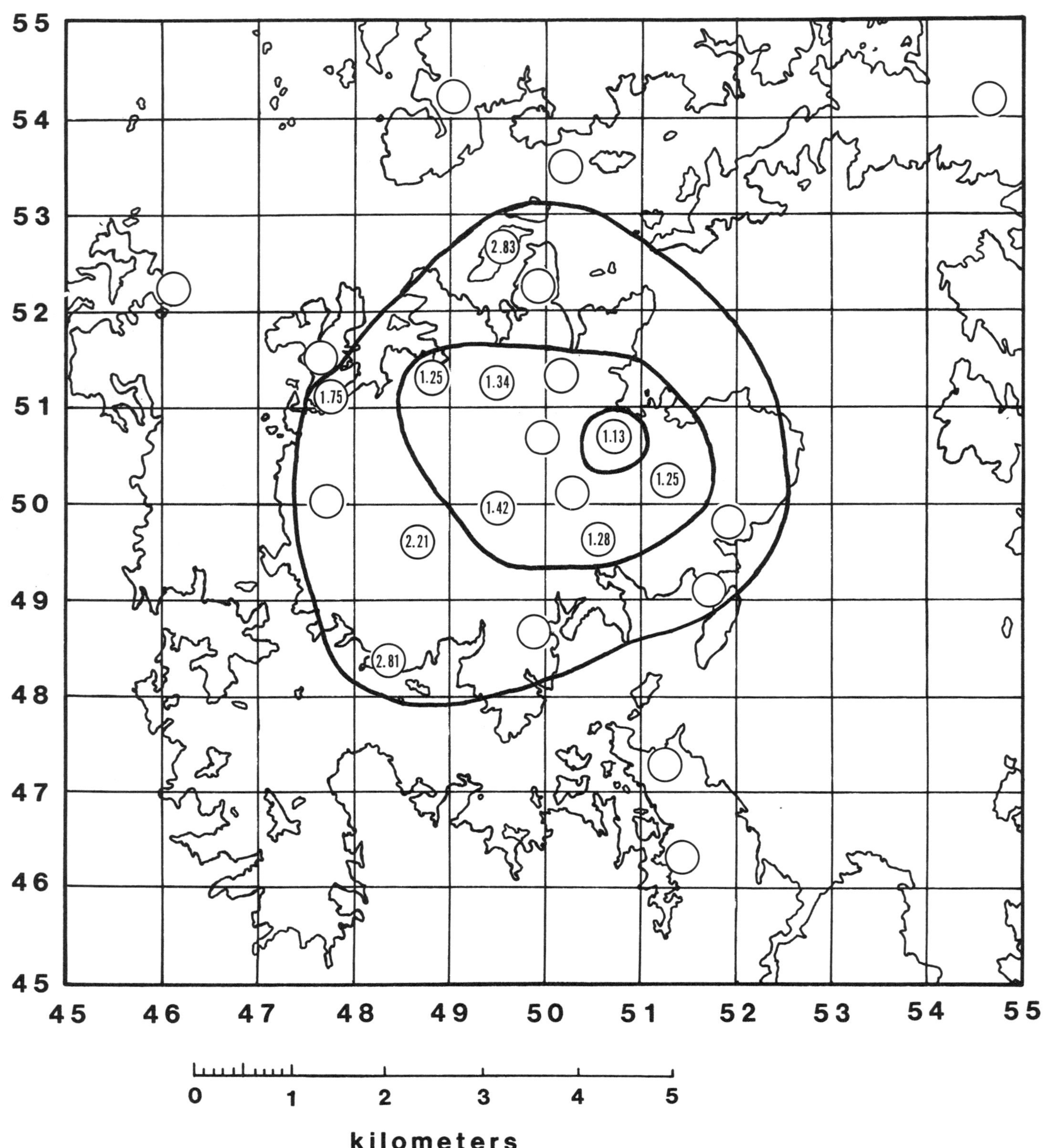

FIGURE 7-26.—Mean distance in kilometers between captures of juvenile male *Artibeus jamaicensis* on and near BCI. Isolines enclose locality groups with similar mean distances and reflect distances between enclosed capture stations and all other capture stations. See text for further explanation.

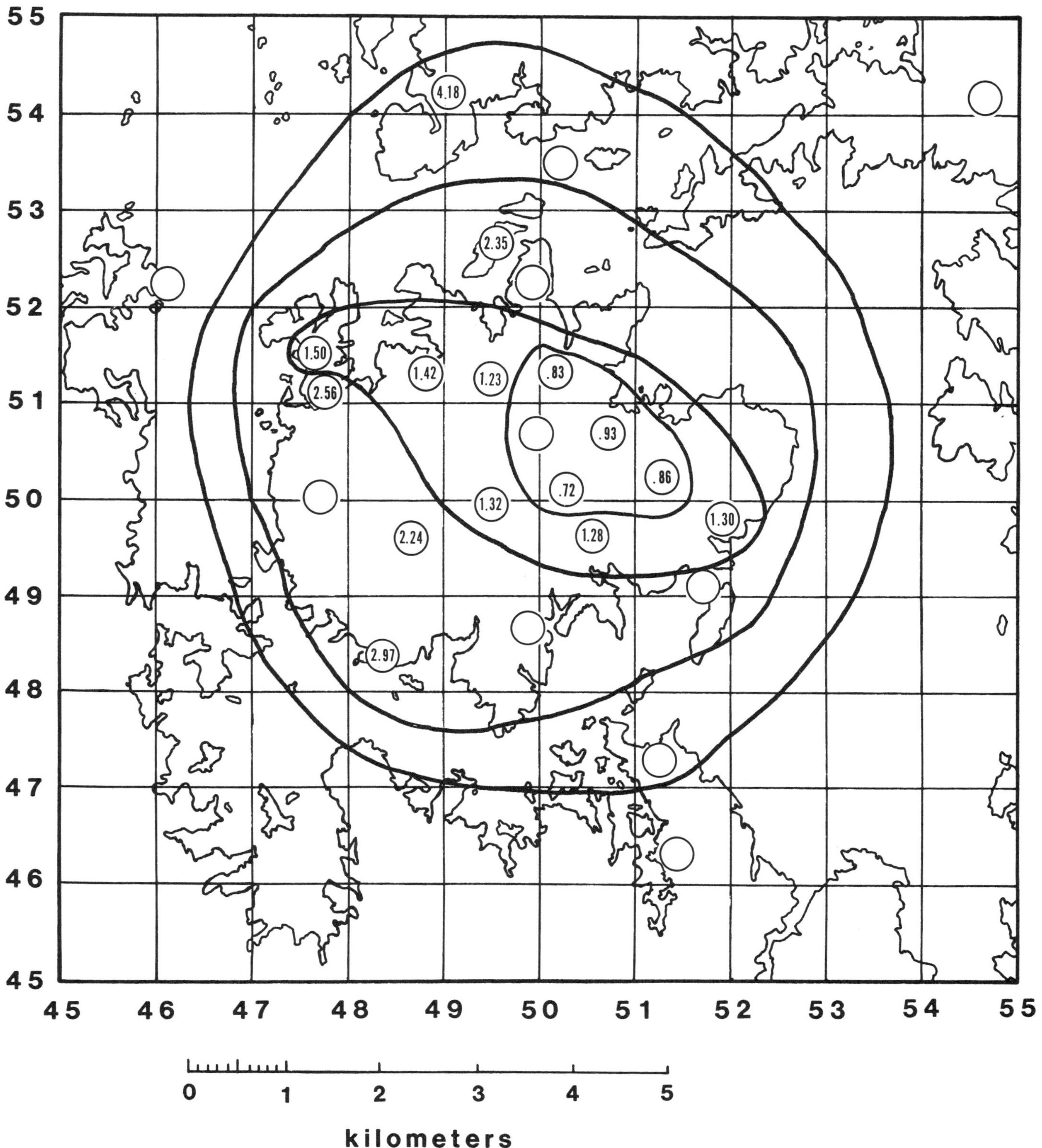

FIGURE 7-27.—Mean distance in kilometers between captures of juvenile female *Artibeus jamaicensis* on and near BCI. Isolines enclose locality groups with similar mean distances and reflect distances between enclosed capture stations and all other capture stations. See text for further explanation.

TABLE 7-12.—Catches of *Artibeus jamaicensis* and other bats during episodes of netting at two or more stations on the same night on BCI.

Date	Locality	*Artibeus jamaicensis*	Other bats
29 Nov 78	Lutz Creek	26	15
	Snyder–Molino 0–2	56	27
11 Jun 79	Barbour–Lathrop 16	109	33
	Snyder–Molino 0–2	6	4
7 Nov 79	Miller 9	77*	47
	Armour–Zetek Jct.	83*	16
8 Nov 79	Armour–Zetek Jct.	2	3
	Armour–Conrad Jct.	27	27
	Miller 9	11	18
9 Nov 79	Donato 3–4	47	54
	Miller 1–2	19	25
	Barbour 3	24	14
11 Nov 79	Donato 3–4	42	47
	Barbour 3	27	22
12 Nov 79	Snyder–Molino–Lutz	5	3
	Standley 15	10	13
	Standley 21	6	10
13 Nov 79	Miller 4–5	2	4
	Standley 21	10	26
	Lutz Creek	1	10
14 Nov 79	Donato 3–4	94	136
	Standley 16	103	82
15 Nov 79	Donato 3–4	14	27
	Standley 16	68	100
16 Nov 79	Standley 16	23	54
	Donato 3–4	42	45
18 Nov 79	Miller 13–14	56	40
	Miller 15	30	32
12 Oct 80	Miller 8–10	101	12
	Chapman 6–9	49	12
13 nights	30 sites	1170	958

* One double capture.

60 for 128 nights; see Table 7-13), only 2% (range, 0%–6%) of individual *A. jamaicensis* were captured more than once at the same tree on successive nights. These recaptured bats were 5% (range, 0%–13%) of all captures of *A. jamaicensis*, which represents a pool of 7427 individuals amassed over 128 nights. In all, only 181 were caught at the same tree on two successive nights, and seven were captured on three successive nights. Individual *A. jamaicensis* were caught more than once in 35 of 46 multinight netting episodes at single ripe trees.

The frequency of repeats (bats caught more than once in a night at the same station) is similar to the frequency of recaptures on successive nights. At least some *A. jamaicensis* reentered a net within minutes of release, probably because of disorientation resulting from the handling process and because of proximity of the processing station to nets. Nevertheless, the impression was that, once captured, these bats avoided the nets. During sampling periods totaling 133 nights in 1979 and 1980 we had 158 repeats of *A. jamaicensis* on 59 nights; an average of 2.7 repeats on the nights with repeats. On the nights with repeats, we caught 4022 *A. jamaicensis,* 3.9% of which represented repeated captures. Repeated captures represented 2.7% of all *A. jamaicensis* caught on all nights repeated (133 nights, 5946 bats).

If we could estimate the proportion of marked and unmarked bats in a foraging aggregation we might be able to estimate the number of *A. jamaicensis* that come to a fruiting tree. We know that the recapture rate is low (<2%), but we do not know what our success rate is in capturing unmarked bats. To judge from the nightly ratio of unmarked to recaptured bats (Table 7-13), these variables must be changing from night to night.

If we assume that many of the same bats return night after night to a choice tree as long as fruit remains, and that we recapture only a small fraction of marked bats, then a tally of

TABLE 7-13.—Results of netting episodes involving consecutive nights at single localities on BCI and the adjacent mainland. The columns of capture records, multiple captures, and mean captures per night pertain to *Artibeus jamaicensis* alone. The last column, total captures all species, includes *A. jamaicensis* and all other bats.

Date	Nights N	Capture records N	Marks	Recaps	Total	Multiple captures* Individuals N	%	Records N	%	Mean captures per night	Total captures all species
Miller Ridge (9 multinight episodes)											
Jan 78	3	291	179	117	296	5	2	10	3	99	529
Apr 78	4	97	37	63	100	3	3	6	6	25	148
Jun 78	3	85	72	16	88	3	4	6	7	29	132
Nov 78	2	106	75	32	107	1	1	2	2	54	150
May 79	2	79	48	32	80	1	1	2	3	40	140
Nov 79	2	94	27	67	94	0	0	0	0	47	179
Dec 79	4	253	76	182	258	5	2	10	4	65	321
Apr 80	4	145	38	109	147	2	1	4	3	37	273
Oct 80	3	319	120	208	328	9	3	18	6	109	402
Total	27	1469	672	826	1498	29	2	58	4	56	2274
Standley Ridge (3 multinight episodes)											
Feb 78	3	99	102	201	12	25	12	188	6	67	274†
Sep 79	2	83	58	141	4	8	6	137	3	71	205
Nov 79	4	109	87	196	4	8	4	192	2	49	448
Total	9	291	247	538	20	41	8	517	4	60	927

TABLE 7-13.—Continued.

Date	Nights N	Capture records				Multiple captures*				Mean captures per night	Total captures all species
						Individuals		Records			
		N	Marks	Recaps	Total	N	%	N	%		
Bohio (4 multinight episodes)											
Dec 78	2	76	64	13	77	1	1	2	3	39	160
Apr 79	2	149	106	46	152	3	2	6	4	76	177
Sep 80	2	101	88	18	106	5	5	10	9	53	156
Oct 80	3	121	56	69	125	3	3	7	6	42	176†
Total	9	447	314	146	460	12	3	25	5	51	669
Fuertes (3 multinight episodes)											
Jan 80	2	50	11	39	50	0	0	0	0	25	119
Aug 80	4	270	176	102	278	8	3	16	6	70	325
Oct 80	2	285	142	146	288	3	1	6	2	144	342
Total	8	605	329	287	616	11	2	22	4	77	786
Plateau (5 multinight episodes)											
Sep 79	2	138	89	51	140	2	1	4	3	70	181
Oct 79	2	105	65	41	106	1	1	2	2	53	393
Nov 79	2	82	19	63	82	0	0	0	0	41	101
Aug 80	3	107	75	35	110	3	3	6	6	37	131
Aug 80	2	98	64	34	98	0	0	0	0	49	112
Total	11	530	312	224	536	6	1	12	2	49	918
Lutz (13 multinight episodes)											
Mar 78	2	67	16	52	68	1	2	2	3	34	159
Mar 78	2	63	19	44	63	0	0	0	0	32	143
Nov–Dec 78	3	120	88	35	123	3	3	6	5	41	178
Mar 79	3	145	54	99	153	8	6	16	11	51	260
Oct 79	3	332	183	161	344	12	4	24	7	115	419
Oct 79	2	108	45	64	109	1	1	2	2	55	149
Nov 79	2	89	34	55	89	0	0	0	0	45	189
Nov 79	2	54	15	39	54	0	0	0	0	27	90
Nov 79	2	108	40	68	108	0	0	0	0	54	271
May 80	13	767	637	157	794	27	4	54	7	61	969
Aug 80	2	147	96	51	147	0	0	0	0	74	171
Sep–Oct 80	3	169	93	80	173	4	2	8	5	58	230
Oct 80	2	104	49	55	104	0	0	0	0	52	172
Total	41	2273	1369	960	2329	56	3	112	5	57	3400
Miscellaneous (9 multinight episodes)											
Barbour-Hood											
Aug 80	2	147	93	56	149	2	1	4	3	75	170
Chapman											
Dec 78	2	99	52	52	104	5	5	10	10	52	134
RCS-AMNH											
Sep 79	4	326	225	119	344	15	5	33	10	86	419‡
Lake-WMW											
Nov 78	3	86	70	19	89	3	4	6	7	30	240
Conrad											
Sep 80	2	199	100	105	205	6	3	12	6	103	234
Armour End											
Sep 80	3	283	203	86	289	6	2	12	4	96	351
Orchid Island											
Apr 79	2	165	163	2	165	0	0	0	0	83	196
Frijoles Road											
Apr 79	2	151	147	6	153	2	1	4	3	72	220
Gigante											
Apr 80	3	130	113	27	140	8	6	18	13	47	224♣
Total	23	1586	1166	472	1638	47	3	99	6	71	2188♦
Total for table	128	7427	4453	3162	7615	181	2	369	5	60	11162

* Individual *A. jamaicensis* captured two or more times during a multinight netting episode.

† One caught three times.

‡ Three caught three times.

♣ Two caught three times.

♦ Seven caught three times.

TABLE 7-14.—Large concentrations of *Artibeus jamaicensis* at single netting stations on BCI on successive nights.

Locality	Date	Nights *N*	*Artibeus jamaicensis* *N*	Average number per night
Miller Ridge	Jan 78	3	291	99
Shannon–AMNH	Sep 79	4	326	86
Lutz	Oct 79	3	332	115
Miller Ridge	Dec 79	4	253	65
Lutz	May 80	13	767	61
Fuertes	Aug 80	4	270	70
Armour End	Sep 80	3	283	96
Miller Ridge	Oct 80	3	319	109
Fuertes	Oct 80	2	285	144

the number of individuals captured during a multinight netting sequence should suggest the number of bats that flock to a tree (Table 7-14). We believe that, on occasion, these numbers must be large, and even if we are catching only a third of the bats, the total at a fruiting tree may exceed a thousand. Our usual impression at the nets during a highly successful night is that while the air seems full of bats, relatively few are getting into the nets.

Group Movements

The number of bats at a netting station, especially if it is at an active feeding roost near a fruiting tree, varies during the night, usually abruptly. This is evidence that the bats are moving about in groups. Whether these groups are associations of individuals that forage together night after night (and could be considered true flocks), or are merely aggregations of bats that form at a feeding roost and shift together from one feeding site to another during the night and then disband for indefinite periods of time is not known.

There is no doubt about the existence of aggregations. Anyone who has much experience working with mist nets in the American tropics has to be impressed with abrupt and dramatic shifts in numbers of bats during the night, signaling the arrival and departure of groups. Heithaus et al. (1974) commented that large bats appeared to arrive in groups to feed and that visits by groups to food sources was pulsed. We refer to these pulses as surges, and we define a surge as an abrupt increase in numbers of bats captured, followed by an abrupt decrease. When and whether there are surges depend on such influences as rain, presence of ripe fruit, the amount of light, and the phase of the moon. Our data on surges were compiled for three-month periods at the height of the rainy season (August–November) in 1979 and 1980 (*n* = 133 nights; means summarized in Table 7-15).

Admittedly these data are biased and incomplete. On a night with few bats we might stop netting after a few hours, thus

recording a night without a surge. On similar nights we persisted, perhaps waiting for the moon to set and still did not record a surge; or, had the first surge commence as late as 2300 h. We undoubtedly missed some second surges by terminating too soon after the first. During the weeks of a full and new moon, moonrise and moonset often coincided with the beginning and end of surges.

Bats are least active in the week following the first quarter of the moon (the week with most light in early evening). Captures were few (average 35.4 per night), surges infrequent and short (2.5 h), and only half of the nights had any surge at all (Table 7-15). Only 4% (one in 28 nights) had a second surge. The first surge was early, beginning at approximately 1915 h and ending at about 2045 h. Presence of fruit and absence of rain seemed to be of relatively little consequence during nights with bright moonlight.

In spite of increasing light in early evening, *A. jamaicensis* was most active in the week following the new moon. Capture rate was high (averaging 53.8 per night) and surges occurred on 80% of the nights. Probably because of the early evening light, second surges occurred on 20% of the nights. The first surges were of long duration, lasting about 3.5 h, and they began about 1945 h and ended near 2215 h. Morrison (1978a) used the term lunar phobia to describe these correlations of activity and moon phase.

Summary

The recapture rate in our 8907 *A. jamaicensis* averaged 1.8 per bat, indicating that less than 50% were recaptured at all, and less than 20% were recaptured three or more times. The low recapture rate, long distances between recapture sites, and potential home ranges that include the entire island and parts of the adjacent mainland complicate movement analyses. We adjusted for some of these difficulties by pooling the data for each locality.

We tabulated captures by species and by half years on a matrix of locality groups and then summarized the entire marking interval (October 1976 through October 1980). Although neither factor completely explains the variation, proximity of netting localities and differences in netting effort led to recapture rates varying from 1% to 43%. Localities in the center of the island consistently yielded better recapture rates than did marginal localities, suggesting that bats living near the edge of the island also forage on the surrounding mainland. It proved impractical to use netting sites randomly on a grid on BCI. Capture success was influenced by rain, wind, cloud cover, moon phase, topography, and the distribution and abundance of preferred foods. Thus, we had to try to standardize capture effort by the use of such measures as recaptures per net-night, or net-hour. We settled on using the number of *A. jamaicensis* caught per locality as the most useful and effective measure of capture effort.

The population of *A. jamaicensis* on BCI is not an

TABLE 7-15.—Surges of *Artibeus jamaicensis* during the rainy seasons (August–November) of 1979 and 1980 on BCI.

Moon	First surge							Second surge							Fruit		Rain	
	Number of nights	%	Mean hour Start	Mean hour End	Mean duration	Mean number caught	Rate per hour	Number of nights	%	Mean hour Start	Mean hour End	Mean duration	Mean number caught	Rate per hour	Yes	No	Yes	No
1st Q	28																	
Surge	14	50	1918	2048	2.5	35.4	16.6	1	4	2300	2300	1.0	18	18	10	4	6	8
No surge	14	50													5	7	6	6
Full	34																	
Surge	25	74	1900	2048	2.8	47.7	19.1	4	12	2212	2242	1.5	14.5	9.7	22	3	6	19
No surge	9	26													7	1	2	6
3rd Q	41																	
Surge	27	66	1924	2130	3.0	49.6	19.5	8	20	2236	2330	1.9	32.7	17.5	22	6	9	19
No surge	14	34													3	10	9	4
New	30																	
Surge	24	80	1948	2218	3.5	53.8	16.6	6	20	2248	2336	1.8	38.2	27.0	19	5	10	14
No surge	6	20													3	2	3	2
Total	133																	
Surge	90	68	1924	2124	3.0	49.1	18.3	19	14	2236	2318	1.7	30.0	16.2	73	18	31	60
No surge	43	32													18	20	20	18
Range			1800–2300	1900–0400	1–6	11–137	6–63			2100–0200	2200–0200	1–4	9–96	7–65				
Percent with surge															80	47	61	77

amorphous unstructured mass of mobile bats flying randomly throughout the island. Individuals are more likely to be recaptured close to where they were marked. Bats netted at a particular locality include those that have day roosts in the area as well as those coming to forage at fruiting trees, and those caught as they travelled between day roosts and feeding sites elsewhere. We believe that the distribution of fig trees is a major influence on the relative number of recaptures at any locality.

We examined fidelity to a particular area (locality group) by analyzing suites of multiple capture records of the 3846 bats recaptured and found three patterns. Highest fidelity occured where we sampled day roosts, rather than foraging sites. Fidelity is at least partially dependent upon netting effort. In locality groups with unusually high fidelity, the sex ratio was skewed toward males. In one such group, most of the males were bachelors, and most of the females were members of harems.

Data on movements of individual bats was restricted by the relatively small number of multiple recaptures. The average bat flies on about 700 nights during its lifetime, yet we captured a bat only an average of 1.8 times; a very narrow window through which to examine its movements. Radio tracking and capture of bats with known day roosts showed average movements of 1–4 km between day roost and feeding site. Mean distances between captures show no correlation with sex or age. Proportionately, records of movements by juveniles were much more frequent than expected and records of adults are about the same or a little less than expected, perhaps reflecting the relative difficulty in recapturing older adults. The longest distance we recorded between captures was about 6 km for three individuals. However, these were not single-night movements; therefore, these distances may represent dispersion instead of actual foraging distances.

Artibeus jamaicensis often visit more than one tree in an evening. We recaptured one individual at a fruiting tree 1.2 km away from another tree where she had been captured earlier the same evening. Recapture rates were only 2% for bats captured more than once at the same tree on successive nights, and similarly low for bats recaptured the same night at the same station.

The number of bats at a netting station shifted during the night, usually abruptly, suggesting that the bats are moving and foraging in groups. These surges of activity seemed influenced by rain, presence of ripe fruit, and phase of the moon. Clearly, bats were least active in the week following the first quarter of the moon (most light early in the evening), and most active in the week following the new moon.

8. Roosting Behavior

Douglas W. Morrison and Charles O. Handley, Jr.

In this section we present a synthesis of our mark-recapture, radio-tracking, and night-viewing-scope observations pertinent to day roosts of *Artibeus jamaicensis* on Barro Colorado Island (BCI). For comparison, we have included data on other species of bats where appropriate. We present a similar synthesis of observations on night roosts as part of the discussion in Section 9, Foraging Behavior.

Mark-recapture and radio-tracking techniques have inherent strengths and weaknesses. With radio-tagging, the behavior of individuals can be intensely monitored for days or weeks. However, sample sizes tend to be small and there is no independent way to measure how much the behavior of an individual has been altered by the transmitter. In contrast, long-term netting studies can generate huge samples, but the data are sometimes difficult to interpret.

Taken together, the two techniques complement each other in important ways. Questions provoked by radio-tracking observations can provide a framework around which to analyze mark-recapture data and netting data provide independent confirmation that the behavior of radio-tagged individuals is typical of the population as a whole. Using the two techniques in concert has increased our confidence in the accuracy of the picture that has emerged.

Bats show diverse roosting behaviors, often using different kinds of roosts for different periods of their daily and annual cycles (Kunz, 1982). Day roosts offer protection from predators and the elements and are generally used for extended periods. In contrast, night roosts may be less protected, temporary sites chosen for their proximity to food sources. Individual bats also may change roosts on a seasonal basis, with certain types of roosts favored for mating, rearing young, or other activities.

Day-roosting Sites

A. jamaicensis apparently is opportunistic in its selection of day-roosting sites. Where caves are available, it roosts in large groups (Dalquest, 1953; Kunz, 1982; Tuttle, 1968). In forested habitats, the largest groups roost by day in tree hollows, while smaller groups and individuals are found in foliage (Morrison, 1979). In the moist tropical forest of BCI, we located 25 day roosts used by 18 radio-tagged *A. jamaicensis*. Sixteen were in foliage and nine in tree hollows.

DAY ROOSTS IN FOLIAGE.—Foliage roosts were used primarily by males. Seven radio-tagged males made transient use of a variety of sites in the foliage, typically occupying a site for 3–5 days (range 1–13 days) before moving to another foliage site 100 m or farther away. In contrast, females normally roosted in tree hollows and rarely used foliage sites. Of 11 radio-tagged females, two roosted in foliage but only after being captured and radio-tagged as they emerged from their day roosts in tree hollows.

The kinds of foliage roosts used by *A. jamaicensis* (a 45 g bat) are typical of those used by canopy-foraging fruit bats (Goodwin and Greenhall, 1961) and were indistinguishable from those used by *A. lituratus* (70 g) and *Vampyrodes caraccioli* (36 g) on BCI (Morrison, 1979, 1980a). Foliage roosts used by radio-tagged *A. jamaicensis* on BCI included shelters under a long, arching frond of a palm, *Oenocarpus panamanus;* under an "umbrella" formed by a single, wilted leaf of a broadleaf epiphyte; in the crown of a spiny black palm, *Astrocaryum standleyanum;* and in a shallow shelter formed by the forest canopy.

Radio-tagged *V. caraccioli* (nine sightings) invariably roosted in groups of three or four adults, 7–12 m above the ground, under the umbrella-like crowns of understory trees (12–20 cm Diameter at Breast Height (DBH)). The day roosts of radio-tagged *A. lituratus* (26 sightings) were more variable, from 2.7–28 m above the ground in a variety of situations, such as under broken or crossed fronds of *Oenocarpus panamanus,* in vine-tangled crowns of subcanopy trees, in cavelike recesses on the underside of the crowns of canopy-height trees, and in branches overhanging the water along the lakeshore.

As variable as these roosts might seem, all had two important features in common: they were difficult to see from the ground and they had unstable supporting structures. Despite their open appearance, foliage roosts may be almost as effective as tree hollows for reducing the exposure of adult bats to predation and rain. The dense leaf cover forming the roof of these recesses is

Douglas W. Morrison, Department of Zoology and Physiology, Rutgers University, 195 University Ave., Newark, N.J. 07102.
Charles O. Handley, Jr., National Museum of Natural History, Smithsonian Institution, Washington, D.C. 20560.

waterproof and provides heavy shade. Contrary to our initial expectations, the availability of the dark recesses these bats prefer as day roosts may be limited, and new ones may be difficult to find. To find a radio-tagged bat often required searching through the foliage for hours from every possible angle with the aid of binoculars aimed at the source of the radio signal. When finally located, the bat was always in the darkest of the myriad recesses overhead.

Foliage roosts would probably frustrate any attack by an aerial predator. A terrestrial predator, even if it did see a roosting group from the ground, would almost certainly vibrate the highly flexible branches of these roosts and be detected long before it could reach the group. Some arboreal predators, such as snakes, may be stealthy enough to make their way across the shaky fronds and branches of foliage roosts without being detected. We observed a large green snake (*Leptophis ahaetulla*) ascending along the rib of an arching frond of *Oenocarpus panamanus* without obvious disturbance to the frond. This snake probably eats lizards and frogs (A.S. Rand, pers. comm.), but might take small bats. On the other hand, fronds and branches sagged under the weight of a 3-m-long rat snake, *Spilotes pullatus,* a known predator of bats.

Changing their roosts almost daily might make foliage-roosting *A. jamaicensis, A. lituratus,* and *V. caraccioli* more difficult to find. The hypothesis that day-roosts are changed to sites closer to the fruiting trees in current use was not supported by our observations (Morrison, 1980a). Increasing the distance between consecutively used sites should decrease the risk from predators attracted to an earlier site. However, foliage-roosting bats do not range freely, but rather remain faithful to a particular ridge or shoreline (Morrison, 1980a). Searching in a familiar area may significantly reduce the cost of finding a new roosting site.

DAY ROOSTS IN TREE HOLLOWS.—Naturally occurring hollows in trees provide shelters that are more permanent. Access to the hollows used by radio-tagged *A. jamaicensis* on BCI was invariably through a single small hole or slit, about 6 cm wide, and 2–15 m above the ground. The hollows typically were long cylinders, 12–25 cm internal diameter, extending 1.5–2.0 m above the entrance hole.

Judged by the size and location of the entrances, hole roosts should give better protection from predators than foliage roosts, but only if their entrance holes are neither too close to the ground nor accessible from nearby branches. We observed predation by a snake on a day-roosting group of *A. jamaicensis* in the hollow trunk of a 1.1-m-DBH fig tree (*Ficus yoponensis*) east of Donato 2. The entrance of the roost, 13 m above the ground, had probably been inaccessible to snakes for many years, until the crown of a nearby sapling (4 cm DBH) grew up to within 0.5 m of it. On a July 1979 afternoon, during a routine check of the roost, we watched a 1.7-m-long yellow-and-black snake (believed to be a rat snake, *Spilotes pullatus*) passing from the roost entrance into the crown of the sapling. The departing snake had five delta-shaped bulges along its length.

Unfortunately, the belief that the bulges were bats could not be confirmed because the snake never came lower than 12 m above the ground before it disappeared in the subcanopy.

Individual *A. jamaicensis* apparently are aware of the location of several holes suitable for roosts, but favor a particular hole for extended periods. Eight of 11 radio-tagged females used the same hollow or pair of hollows for the 4–22 day life of their transmitters (Morrison, 1979). The male and four of six females trapped at a roost near Barbour-Lathrop 1 were still in residence when the roost was netted again three months later. The movement of roosting females between tree hollows was not completely random. For example, a female that usually roosted near Barbour-Lathrop 1 twice used a second hollow 100 m away. This same pair of hollows was used by a second female two months later. Similarly, a female that favored a hollow near Wheeler 3 also roosted in a *Quararibea asterolepis* 300 m away. Three months later another female, captured as she left the Wheeler 3 hollow, took up residence in the *Quararibea*. These observations are more likely a reflection of a shortage of preferred roosting holes than a reflection of cohesiveness within female groups. Female roostmates showed no cohesiveness when foraging (see Section 9, Foraging Behavior). Further, the few females that temporarily abandoned their tree hollows after being netted there typically went to different alternative sites.

DAY ROOSTS IN MAN-MADE STRUCTURES.—Groups of day-roosting *A. jamaicensis* were found inside the hollow pyramidal concrete markers used as navigational aids by ships transiting the Panamá Canal. Each marker stands 7.6 m high at the peak, is 1.5 m on each side, is open at the bottom, and resembles a hollow pyramid on long stilts. Two such markers, one on Buena Vista Island and the other near Palenguilla Point (= Peña Blanca), have been occupied by *A. jamaicensis* for many years. Each of these roosts was censused ten times over a three-year period beginning in October 1977. Of 83 adult females trapped in these roosts at least once, 44 (53%) were recaptured in the same roost 1–6 times. The average interval between first and last capture was 14 ± 9 months, with 14 females in residence for over two years.

The recapture records from the Buena Vista and Peña Blanca markers provide a good estimate of long-term fidelity, despite the artificial nature of the roosts themselves and disturbance by Bat Project personnel. Actually, the average tenure of females may be longer at hole roosts that are not so frequently disturbed. The number of *A. jamaicensis* roosting in the two markers declined steadily over the study period, probably due to the disturbance of repeated capture. At Buena Vista, the number of adult females went from 20–30 in 1977 to 10–15 in 1978 and 1979. By the end of 1980, only four adult females remained, three of which were first-time captures. At Peña Blanca, adult females declined from 10–30 in 1977 to 10–15 in 1978 and 1979. By the end of 1980, the only two bats in the roost were subadult females, both first-time captures.

EXPERIMENTS WITH "ARTIFICIAL" TREE HOLLOWS.—

Walking through a tropical forest, one can get the impression that holes in the trunks of trees are almost as numerous as foliage umbrellas. However, closer inspection reveals that these holes usually open into cavities too shallow for even one *A. jamaicensis,* much less a group, and many of the larger cavities either extend down from the entrance hole (the preferred direction is up), have an entrance hole that is too large (wider than 6 cm), or have more than one entrance hole.

We suspected that the availability of suitable tree hollows might be a factor limiting the population of *Artibeus jamaicensis* on BCI. Finding *A. jamaicensis* roosting inside concrete pyramids encouraged us to put up artificial tree hollows to determine whether the demand for preferred roost holes was greater than the supply and what the pattern of colonization of new roost holes might be. From balsa trees (*Ochroma pyramidale*) that we felled on Delesseps Island, a regularly cleared area adjacent to a dangerous bend in the Canal, we constructed 40 artificial tree hollows. The trees were cut into 2 m lengths, 0.35 to 0.45 m in diameter, sawed in half lengthwise and hollowed out with an ax. Internal cavities approximated the "ideal" hole roost: 15 cm in diameter and extending 1.1 m above a 6 cm × 40 cm entrance hole. To facilitate inspection of roost contents with a flashlight and mirror, a small trap door was cut in the bottom of the back half.

During July 1978, we placed the bat houses all over BCI in habitats structurally similar to those of known hole roosts. The hollows were lashed high enough on understory trees for the entrance holes to be at least 1.7 m above the ground. Sixteen roosts were placed within 800 m of the laboratory clearing, eight within 500 m of the Tower Clearing in the center of BCI, and four each along Miller Ridge, Chapman End, Drayton End, and Armour End. In an attempt to make some of the new roosts more conspicuous, ripe figs, banana, and *A. jamaicensis* feces were placed inside the entrances of eight of the roosts near the lab clearing.

We hoped that regular inspection of the roosts would reveal some pattern in the colonization of the newly available tree hollows and shed light on the dynamics of harem formation. Do females aggregate at a tree hollow first, with a male appending himself later? Or is the roost first discovered and used by a male who then "recruits" females?

These questions remain unanswered. The hollows were a disappointing failure. Although used by a variety of invertebrates and a few vertebrates, none was ever found to have been used by bats. A thorough census in July 1979, after the hollows had been in place for a year, revealed the most common occupants to be termites (16 hollows), crickets (13), spiders (11), and ants (6). Other inhabitants were whip scorpions (3), earwigs (3), roaches (2), and geckos (2). A few contained mammalian (2) and avian (1) nesting material. Two hollows were completely empty. All the hollows were taken down and destroyed in January 1980. Unfortunately, negative data cannot be used to argue either for or against a shortage of preferred roost hollows.

Composition of Day-roosting Groups

GROUPS IN FOLIAGE.—*A. jamaicensis* has been reported roosting in large groups containing both sexes in foliage in Trinidad (Goodwin and Greenhall, 1961) and Brazil (Jimbo and Schwassmann, 1967). On BCI, groups of *A. jamaicensis* roosting in foliage were small (1–3 individuals) and primarily male (Morrison, 1979, table 1). Five of eight foliage roosts were occupied by solitary adult males and another by two adult males. The only adult females found roosting in foliage were two that had been radio-tagged when they emerged from hole roosts. They used foliage roosts for only two to five days before returning to a hole roost.

GROUPS IN TREE HOLLOWS.—Hole roosts were used by groups of adult females, their nursing young, and a single adult male. Nine of 11 radio-tagged females used hole roosts exclusively. The two exceptions had been captured at hole roosts, subsequently used foliage roosts, but returned to hole roosting a few days later.

The composition of hole-roosting groups was determined by capturing the occupants in a "laundry-chute" trap as they emerged in the evening. Each of six hole roosts censused this way contained only one adult male, 3–14 adult females, and 0–6 juveniles (Morrison, 1979; Morrison and Morrison, 1981).

GROUPS IN MAN-MADE STRUCTURES.—The pattern of sexual segregation at day roosts suggests that *A. jamaicensis* has a harem mating system. This hypothesis is supported by the long-term records of *A. jamaicensis* roosting inside the navigation markers at Buena Vista and Peña Blanca. In 18 of 20 capture episodes when adult females were present in the roosts, only one adult male was captured or some bats escaped (among which there might have been an adult male). We never captured more than one adult male in a roost.

Of 46 *A. jamaicensis* captured as adults at the Buena Vista marker, only three were males. The first male was captured three times over a 17-month period, first in December 1977 and last in April 1980. The second was captured at the marker once, in November 1979, after it had been marked as a subadult at Snyder-Molino 2 on BCI in December 1977. A third was captured in March 1980. At no time was there more than one adult male at the roost. This capture sequence suggests that the second male had replaced the first as the harem male at Buena Vista and was himself replaced a few months later.

Displaced harem males may be able to regain their former position. Of 42 *A. jamaicensis* captured as adults at Peña Blanca, four were males but no more than one was found at a time. The first male was captured only once, in October 1977. The second male was captured three times between April and July 1978. The third male was captured in December 1978 and the fourth in April 1979. The second male reappeared as the only male present in November 1979.

Many adult females showed long-term fidelity to marker roosts that extended far beyond the tenure of any of the harem males. This suggests that females are attracted by the roost site rather than by a particular male.

Juvenile females do not stay to become members of their mother's harem. Of 31 females captured as juveniles at Buena Vista and Peña Blanca, only one was ever recaptured at either (Peña Blanca) as an adult. Instead, females leave their natal roosting groups and join other harems as subadults. Of 21 females present at the two marker roosts as subadults, seven (33%) were recaptured there as adults. The harems of *A. jamaicensis* contain adult females of all ages, suggesting that subadult females join established harems. In contrast, harems of the greater spear-nosed bat, *Phyllostomus hastatus,* are formed from a single generation of subadult females (McCracken and Bradbury, 1977). Young females of the short-tailed fruit bat, *Carollia perspicillata,* are more likely to disperse away from their natal roosts than are young males, but even so, 42% remain in the natal roost. Those roosts, however, tend to be larger colonies in caves or hollow trees and contain more than one harem (Fleming, 1988).

Males also leave their natal roosting groups by the time they are subadults. None of the 21 males marked as juveniles at Buena Vista and Peña Blanca was ever captured again at these roosts. The single male marked at Peña Blanca as a subadult was never recaptured.

Male Defense of Tree Hollows

The sexual segregation apparent at day roosts suggests that males gain harems by defending hole roosts used by females. This hypothesis is supported by Morrison and Morrison's (1981) observations on the nocturnal interactions of harem group members, monitored with an infrared viewing device in conjunction with the radio-telemetry of individuals.

During the breeding season of June–July 1979, a tripod-mounted, 1.5 power night-viewing device (Javelin 221) fitted with an infrared light source was used on six nights to observe the activity around two roost hollows. Transmitters were fitted to each harem male and three and four of the three and 14 females in their respective harems. The harem male at a third hole roost was also radio-tagged.

The transmitters enabled us to monitor the foraging activities of individuals on 27 10-hour nights and to identify individuals within the field of view of the night-viewing device. In addition, because the relative intensities of the transmitter pulses became distinctively modulated whenever the bats were flying, flying times could be measured even when the bats were not in sight.

Continuous radio contact was maintained on the three harem males during 12 10-hour nights. All three harem males spent over 90% of the night within 50–100 m of their respective roost hollows. Unlike the females, who carried fruits only 25–200 m from the fruit tree to a night roost before eating them, harem males brought each fruit back to the day roost area. On several nights this meant the male was making feeding passes of over 1.5 km on each roundtrip.

The foraging activities of the two males with radio-tagged harems were closely correlated with those of their females. The male was invariably the first out of the roost at dusk. He began making feeding passes before most of his harem had left the roost. Each male suspended feeding passes as soon as the first female returned to suckle the juvenile she had left behind inside the hollow. Females remained inside the hollow for 1–4.5 hours, depending on the phase of the moon. Both sexes minimized flying in bright moonlight, probably to reduce the risk of predation from visually orienting owls and opossums (Morrison, 1978a). During these periods of inactivity, the harem male stayed outside the roost, but did not resume feeding passes until his females began to leave again for a second bout of feeding. The harem male was always the last to reenter the roost at dawn.

Harem males did an extraordinary amount of flying in the immediate vicinity of their day-roost trees. While their females were leaving the roost, the male preceded each of his feeding passes with 1–4 minutes of local flying, during which he crossed in front of the roost entrance one to five times, occasionally hovering there for less than a second. In contrast, when all his females were out foraging, most local flying was done away from the roost entrance out of view of the night-viewing scope.

Local flying probably keeps the nearby area free of potential intruders. On two occasions the harem male chased away an unidentified bat that repeatedly tried to enter the roost. The unidentified bats may have been rival adult males or they may have been young of the previous generation that were no longer allowed to roost with their mothers. On the two nights when we were trapping emerging harem members for radio-tagging, we also netted two adult males and seven juveniles trying to enter the roost.

Local flying by the harem male might also serve to advertise the roost to females. During one predawn return, the harem male was seen closely following (escorting?) five of his 14 females as they approached the roost entrance, circling back to inspect the entrance after each female had entered. The frequency and duration of local flying bouts increased during both periods of the night when females were returning. In the absence of bright moonlight, harem males were typically in flight for 60 minutes of the final 90 minutes before dawn.

These observations provide additional support for the hypothesis that *A. jamaicensis* males maintain harems by defending roost holes. Although it would not seem feasible to defend more than one female at a time when they are foraging away from the roost, the defense of a roost hollow could significantly increase the harem male's unimpeded access to females in postpartum estrus. In the 18 g frugivorous short-tailed fruit bat (*Carollia perspicillata*), males defend roosting sites both in hollow trees and within the confines of larger colonies roosting in caves (Fleming, 1988).

In two other species of Neotropical bats known to have harems, males have evolved other strategies for harem maintenance. In the 8 g insectivorous white-lined bat (*Saccopteryx bilineata*), males gain harems by defending prime feeding territories (Bradbury and Vehrencamp, 1977a). In the 115 g omnivorous greater spear-nosed bat (*Phyllostomus hastatus*), males defend females clustered on cave ceilings (McCracken and Bradbury, 1977). These behaviors are consistent with the theory that harem mating systems evolve only where it is feasible for one male either to defend a group of females directly or to defend some limited resource used by more than one female (Emlen and Oring, 1977).

Summary

Data from 18 radio-tagged *A. jamaicensis* yielded information on 25 day roosts on Barro Colorado Island. Day roosts in foliage are more commonly used by males than females, and the bats typically change roosts every three to five days. Foliage roosts occur in a variety of species of plants, but all are difficult to see, have unstable support, and yet provide good protection from rain and predators. These roosts probably are used primarily by bachelor males, either singly or in small groups.

Day roosts in tree hollows provide more permanent shelter and probably better protection from predators. Tree holes are occupied by harems and are defended by males. Each of six hole roosts, censused by capturing all occupants, contained 3–14 adult females, 0–6 young, and a single adult male. Females favor a particular hole for extended periods. Harem males spend over 90% of each night within 50–100 m of their hollows. Even when females are not present, the male carries individual fruits back to the hollow to eat. Harem males do an extraordinary amount of flying near the hole roost, chasing away intruders and escorting returning females.

Two man-made navigational markers housed harems of bats for extended periods of time. These roosts never contained more than one adult male. Adult females frequently outlasted harem males at these sites, suggesting that females are faithful to the roost site rather than to the male. All juvenile males leave their natal roosts by the time they are subadults. Most juvenile females also leave their natal roosts and join other harems as subadults.

In an effort to determine how new harems are established, we constructed artificial tree hole roosts from pieces of tree trunk. These artificial roosts failed to attract any bats during the course of a one-year trial.

9. Foraging Behavior

Charles O. Handley, Jr., and Douglas W. Morrison

Enough is known about the foraging behavior of *Artibeus jamaicensis* to permit a quantitative description of its daily routine and energy budget (Morrison, 1978d; Section 2, Physiology). However, many questions remain, especially concerning the social aspects of foraging. Here we present a more qualitative, at times speculative, discussion of behaviors related to foraging, based on our radio-telemetry and mark-recapture studies of *A. jamaicensis* on Barro Colorado Island (BCI).

The Basic Feeding Pattern

From radio-telemetry of 32 individuals we know that an *A. jamaicensis* typically leaves the day roost 0.5 hours after sunset and flies directly to one of the fruit trees visited on a previous night. Individuals may return to the same tree for up to eight consecutive nights (4.3 ± 1.8, $n = 27$ trees). A bat does not roost in the fruiting tree to eat, but instead carries the fruit to a feeding roost 25–200 m away. Small bites of fruit are crushed, sucked dry, and dropped as pellets ("bat chop") under the roost. From its feeding roost, the bat makes numerous passes to the fruiting tree, returning with a fruit carried in its mouth each time. At 0.5 hours before sunrise the bats return to their day roosts.

The basic feeding pattern is affected by the phase of the moon (Morrison, 1978a). On "dark moon" nights (one week either side of the new moon) *A. jamaicensis* is active from dusk to dawn and visits as many as five separate feeding areas (3.0 ± 1.5, $n = 14$ nights). On "bright moon" nights (one week either side of the full moon) the bats interrupt foraging to return to their day roosts for one to seven hours when the moon is nearest its zenith. On these nights, the bats usually visit no more than two feeding areas (2.1 ± 0.7, $n = 23$ nights), with the more distant one visited either before moonrise or after moonset.

"Lunar phobia" (Morrison, 1978a) and the use of feeding roosts probably reduce exposure to predation. Bright moonlight would make a bat hovering to select a fruit more conspicuous to visually-orienting predators such as owls and opossums. Anecdotal evidence suggests that potential predators of bats are attracted to fruiting trees. Judging from the number of vocalizations heard, there seem to be more owls around trees with ripe fruit than in other parts of the forest. Furthermore, opossums (e.g., *Didelphis marsupialis*) are often seen in the branches of fig trees containing ripe fruit, and we once saw a gray four-eyed opossum (*Philander opossum*) almost capture a bat that had chanced to roost nearby. This helps explain why *A. jamaicensis* does not hang in fruit trees to feed.

Lunar phobia is present, to a lesser extent, in *A. lituratus* and *Vampyrodes caraccioli*, two other canopy fruit bats of BCI (Morrison, 1980a). Aspects of lunar phobia that we observed in these species may apply to *A. jamaicensis* as well. Perhaps because their day roosts are in foliage rather than in tree hollows, radio-tagged *A. lituratus* and *V. caraccioli* did not return to them during bright moonlight, even when the day roost was relatively close (within 350 m) to the feeding area. However, total flying activity was reduced significantly on bright moonlit nights.

To simplify studying activity patterns of bats we divided the night into 15-minute intervals. On bright moon nights, *A. lituratus* did not fly in 55% of the intervals, in contrast to flying in 80% on dark moon nights. *V. caraccioli* showed a similar pattern, sometimes remaining inactive (not flying) for periods of up to four hours. Although feeding passes were suspended or greatly reduced in moonlight, both species flew between feeding areas and conducted prolonged search flights even in the brightest moonlight. These behaviors suggest that the risk of predation is greatest near fruiting trees.

Various workers (Crespo et al., 1972; Schmidt et al., 1971; Tamsitt and Valdivieso, 1961; Villa-R, 1966; Wimsatt, 1969) have reported catching fewer foraging bats (*Desmodus rotundus*, *A. lituratus*, *Phyllostomus discolor*, and *Glossophaga soricina*) in their nets during periods of moonlight. Eckert (1974) painstakingly demonstrated the inhibitory effect of moonlight on *A. lituratus* and *P. hastatus* in captivity and Häussler and Eckert (1978) confirmed the effect on *P. discolor* with simulation experiments. Several African insectivorous microchiropterans show the same effect as evidenced by bat

Charles O. Handley, Jr., National Museum of Natural History, Smithsonian Institution, Washington, D.C. 20560.
Douglas W. Morrison, Department of Zoology and Physiology, Rutgers University, 195 University Ave., Newark, N.J. 07102.

detector recordings of activity (Fenton, Boyle et al., 1977). Eckert (1982) argued persuasively that all such effects are adaptations for the avoidance of visually orienting predators.

Longer-term Foraging Patterns

Patterns of fruit tree use were most apparent on bright moonlit nights because flying was limited and feeding passes seemed especially regimented. Occasionally on such nights, a radio-tagged *A. jamaicensis* made what appeared to be a feeding pass in the wrong direction. In some cases, these flights were simply feeding passes to another fruiting tree, demonstrating that bats sometimes use a single night roost while feeding on more than one tree. Or, if these were nonfeeding passes they may have been reconnaissance flights to gain information about future feeding sites.

Although bats returned to feed on the same fruit tree for as many as eight consecutive nights, they discontinued feeding visits when the density of the figs declined to less than one or two per square meter ($n = 6$), unless a fig tree was less than 150 m from the day roost. In switching fruit trees, a bat typically made no more than the first feeding pass of the night to the previously used tree and then immediately, without search flying, switched its feeding roost or feeding passes to a new tree. In 11 instances this shift was to an area that the bat had visited from 1–8 (4.2 ± 2.2) nights previously. In five of these 11 cases, the shift was to a fig tree that had almost certainly been the object of earlier reconnaissance flights.

Flying by radio-tagged *A. jamaicensis* during the bright half of the lunar month was reduced to the minimum length of time required to get to a feeding roost (usually from one to five minutes). Even on moonless nights, the bats often made only short flights and spent most of their time hanging at the feeding roosts. Occasionally, however, bouts of sustained flying that lasted 10–45 minutes were recorded. Six of 40 all-night tracking sessions contained at least one such period of prolonged flying. All but one of those longer flights occurred during the dark half of the lunar month. The one exception occurred in the evening before a late (2345 hours) moonrise.

Prolonged flights began either from night feeding roosts ($n = 3$) or upon emergence at dusk from day roosts ($n = 3$). The area covered by these flights could only be approximated, as it was not possible to triangulate the bats' constantly changing position. Usually the flying was confined to the watershed in which it began ($n = 4$), but once it shifted into an adjacent watershed and another time into a nonadjacent watershed. These long search flights undoubtedly increase the chances of encountering newly fruiting trees and might have been initiated for that purpose.

Feeding Roosts

Triangulation by radio of foraging *A. jamaicensis* suggests that feeding roosts usually are in one or a few favored trees, within a 0.25 to 0.5 ha area, 25 to 200 m from a fruiting tree. Morrison (1975) was able to pinpoint and confirm, by on-site inspection, two feeding roosts used by his radio-tagged *A. jamaicensis*. One was in the crown of a 13-m-tall, spiny black palm (*Astrocaryum standleyanum*). The other was about 3 m above the ground under the long, arching frond of an *Oenocarpus panamanus*. Both sites could have been selected for their reduced accessibility to terrestrial predators (see Section 8, Roosting Behavior).

After Alfred Gardner discovered that bats drop pellets of chewed pulp when feeding on figs, Bat Project personnel routinely used accumulations of pellets to recognize sites used as feeding roosts. Depending on the height from which they are dropped, the pellets may be in small piles, distinct clusters, or broadly dispersed. The area covered can be quite extensive, sometimes exceeding 100 m^2. These feeding roosts are often located along ridges above fruiting trees, but may be found elsewhere, especially where young palms or subcanopy trees are numerous. The single common feature is an open understory that does not restrict flight. This may explain why we have found so many feeding roosts along or adjacent to trails, which also function as open flight paths for bats.

Radio-tagged *A. jamaicensis* were relatively faithful to feeding roosts. Invariably the same feeding roost area was used night after night for as long as a bat fed from the same fruit tree ($n = 14$ fruiting trees). Four radio-tagged bats made foraging passes from the same feeding roost to two simultaneously fruiting trees 35–220 m apart, and two bats used the same feeding roost areas when they returned after three and eight nights, respectively, to get fruit from another tree in the neighborhood.

There are several possible, but untested, explanations for this apparent fidelity to feeding roost areas. It might simply be the result of the patchy distribution of vegetation preferred for feeding roosts. Or it might be that *A. jamaicensis* is a creature of habit, preferring to use familiar sites along well-known flight paths. The latter hypothesis is supported by our observation that radio-tagged *A. jamaicensis* frequently did not use (or failed to find) a fig tree producing ripe fruit close to their day roosts. Instead they commuted two to three times farther to another tree. This suggests that many fruiting trees are found because they stand along familiar foraging routes.

Feeding roosts may have other functions not related to feeding. For example, if they are closer to a fruiting tree than the day roost, they may serve as convenient havens during bright moonlight. This at least seems to be the case with *A. lituratus* and *V. caraccioli,* two fruit bats that have day roosts in foliage (Morrison, 1980a). Feeding roosts also may be sites for social interaction. Adult bachelor males may try to copulate with females at feeding roosts away from the day roost and its defending harem male because they probably lack other access to females (Morrison and Morrison, 1981). However, females may be less receptive to males encountered outside the day-roost hollow if competition for hollows selects for males who sire stronger, more aggressive, or otherwise fitter young. Also, day roosts are probably safer sites for copulating bats than are night roosts.

Group Foraging

Feeding roosts might be "information centers" for food finding (sensu Ward and Zahavi, 1973). In theory, *A. jamaicensis* is a likely candidate for group foraging information sharing. A prerequisite for the evolution of information centers is that food be found in short-lived, locally superabundant patches. This sort of food distribution means that users must continually find new food patches. However, because the patches contain so much food (and for so short a time), nothing is gained by trying to conceal or defend a food patch once it is found. Fig trees fit this description. A single fig tree on BCI may bear 40,000 figs, of which a single *A. jamaicensis* would likely carry away no more than 100 (Morrison, 1978d) before the fruits were gone or spoiled.

A second characteristic that makes *A. jamaicensis* a potential user of information centers is that foraging bats aggregate at night (feeding) roosts. A night roost may be occupied by dozens, if not hundreds, of bats, and they may be making feeding passes to several different fruiting trees.

A simple mechanism for information exchange is plausible. If an *A. jamaicensis,* which has been successful in finding a fig tree, returns to its feeding roost with a fig or the odor of figs, any roost mates who had been unsuccessful in finding food could follow the successful bat out on its next feeding flight. The information sharing would not need to be intentional, but the information would greatly benefit the unsuccessful bat and would not significantly deplete the food at the tree for the successful bat. In a group that remained together over a period of time, it is possible that tonight's successful bat could be tomorrow night's unsuccessful bat, and vice versa, so all would gain over the long term. Note also in this context the probability of food finding reconnaissance flights initiated in advance of need.

Despite the theoretical potential, it is still unknown whether *A. jamaicensis* on BCI forages in cohesive groups. Certainly this bat does forage in groups, as the nightly ebb and flow of bats at a fruit tree demonstrates (see Section 7, Movements). However, whether the flocks that swirl about a fruit tree stay together for hours, all night, several nights, or for long periods of time is unknown.

Morrison and Morrison (1981) tracked a group of three and another of four radio-tagged female *A. jamaicensis* from two different harems for eight complete nights and found no evidence of cohesiveness of harem members away from the day roost. These females left and reentered their day roost hole individually. Because emergences were typically more than a minute apart, each female emerged after the previous one had disappeared from the day roost area. On three of the eight nights, three of four females did visit the same ripe-fruit-bearing *Ficus insipida* in the course of the night, but they moved between fruit trees independently and did not roost anywhere near each other while feeding. This radio-tracking study suggests that harem females do not forage in groups, at least not during the circumstances operant when our observations were made.

Nevertheless, it is still possible that other sex and age classes forage in groups, or that *A. jamaicensis* forage together in habitats where fruiting trees are not as abundant and easy to find as they are on BCI. In México and Costa Rica, for example, clumped mist-netting capture times have been interpreted to mean that the *A. jamaicensis* there forage in groups (Dalquest, 1953; Heithaus et al., 1975). Group foraging has been reported in flower-feeding bats in Brazil (Sazima and Sazima, 1977) and Arizona (Howell, 1979).

It should be possible to detect foraging-group associations by analysis of the occurrence of "double recapture pairs" in the mark-recapture data. A "double recapture pair" consists of two bats that were captured at the same netting site on the same date and were subsequently recaptured together at another site some days, months, or years later. Morrison (1975) found 22 such pairs among the 259 recaptures of 1472 *A. jamaicensis* he and Bonaccorso marked over a 14-month period in 1972 and 1973.

What is the probability that this number of double recapture pairs might occur simply as a result of chance associations in a population of independently foraging bats? Given that two or more recaptures are made on night n, let P_n stand for the conditional probability that any two of these recaptures were captured together previously (Morrison, 1975). If $(n-1)$ = the number of previous capture nights, x_i = the number of bats captured on previous capture night i, where $1<i<n-1$, and y = the total number of bats banded to night *n,* then

$$P_n = \sum_{i=1}^{n-1} \frac{x_i\,(x_i-1)}{y\,(y-1)}$$

The number of double recapture pairs c_2 that could be expected to occur by chance on any night *n* is simply P_n times the total number of pair combinations of recaptures on night *n,* or P_n times $c_2{}^z$, where z_n is the number of recaptures on night *n.* The total numer of recaptures expected to occur by chance during the entire 14-month netting program is the sum of the $P_n \cdot c_2{}^z$ values of all 60 of the 131 sampling nights in which two or more recaptures were made. This sum was calculated and found to equal 14.2, the number of double recapture pairs expected to occur simply by chance. The observed value of 22 was not significantly greater than that expected from random assortment as determined by a one-tailed binomial test, $P = 0.195$ (the *a priori* expectation for higher than random assortment, justified the use of a one-tailed test). Thus, Morrison concluded that *A. jamaicensis* does not forage in groups, at least not in groups with memberships sufficiently constant over the long term to be detected by this method.

However, this conclusion was based on the unrealistic

assumption that there is no mortality of marked bats. The chance that a marked bat will die reduces the expected number of double recapture pairs and so increases the significance level of the observed value. To reflect this reality we used the equation

$$P_n = \sum_{n=1}^{n-1} \frac{x_i\,(x_i-1)}{y\,(y-1)}\ s^2 t_i$$

Here it is reasonable to make s equal to 0.57, the annual survivorship for adult *A. jamaicensis* (see Section 6, Population Estimates), and for t_i we use 0.32 years, the average interval between recaptures for Morrison's (1975, table 1) data. The expected value now drops from 14.2 to 9.9 [$14.2 \times (0.57)^{0.64}$]. The observed value of 22 double recapture pairs is significantly greater than this revised expected value (one-tailed binomial test, $P < 0.05$).

Unfortunately, even if these revised calculations reveal an incidence of double recapture pairs significantly greater than expected, we still cannot be sure that this difference is due to the existence of foraging groups. Other factors could produce the same result. For example, the probability of observing a double recapture pair would be increased if independently foraging bats frequented the same feeding area. Two bats might frequent the same area because it is especially attractive (e.g., the great Lutz fig patch) or because it is near the day roosts of both bats. The calculations of site fidelity (see Section 7, Movements) indicate that the capture sites were not in equally attractive areas and that individual bats were not equally likely to be captured at all sites. Furthermore, we suspect that after the first capture, bats tend to avoid mist nets because they are seldom recaptured (see Section 6, Population Estimates). Recapture pairs might be together at fruiting trees many times before one or both are recaptured again.

These inequalities tend to increase the expected number of double recapture pairs. Without some estimate of the magnitude of these biases, any test for foraging groups based on the incidence of double recapture pairs will be inconclusive.

Thus, although clustered capture times and double recapture pairs suggest group foraging, we lack sufficient evidence to show that these associations are anything more than coincidental. A powerful test for group foraging could be based on direct observation of the foraging movements of a large number of radio-tagged bats captured together at the day roost or captured in "clusters" while foraging. This remains to be done.

Summary

The usual pattern for an *A. jamaicensis* is to leave the day roost half an hour after sunset, fly to a fruit tree, take a fruit to a nearby feeding roost, consume it (dropping dry pellets to the ground beneath), make several more feeding passes to the fruit tree, then return to the day roost half an hour before sunrise. On dark moon nights, a bat may visit as many as five feeding areas. On nights with bright moonlight, bats visit only one or two feeding areas, and several hours may be spent back in the day roost during the brightest hours. Both lunar phobia and the use of feeding roosts separate from the fruiting tree may reduce exposure to predation.

Radio-tracking observations suggest that bats sometimes use the same feeding roost to make feeding passes to more than one fruiting tree. The bats also appear to make brief reconnaissance flights to assess the condition of previously located fruit trees. The bats spend most of their time hanging in the feeding roost and seem to minimize feeding time even on dark moon nights. Prolonged flights of 10–45 minutes (searching for new trees?) were rare and occurred only in the absence of bright moonlight.

Feeding roosts tend to be groups of favored trees 25–200 m from a fruit source, and most are located in areas of open understory. Bats are faithful to individual feeding roosts as long as the nearby fruiting tree remains productive. This fidelity may result from patchy distribution of preferred roosting trees, or may simply reflect the habitual use of a familiar area.

The bats clearly congregate at food sources, but whether cohesive, long-term group foraging occurs is unknown. Clustered capture times and double recapture pairs suggest group foraging, but direct observation of the phenomenon is lacking. Radio-tracking data on females from the same harem show no evidence of such flocking.

Whether or not *A. jamaicensis* forms foraging groups, the potential exists for an exchange of food location information among individuals at feeding roosts. Foraging theory suggests that feeding roosts could be information centers for finding trees with ripe fruit (short-lived, locally superabundant food patches), especially if the bats using a night roost are making feeding passes to different fruiting trees.

10. Food Habits

*Charles O. Handley, Jr., Alfred L. Gardner,
and Don E. Wilson*

Initially, the Bat Project focused entirely on capturing and marking bats, and little attention was given to food habits and other aspects of natural history. Gradually it became apparent that capture rate usually was linked to the foraging behavior of the bats and to the location of nets in relation to food sources and feeding roosts. Consequently, Bat Project personnel took increasing interest in the content of the bats' feces, in evidence and location of feeding roosts, in food items carried by bats into nets, and in the location and phenology of fruiting trees. As a result, capture effort became more productive in number of bats marked and recaptured. Although our data on food habits lack the precision of the smaller-scale studies of Bonaccorso (1979) and Heithaus et al. (1975), they do offer insights on food preferences, seasonality of food availability, and the foraging habits of *Artibeus jamaicensis* on Barro Colorado Island (BCI).

Project personnel eventually established a protocol that included searching for evidence of bat activity before selecting sites as capture stations. Along the trails we watched for pellets ("bat chop") that stenodermatine bats spit out after chewing and pressing the juice from the pulp of fruits such as *Ficus insipida, F. yoponensis,* and *Spondias radlkoferi.* The pellets appear in discrete piles or loose clusters on the ground when the bats drop them from roosts low in the subcanopy, or they may be scattered, as though broadcast, when they are dropped from the canopy. Because the bats return again and again during the night to the same perch, piles of pellets may represent many fruits.

We also watched for fragments of partly eaten fruits; large seeds such as those of *Spondias mombin, S. radlkoferi,* and *Dipteryx panamensis* discarded by bats after they had scraped off the pulp; cast off skins of fruit such as *Quararibea asterolepis;* and trees dropping ripe fruit of kinds known to be favored by bats. Congregations of noisy diurnal frugivores such as monkeys (*Alouatta palliata* and *Cebus capucinus*), guans

(*Penelope purpurascens*), and parrots (*Amazona* spp.) often led to the discovery of trees with ripening fruit.

We left the trails to search under certain trees (e.g., *Oenocarpus panamanus* Bailey and *Gustavia superba* (Humboldt, Bonpland, and Kunth) (Berg) particularly favored by bats as feeding roosts. We found that patches of these trees associated with several fruit trees were used repeatedly by *A. jamaicensis* as feeding roosts. Thus, in the course of a year, the bats may use the same feeding roost many times. Taking this into account, we routinely inspected known feeding roosts in addition to walking the trails in search of new sites.

Fruits Used as Food

Our observations on fruits eaten by *A. jamaicensis,* mostly late in the rainy seasons (August to November) of 1979 and 1980, are summarized here (Figure 10-1 and Table 10-1).

Ficus insipida Willdenow: Bonaccorso (1979), Fleming (1971), Morrison (1978d), and our own observations agree that in central Panamá *F. insipida* is the favorite food of *A. jamaicensis* (Figure 10-1), *A. lituratus,* and perhaps *Vampyrodes caraccioli,* and is eaten in lesser amounts by several other bats. Its fruit production is aseasonal, but the crop is limited in October and from January through March (Table 10-1). Few fruits reach maturity in August and September, two months when *F. insipida* does not seem to be a significant food for bats. Large concentrations of bats, mostly large stenodermatines, were found near *F. insipida* bearing ripe fruit in March, April (four sites), May, June, October (two sites), November (two sites) and December. Normally the bats pick soft, fragrant, fully mature fruit, but occasionally (as was noted on 11 January, 8 February, 21 October, and 7–9 November when few suitable *F. insipida* were available) *A. jamaicensis* carried partly eaten small, hard, and latex-laden unripe fruits into the nets.

Ficus yoponensis Desvaux: Similar to *F. insipida* in abundance and seasonal availability on BCI (Table 10-1), the small-fruited *F. yoponensis* is a favorite of smaller frugivores such as *Uroderma bilobatum, Vampyressa pusilla, V. nymphaea,* and *Vampyrops helleri* and is consumed in great

*Charles O. Handley, Jr., and Don E. Wilson, National Museum of Natural History, Smithsonian Institution, Washington, D.C. 20560.
Alfred L. Gardner, NERC, U.S. Fish and Wildlife Service, National Museum of Natural History, Washington, D.C. 20560.*

	JAN	FEB	MAR	APR	MAY	AUG	SEP	OCT	NOV	DEC
Ficus insipida	□1 △9	△1	△1		△1	□55 △9	□8 ○1 △1	□67 △1	□4	□11 △5
Ficus yoponensis						□8 ○1	□15	□38	□4	
Ficus sp.	□8	□2	○1	□1			□1 △1	□7 ○2 △1	□14	□8 ○1 △1
Ficus trigonata	△4					□10	□46	□6		
Ficus obtusifolia	△2		△2			□6	□14	□1		
Ficus costaricana							□4			
Ficus popenoei							□1			
Dipteryx panamensis	□1 △1	□1 △1					□3			
Quararibea asterolepis							□14 △1	□4		
Spondias mombin						□4	□17	□20		
Spondias radlkoferi						□2	□5	□8 △2	□1	
Poulsenia armata									△1	
Anacardium excelsum			△1							
Calophyllum longifolium							□1			
Cecropia sp.							□5			
Unidentified fruit pulp			□8	□8		□5	□23	□9	□2	
Unidentified seed		○1		○5			○3			
Pollen on fur	8								7	2

FIGURE 10-1.—Seasonal distribution of foods of *Artibeus jamaicensis* on BCI observed during the period 1977–1980. Squares represent seedless feces, circles represent feces with seeds, and triangles represent fruit carried into nets. Numbers represent observations per month.

quantities by larger stenodermatines when *F. insipida* is scarce. Although we caught large numbers of *A. jamaicensis* at fruiting *F. yoponensis*, we often had the experience of netting under a tree dropping ripe figs and finding the area devoid of bats, or of catching *A. jamaicensis* carrying *F. insipida* from some more distant tree. Bats with feces containing seeds of *F. yoponensis* were outnumbered by about two to one by bats with feces containing *F. insipida*. However, large concentrations of bats were found at ripe *F. yoponensis* in May, October (two sites), and November (three sites).

Ficus dugandii Standley: On the breezy nights of 25 and 26 October 1979, a giant *F. dugandii*, with a crown emerging above the canopy near the highest point of the island, attracted (perhaps by wind-wafted odor) large flocks of unmarked stenodermatines, presumably from the mainland. In mist nets beneath the tree, which was dropping many ripe fruits, we captured only 106 *A. jamaicensis*, 7 *A. lituratus*, and 2 *V.* *caraccioli*, while the catch of smaller stenodermatines totalled an astonishing 265, including 138 *Uroderma bilobatum*, 90 *Chiroderma villosum*, 15 *Vampyressa nymphaea*, 6 *V. pusilla*, 7 *Vampyrops helleri*, 7 *Artibeus phaeotis*, 1 *A. watsoni,* and the only *A. hartii* ever taken on BCI.

Ficus trigonata Linnaeus: The red-spotted, small-seeded fruit of *Ficus trigonata* was an important food of *A. jamaicensis* and *C. villosum* at the height of the rainy season in September when fruit of *F. insipida* and *F. yoponensis* was scarce (Table 10-1). Many fruits of *F. trigonata* were carried into the nets and its seeds were common in feces (Figure 10-1). *F. trigonata* also occasionally was found in the feces of *Carollia perspicillata* and *Phyllostomus hastatus*. Although commonly seen in September, the lean month, the fruit of *F. trigonata* seemed not to dominate the diet of any bat in that month save possibly *Chiroderma villosum*. *Spondias mombin* and *Quararibea asterolepis* were eaten more commonly. One *F. trigonata* was

TABLE 10-1.—Number of trees bearing ripe figs in successive fortnights on a 25-hectare plot surrounding Lutz Creek, BCI, 1976–1979. Numbers represent individual trees of *Ficus insipida* (i), *F. yoponensis* (y), *F. obtusifolia* (ob), and of other species of fig (o), fruiting in the fortnight prior to the date given (D). Trees had full crops, unless expressed as a/b, where a is the number with full crops and b is the number with reduced crops. Data from M. Estribí.

Month	1976					1977					1978					1979				
	D	i	y	ob	o	D	i	y	ob	o	D	i	y	ob	o	D	i	y	ob	o
January						8	3	3	0	0	8	3/1	4/2	0	0	7	2	7/2	0	0
	24	2	0/1	0	0	23	2	8	0	0	21	0	1/3	0	0	20	1	5/1	0	0
February	7	1	2	0	0	6	2	6	0	0	4	1	1	0	0	4	1	2/1	0	0
	21	1	0/1	0	0	19	0	1	0	0	18	1	2	0	0	17	1	1/1	0	0
March	6	1	5	0	0	6	0	3	0	1	4	3/1	2	0	0	4	2	0	0	1
	20	0	4/1	0	0	19	1	1	0	0	19	0	3/1	0	1	18	4	1	1	0
April	3	0	1	1/1	0	2	0	1	1	0	1	1	2	0	0	1	1	2	0	0
	17	2	2	0	0	16	1	2	0	0	15	3	3/3	0	0	14	1	2	1	0
						30	1	3	0	0	30	2	3	0	0	28	2	0	0	0
May	1	2	3	0	0															
	15	4	6/1	0	0	14	1	1	0	0	14	0	4	0	0	13	2	2	0	0
	29	2	0	0	0	29	0	0	0	0	28	1	4	0	0	27	2	0	0	0
June	12	1	1	0	0	11	3	2	0	0	10	0	3/1	0	0	10	2	0	0	0
	26	0	5	0	0	25	3	3	0	0	24	0	2	0	0	24	1	9	0	0
July	10	0	5/1	0	0	9	4	2	0	0	8	0	0/1	0	0	7	2	4	0	0
	24	0	0	0	0	23	1	7/1	0	0	23	1	7	0	0	21	0	1	0	0
August	6	0	0/1	0	0	6	3	8	0	0	5	2	0	0	0	4	2	1	0	0
	21	0	0/1	0	0	20	2	2	0	0	19	3/1	0	0	0	18	2/1	1	0	0
September	4	0	1	0	0	3	0	2	1	0	2	1	0/2	0	0	1	2	2	0	0
	18	2	0	0	0	17	2/1	3	0	0	16	2	2	0	1	15	1	0	0	0
											30	2	3	0	0	29	0	0	0	0
October	4	0	3/1	0	0	2	0	0	0	0										
	16	1	1	0	0	15	0/1	0	0	0	16	2	4	0	0	13	0	3/1	0	0
	30	3/1	2	0	0	29	0	0/2	0	0	29	0	0/2	0	0	27	0	3	0	0
November	13	2	6	0	0	12	0	1	0	0	11	0	0	0	0	11	0	1	0	0
	27	2	1	0	0	26	0/1	0/2	0	0	26	1	0	0	0	24	2	0	0	0
December	11	6	3	0	0	10	1	3	0	0	9	0	6	0	0	9	1	5	0	0
	26	1/1	6	0	0	24	2	9/1	0	0	26	3	10/2	0	0	22	2/1	7/1	0	0

carried into a net by an *A. phaeotis* in March.

Ficus obtusifolia Humboldt, Bonpland, and Kunth: The large velvet-skinned fruits of *F. obtusifolia* were carried into nets occasionally by *A. jamaicensis*, *A. lituratus*, and *C. villosum*. The fruits seemed to be more attractive to the smaller frugivores such as *C. villosum*. We found bats concentrated at an *F. obtusifolia* with ripe fruit in March.

Ficus popenoei Standley: The velvety, oblong fruits of *F. popenoei* are uncommon on BCI, and we found only a few trees with fruit. We netted near one that was dropping great quantities of ripe fruit, and although we caught a number of small stenodermatines of various species, the tree was ignored by *A. jamaicensis* and the other larger fruit eaters.

Spondias mombin Linnaeus: An important food of bats (Gardner, 1977), *S. mombin*, ripens in August and September when fig production is low. It is found in scattered patches in the young forest on the lower-lying areas of the island. Apparently *A. jamaicensis* ingests the pulp of this fruit without trying to extract the juice and, thus, does not make pellets. Although *S. mombin* fruit seldom was carried into the nets, those nets set nearest to trees with ripe fruit invariably caught the most *A. jamaicensis* and *C. perspicillata*. A large concentration of bats, mostly *A. jamaicensis*, was found in a patch of *S. mombin* between 12 and 17 September.

Spondias radlkoferi Donnell Smith: Widespread on BCI, but neither forming patches nor producing such great quantities of fruit as *S. mombin* and *Q. asterolepis*, *S. radlkoferi* is an important source of food for bats in the season of fig scarcity at the height of the rains. Fruit of this tree is ripe from September to November and is much sought by *A. jamaicensis*, *A. lituratus*, *A. phaeotis*, *A. watsoni*, and *V. caraccioli*. In 1979, ripe fruit began to fall on 19 September. Many fruits were carried into the nets on 26 September, and by 30 September most *A. jamaicensis* and *A. lituratus* that we captured were eating it. A large concentration of bats was associated with a *S. radlkoferi* with ripe fruit on 7 October. The last time we noted feces containing fruit pulp of this species was on 7 November. Seeds of *S. radlkoferi* with much or all of the pulp scraped off by bats were common under feeding roosts throughout October.

Quararibea asterolepis Pittier: The beautiful *Q. asterolepis* grows in patches, which frequently contain many individuals, mostly in the old forest on the elevated interior of the island. We found it at lower elevations at Snyder-Molino 2, Shannon 1, mouth of Barbour Creek, Standley 16, and Wheeler 26.5. When its fruit ripens in August, September, and October it is the most abundant tree on the Plateau with fruit eaten by *A. jamaicensis*. Individual trees drop large quantities of ripe fruit over a period of two or three weeks or more. We found concentrations of bats at *Q. asterolepis* with ripe fruit on three occasions in September, and *A. jamaicensis* and *C. perspicillata* frequently carried the fruit into our nets. The fibrous fruit shell remains for weeks on the ground under feeding roosts, persisting long after the fruiting season has passed. *Q.*

asterolepis and *S. mombin*, together with the much less abundant *F. trigonata*, are the fruits of choice for *A. jamaicensis* in September, and possibly in August as well, when fruits of *F. insipida* and *F. yoponensis* are scarce or absent. It is of interest that *Q. asterolepis* and *S. mombin*, which fruit simultaneously and are visited by the same bats, have complementary, nearly nonoverlapping distributions on BCI. Both occur in patches and produce much more fruit than the vertebrate frugivores can consume. Near fruiting *Quararibea astrolepis*, *A. jamaicensis* usually was caught in small groups rather than in such large concentrations as often appeared at favored fig trees at other seasons. Some *Q. asterolepis*, even when dropping quantities of fruit, almost seemed to be ignored by bats on some nights when we netted near them.

Dipteryx panamensis (Pittier) Record and Mell: During the dry season when fig productivity is low, the fruit of *D. panamensis* is an important food of *A. jamaicensis*. On BCI *Dipteryx* is a widespread and abundant tree, sometimes forming large patches (as at Zetek 22 and east of Standley 19). Its fruit is ripe from December to March and occasionally at other seasons of the year (fruit was carried into a net in mid-November). We found large concentrations of bats, principally *A. jamaicensis*, at ripe *D. panamensis* in January 1979 and February 1978, when numerous fruits were carried into the nets.

Brosimum alicastrum Swartz: Robin Foster (pers. comm.) regards bats as the principal dispersers of *B. alicastrum* seeds, and Croat (1978) believed that *B. alicastrum* is second only to figs in importance as a food for forest animals. We found a large concentration of bats, including almost 100 *A. jamaicensis*, at a *Brosimum* bearing ripe fruit in May. However, a large tree raining ripe fruit in October, near the end of the fruiting season, was attractive to kinkajous (*Potos flavus*) and peccaries (*Tayassu tajacu*), but apparently was of little or no interest to bats.

Anacardium excelsum Bertero and Balbis: Abundant and widespread, and fruiting from March to May, *A. excelsum* is an important food of *C. perspicillata*. We often caught this bat, and *A. phaeotis* on one occasion, carrying the fruit of *A. excelsum*. Except for one fruit carried into a net in March, we have no evidence that *A. jamaicensis* eats the fruit of this species.

Poulsenia armata Miguel: The spiny fruits of *P. armata* occasionally were carried into the nets by *A. phaeotis* (January, September, and October), *A. watsoni* (October), and once by a *A. jamaicensis* (November).

Calophyllum longifolium Willdenow: In September and October we often found the large seeds of *C. longifolium* under dining roosts of *A. jamaicensis*, and occasionally we found remains of pulp and skin in the feces of this bat. The pellets from the fruit of *C. longifolium* dropped by *A. jamaicensis* and *A. lituratus* seem extremely resistant to decay and persist on the ground for several weeks. Although the large round seeds are conspicuous, the pellets are much smaller and darker than those dropped by *A. jamaicensis* when feeding on figs, and consequently we often overlooked them. Fruits of *C. longifolium* that were carried into nets by bats were offered to *A. jamaicensis, A. lituratus,* and *V. caraccioli* temporarily held in captivity on BCI. Both species of *Artibeus* ate the fruit, but *V. caraccioli* did not, although it readily ate figs when those were offered.

Cecropia spp.: Occasionally we found seeds of unidentified species of *Cecropia* in feces and under feeding roosts of *A. jamaicensis.*

Guettarda foliacia Standley: Partly eaten fruits of *G. foliacia* were found beneath feeding roosts, probably of *A. jamaicensis,* in September 1979 and October 1978.

Hura crepitans Linnaeus: Numerous fleshy staminate flower stems of *Hura crepitans* were carried into nets by *Phyllostomus discolor, A. jamaicensis,* and *A. lituratus* in November.

Unidentified Pollen: In November and December, numerous bats of the species *Phyllostomus hastatus, Glossophaga commissarisi, Uroderma bilobatum, U. magnirostrum, A. jamaicensis,* and *A. lituratus* were stained yellow with unidentified pollen (probably mainly from flowers of the Bombacaceae).

Tree Selection

We know that *A. jamaicensis* will eat a variety of fruits, and sometimes other foods, but most of its meals are figs. On BCI it prefers *F. insipida* over all other figs, and it must be a rare night on the island when there are no ripe fruits of this species available. We have no idea why *A. jamaicensis* occasionally chooses the fruits of *Brosimum, Calophyllum, Dipteryx, Quararibea,* or *Spondias* over *Ficus,* but we have learned much about why and when they choose a particular fig tree.

The normal massive crop of figs of each tree goes through several stages of harvest as it ripens. First come the howler monkeys (*Alouatta palliata*), which begin to eat the fruit long before it is ripe, while the pulp is hard and the skin is still full of latex. The monkeys are followed by the local bats from nearby roosts. They also begin to harvest a crop before it is fully ripe, probably selecting scattered ripe fruit, and sometimes picking unripe fruit. We have caught *A. jamaicensis* carrying unripe figs, but we do not know whether they actually eat such fruit, and if they do, whether they eat all or only part of it. Perhaps they are sampling crops to determine the stage of ripeness. Pellets dropped by local bats may be conspicuous during several nights before enough of the fruit crop is ripe to attract large groups of bats.

When the big crop finally is fully ripened, bats at the tree may number in the hundreds or even thousands. They come in surges, and some may stay in the vicinity all night (see Section 7, Movements and Section 9, Foraging Behavior). Groups will return to a tree that has a big crop for three or four nights. In the final stage of harvest, monkeys, other diurnal frugivores, and

local bats still return to the tree after its crop has been too depleted to continue to attract large flocks of bats. Altogether, local bats may harvest figs from a tree over a period of eight or nine nights.

Some trees produce large crops that ripen a few figs at a time during a prolonged period (up to 10 to 15 days) rather than ripening abruptly as figs normally do. These trees are attractive to local bats and nonchiropteran frugivores, but not to groups of bats, except for brief passes.

The remnants of fig crops aborted because of lack of pollination by fig wasps, or because of infestation with larvae of beetles and flies, as well as the crops produced by small, young trees, might not attract many bats. Size of the fruit crop has an important bearing on its use by bats. Small crops may be harvested by local bats, but usually they are not attractive to groups. Because monkeys (*Alouatta palliata*) begin to harvest a crop before bats do, they may strip a small crop before it is ripe enough to attract bats. Local bats may help finish harvesting a crop that is too small to attract a group. Or, a group may make one or two feeding passes and then leave. Radio-tracking revealed that a bat may visit three to five trees in a night (Morrison, 1978a).

The fruits of figs are subject to destructive processes such as infection with fungi and infestation by insects that make them unattractive to bats. During the rainy season we often saw *F. insipida* infected with a fungus manifested in fully developed ripe fruit that have a bearded appearance while still on the tree. As falling fruit accumulates, the ground beneath the tree becomes carpeted with the fuzzy white fruit. Nothing eats it, in the tree or on the ground, and eventually it rots away. Fungi, as well as invertebrates of the litter, accelerate the breakdown of the pellets of chewed fig pulp dropped to the ground as bats feed. Only the most discerning eye will spot traces of these pellets four or five days after they have been dropped.

The fruit of both *F. insipida* and *F. yoponensis* can be infested by the larvae of flies and beetles. If the infestation is heavy, the tree aborts part or all of its crop at about the sixth week (about three-fourths of the way through the developmental cycle). These hard, latex-laden, and insect-riddled fruits are eagerly eaten by monkeys (both *Alouatta palliata* and *Cebus capucinus*), frugivorous birds, and terrestrial frugivores such as pacas (*Agouti paca*), peccaries (*Tayassu tajacu*), and tapirs (*Tapirus bairdii*). However, although they may be large and appear to be ripe, they are not sweet, and they are ignored by bats. Until we learned the nature of this kind of fruit-crop abortion, we sometimes wasted effort by netting at these trees. We were misled by the abundance of falling fruit and the feverish activity of many frugivores into believing that bats also would flock to the tree.

Summary

Because capture rate is clearly linked to foraging behavior, we gathered data on fecal contents, feeding roosts, food carried into nets, and on the location and phenology of fruiting trees. The bats' habit of dropping small pellets of chewed fruit pulp under favorite feeding roosts frequently dictated our choice of netting sites. Congregations of diurnal frugivores led us to trees with ripening fruit that would be visited at night by bats. Certain species of trees favored as feeding roosts are used repeatedly, especially in the vicinity of fruiting trees.

Fruit of the fig *Ficus insipida* is the favorite food of *A. jamaicensis* as well as several other bats on BCI. Fruits of other species of figs, including *F. yoponensis, F. dugandii, F. trigonata,* and *F. obtusifolia,* are sometimes eaten by *A. jamaicensis,* but the uncommon *F. popenoei* is ignored. *Spondias mombin, S. radlkoferi,* and *Quararibea asterolepis*

are important food sources late in the rainy season when figs are scarce. *Dipteryx panamensis* is used in the dry season as an alternate food source. The fruits of *Anacardium excelsum* and *Poulsenia armata* and the flowers of *Hura crepitans* occasionally are eaten by *A. jamaicensis.*

The normal massive crop of a *F. insipida* is used by local bats for several days before the entire crop is fully ripe, at which time hundreds or even thousands of bats may visit the tree for three or four nights until the crop is depleted. Then local bats continue to use the tree for several nights until there is no more fruit. Trees with small crops and trees producing crops that ripen asynchronously over a period of a week or more are visited by local bats but are largely ignored by groups.

11.   Diet and Food Supply

Charles O. Handley, Jr., and Egbert G. Leigh, Jr.

Bonaccorso (1979) argued that food supply limits populations of bats on Barro Colorado Island (BCI). Late in the rainy season when fruit is scarce, he found more fruit bats with empty stomachs than he did at other times of year. About 83% of the frugivores he netted in October and November had empty stomachs, in contrast to 71% of those netted in March and April when fruit was far more abundant. However, these data must be interpreted with caution. Bonaccorso kept his data free of bias by netting the same localities each month, without regard for presence or absence of fruit near the netting stations. The bats he caught with empty stomachs must have been on their way to someplace else where there was fruit. They could not endure empty stomachs for more than a few hours without starving to death.

Moreover, Bonaccorso (1979) observed that birth (and breeding) of bats is timed to coincide with seasons of fruit abundance, with one birth peak in March and April coinciding with the fruiting peak at the onset of the rainy season and another in July and August coinciding with the fruiting peak of August and September (Figures 4-6 and 4-7). Few frugivorous bats give birth between November and mid-March when fruit is least abundant.

Bonaccorso (1979) also found that the diets of various species of bats that forage for fruit in the canopy differed to the extent one would expect if these animals were food limited. He found that larger bats ate larger figs such as *Ficus insipida* and *F. obtusifolia*, whereas smaller bats concentrated on smaller figs such as *F. popenoei* and *F. yoponensis*. For the three larger frugivores (*Artibeus lituratus, A. jamaicensis,* and *Vampyrodes caraccioli*), he found that the regression of the mass Y of a fruit carried by a bat on the mass X of its carrier was $Y = 0.23X - 3.92$ g ($r^2 = 0.46$, $n = 27$). Presumably, smaller bats carried fruit into the nets so rarely that he could not extend the regression.

Reading the mass of these fruits from Bonaccorso's graph and calculating the mean and standard deviation of the logarithms of the mass of the fruits these bats carried, we

Charles O. Handley, Jr., National Museum of Natural History, Smithsonian Institution, Washington, D.C. 20560.
Egbert G. Leigh, Jr., Smithsonian Tropical Research Institute, Unit 0948, APO AA 34002-0948 or Apartado Box 2072, Balboa, Republic of Panamá.

derived values of 2.30 ± 0.24 ($n = 6$) for *A. lituratus*, 2.04 ± 0.33 ($n = 17$) for *A. jamaicensis*, and 1.54 ± 0.31 ($n = 4$) for *V. caraccioli*. The standard deviations are roughly equal to the differences between neighboring means. May and MacArthur (1972) showed that in an idealized competitive community, species could coexist securely if the sizes of foods eaten by the various species differed to this extent. Excepting *Artibeus phaeotis,* which eats few figs, the ratio of the mass of each species of canopy frugivore to that of its next smaller competitor was roughly the same as in these three species, about 1.4 : 1 (Bonaccorso, 1979).

It is tempting to assume that the relation between the sizes of the smaller stenodermatines and the sizes of fruit they eat is the one Bonaccorso inferred from the few data on his largest frugivores. Furthermore, we could conclude from the elegant theory developed by May (1974) and May and MacArthur (1972) on niche overlap, that the canopy frugivores coexist by virtue of the differences Bonaccorso observed in the sizes of fruit these bats eat.

Ideally, if all sizes of figs were available at all times and were uniformly distributed, the smaller bats would usually take the smaller figs and the larger bats would usually take the larger figs. However, small-fruited and large-fruited species of figs are not uniformly distributed either in time or space, and the bats are adaptable enough to take what is available. Of course, in standardizing their diets, it is also necessary to take into account the foraging behavior of these bats. The energy expended in commuting from dining roost to fruiting tree makes it energetically imprudent for the large bats to routinely feed on small figs from each of which they would extract a comparatively tiny amount of nutrients (see Section 2, Physiology).

Fig Production

When we became concerned about the relation between food availability and the bats' energy requirements (see Section 2, Physiology), we began to gather information on estimates of fig production on BCI. Each individual fig tree bears fruit to its own rhythm (Morrison, 1978d) so that at any season some trees are bearing ripe figs (Table 10-1). Fig production peaks early in the dry season in December and January, when other fruit is scarce, and there seems to be a lower peak at the beginning of the rainy season. Sometime between July and November, a month may elapse during which no more than one of the 130 fig trees in the Lutz Watershed will produce a crop of fruit. This low could very well be an islandwide phenomenon (Morrison, 1978d).

In the 25 ha of high second-growth forest of the Lutz Ravine, the 49 *F. insipida* produced 42 full crops of fruit per year between March 1973 and March 1975 (Morrison, 1978d) and 34 full crops of fruit per year in the 4-year period beginning 1 January 1976 (Table 10-1). The 71 *F. yoponensis* produced 64 full crops of fruit per year during each period.

We have two estimates of the size of a full crop of figs. Hladik and Hladik (1969) counted 8,000 figs on a *F. insipida* that had roughly 100 m^2 of crown, or 80 figs per square meter of crown. They thought the crop measured a little low. Morrison (1978d) recorded 112 figs per square meter in a 1.3-m^2 section of a crown of *F. yoponensis* and 113 figs per square meter in a 3.9-m^2 section of a crown of *F. insipida*. For both species, the average crown area was 345 m^2, so there may be about 40,000 figs in the average full crop. The 25 ha of Lutz Ravine, accordingly, could yield about 100,000 fruits of *F. yoponensis* and about 60,000 fruits of *F. insipida* per hectare per year.

Morrison (1978d) calculated average mass for *F. insipida* fruits as 9 g, 81% of which was water. His average mass for *F. yoponensis* was 3.5 g, again 81% water. Accordingly, we estimated that 100 kg dry mass of *F. insipida* fruit and 70 kg dry mass of *F. yoponensis* fruit are produced per hectare per year in the 25 ha of Lutz Ravine. These figures are probably representative for the other second-growth forest of the island. The old forest has far fewer fig trees and, consequently, much lower production of figs than does the second-growth forest in Lutz Ravine. Older forest covers roughly half of the surface area of BCI. Therefore, we estimate that, islandwide, fig production is roughly 100 kg dry mass of fruit per hectare per year.

Figs Consumed by Bats

We also wanted to know how our estimates of the availability of figs on BCI compared with the capability of the canopy frugivores for consuming figs and their feeding rates. Morrison (1978d) determined that a *A. jamaicensis* ate an average of seven *F. insipida* fruits per night ($n = 7$ nights). He assumed from Bonaccorso (1975) that the fruits of *F. insipida* consumed by *A. jamaicensis* weigh an average of 9.5 g each. However, in December 1979 and January 1980 we found the average fresh mass of fruits of *F. insipida* carried into nets by *A. jamaicensis* and *A. lituratus* to be 6.8 g ($n = 40$). If a 45 g *A. jamaicensis* processes seven figs with a total fresh mass of 49 g, or 9 g dry mass of fruit per night, then the estimated resident population of 3,000 *A. jamaicensis* on BCI might consume an average of 7 kg dry mass of fruit per hectare per year, and the total population of residents and transients together, might consume an average of 9 kg dry mass of fruit per hectare per year. These uncritical evaluations indicate that these bats eat a surprising amount of fruit per year.

Based on the overall data set for the Bat Project, we believe that about one-third of the frugivorous bats that forage in the forest canopy on BCI are *A. jamaicensis,* one-third are *A. lituratus,* and all the other stenodermatines make up the other third. If we assume from Bonaccorso (1979) that 65% of the diet of *A. lituratus,* 76% of the diet of *A. jamaicensis,* 30% of the diet of *A. phaeotis,* and 81% of the diet of the other stenodermatines is figs, then these frugivores process 28 kg dry mass of figs per hectare per year. In all, the frugivorous bats of the canopy of the forest on BCI eat roughly 40 kg dry mass of fruit per hectare per year when all sources are considered.

Competition and Food Limitations

Aside from bats, howler monkeys *Alouatta palliata* are the primary consumers of figs. Howler monkeys on BCI eat an average of 90 kg dry mass of food per hectare per year (Nagy and Milton, 1979), about half of which is fruit (Hladik and Hladik, 1969; Milton, 1978). Howler monkeys near Lutz Ravine spend over 50% of their foraging time in fig trees eating fruits and leaves. Troops in old forest habitats near the center of the island spend more than 25% of their time in fig trees. In all, they may well eat 20 kg dry mass of figs per hectare per year. These monkeys and kinkajous (*Potos flavus*) are important competitors of bats because they eat figs before the fruits are ripe enough to attract bats and because they knock down many figs in the process of foraging. Figs attract many other animals, such as guans (*Penelope purpurascens*), toucans (*Ramphastos* spp.), coatimundis (*Nasua narica*), and other primates such as the night monkey (*Aotus lemurinus*) and the capuchin (*Cebus capucinus*), but considering the year as a whole, these are incidental competitors that depend primarily on other foods.

Despite a remarkably tight fit between numbers of bats, consumption of figs, and supply of figs, the breeding cycle of *A. jamaicensis* seems related to the abundance of fruit in general, not to the rhythm of fig production. Fruiting of figs peaks in December and January when few bats are being born. To be sure, a variety of other trees composing the forest canopy, notably *Anacardium excelsum, Licania platypus, Spondias radlkoferi, Calophyllum longifolium, Symphonia globulifera, Brosimum alicastrum,* and *Poulsenia armata* produce fruits apparently designed to attract bats (Robin Foster, pers. comm.; also see Gardner, 1977). Because all of these trees produce green fruit that is often difficult to see, phenological records are far from precise, but it is clear that their fruiting rhythms will not resolve our puzzle. A few of these species fruit when young bats are being born: *Anacardium excelsum* in April and May, and *Licania platypus* in July and August. The others either fruit at odd times, as does *Calophyllum longifolium,* which may fruit in December and January and again in June and July, and *Spondias radlkoferi,* which fruits in October and November; or they fruit irregularly, perhaps several times a year, as do *Poulsenia armata* and *Brosimum alicastrum.* From August to October bats sometimes eat bright-colored fruits such as *Spondias mombin* and *Quararibea asterolepis,* both of which also attract other animals.

The breeding rhythm of frugivorous bats seems dictated by the availability of edible fruit in general and not of fruit preferred by bats alone. When the fruit supply is short in the forest all over the island, animals of all kinds focus on what is left. If this remainder is mostly fruit preferred by bats, the bats too will suffer shortages, and as Bonaccorso (1979) claimed, it will be a factor limiting population size.

Summary

Seasonality of the food supply controls the reproductive cycle of *A. jamaicensis,* with birth peaks timed to take advantage of succeeding fruiting peaks. Previous studies observed that larger species of bats took larger fruits and smaller species took smaller fruits, suggesting food-limited populations of canopy frugivores. Competition theory also suggests that coexistence is possible if food particle size differs by the ratio of 1.4 : 1 estimated by earlier work. However, small- and large-fruited species of figs are not uniformly distributed in time or space, and the bats are adaptable enough to take what is available.

Although fig trees bear fruit asynchronously, there is a peak early in the dry season, a lower peak at the beginning of the rainy season, and a low late in the rainy season. A 25 ha plot of forest on BCI contained 49 *F. insipida,* which produced 34 full crops per year during the study period, and 71 *F. yoponensis,* which produced 64 full crops per year. If previous estimates of 40,000 fruits in an average full crop of either species are correct, the 25 ha of Lutz Ravine could yield about 60,000 *F. insipida* and about 100,000 *F. yoponensis* fruits per hectare per year. This converts to about 170 kg dry mass of figs per hectare per year for second-growth forest and averaged with the much lower production in older forest, suggests an islandwide production of roughly 100 kg dry mass of fruit per hectare per year. Our calculations suggest that frugivorous bats consume 28 kg dry mass per hectare per year of this production.

Howler monkeys and kinkajous compete with bats for figs, and guans, toucans, coatimundis, and other monkeys are incidental competitors. Despite the good fit between numbers of bats, consumption of figs, and supply of figs, the reproductive cycle of *A. jamaicensis* seems not bound by the cycle of fig production. Rather, reproduction seems geared to take advantage of peaks of production of edible fruit in general, and competition in times of short supply may be a factor limiting size of populations of bats.

12. Appendix

Methods of Capturing and Marking Tropical Bats

The Bat Project on Barro Colorado Island (BCI) developed an array of techniques and methods that facilitate and simplify capturing and marking bats. This protocol evolved throughout the first five years of field effort. To assist future projects dealing with bats, we have outlined our methods in considerable detail. Additional useful tips on methodology relating to many aspects of the study can be found in Kunz (1988).

Finding Bats

We quickly discovered that it was more productive to net for *Artibeus jamaicensis* in the vicinity of trees bearing ripe fruit than in sites selected at random. Fruiting trees often could be located by the calls of noisy frugivores such as guans, parrots, and monkeys by day and kinkajous by night. If the crop was particularly large, the odor of fruit rotting on the ground often carried on the breeze for some distance. We also searched for the piles of pellets of pulp that stenodermatines spit out under their feeding roosts after they have extracted the juices from the fruit. Great numbers of pellets scattered over an extensive area indicated a nearby fruit bearing tree used by large numbers of bats.

Netting Stations

Our standard netting station consisted of ten mist nets, each 10 or 12 m long, strung between 3 m tubular aluminum poles. Rarely, if space for nets was limited, we substituted one or more 5 or 6 m nets. For transport we apportioned the 20 poles into bundles tied securely at each end, and carried them over the shoulder, resting between the horns of a pack frame. Two persons could carry the poles and other paraphernalia required for a capture station over a rough trail without undue stress, but it was easier with three persons.

If possible, we placed the nets along a trail near a fruiting tree, around the base of the tree, or on a ridge overlooking the tree. As far as was possible, the nets were arranged in a zig-zag pattern along trails with ends overlapping about 1 m, thereby blocking as much of an underbrush-free area as possible.

We stretched the net shelf strings tight to minimize tangling of large *Artibeus,* and high enough to keep heavy bats in the lower shelf off the ground. These tactics facilitated the capture of the larger bats, but somewhat diminished our chances of catching smaller kinds.

Our marking station, always near the netting area, consisted of a 3×4 m lightweight tarp stretched over a rope tied between two trees to provide shelter; lightweight folding stools; and a small plastic ground cloth for spare batteries, canteens, packs, and other equipment. All marking stations were temporary and could be set up or taken down in a matter of minutes.

Catching Bats

We usually set nets five nights each week following a schedule of two nights on, one off, three on, one off, but we modified this schedule if circumstances dictated. Normally the netting sessions began at dusk and ended when the bats quit flying, regardless of light conditions. Sometimes we closed the nets after an hour or two if netting was slow, but occasionally we netted until dawn. When possible, we furled the nets during showers.

We kept the bats in 20×30 cm bags made of Monk's Cloth, a soft but strong open-weave fabric that would not chafe the elbows, wrists, and knees of the bats and minimized overheating and suffocation. We kept empty bags handy on rope belts and as we collected bats from the nets and bagged them, they hung from the same belts. We never mixed species in a bag and usually used a separate bag for each individual, although on exceptionally busy nights we put two or three bats in a bag. We hung the bagged bats in the marking station in hourly lots on poles or lines strung close to the processors. Poles laid over pack frames stuck upside down into the ground about a meter apart made handy bag racks.

Because they might not have eaten before we caught them and, therefore, might easily become fatally stressed, we processed bats of the first hour promptly. Small bats, regardless of capture time, were segregated and processed quickly. The energy budget of small stenodermatines is so constrained that they can deplete energy reserves to lethal levels in minutes.

We patrolled the nets constantly and removed tangled bats as soon as possible. If the bats began to come in faster than we could remove them, we closed nets as each was cleared until we

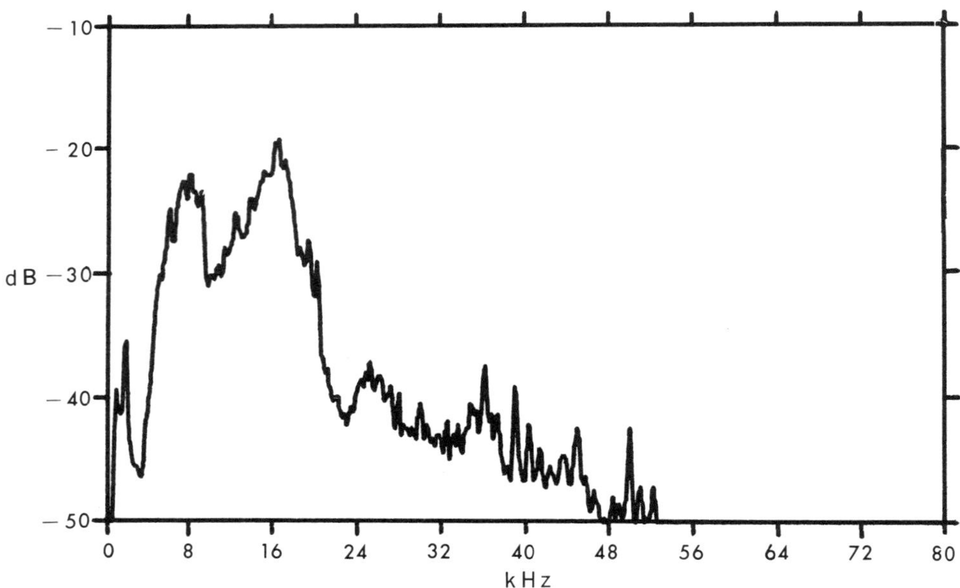

FIGURE 12-1.—Sound spectrum of Audubon Bird Call, used to attract bats. From recording by James Fullard and Jackie Belwood.

had only as many nets open as could be kept cleared. When the catch slowed we reopened nets.

The distress calls of particularly vocal bats were useful for attracting other bats into the nets. In the absence of such bats, the squeaks of an Audubon Bird Call often were effective in attracting bats. James Fullard and Jackie Belwood tested these mechanical squeakers and found all the squeaks were in the audible range, with no ultrasonic components (Figure 12-1). The bird calls put out most of their energy at values between 8 and 16.8 kiloHertz (kHz) with a peak mean of 14.85 kHz. Maximum frequencies ranged between 50 and 52 kHz. The effectiveness of the bird call for bats may be limited by its absolute intensity rather than its frequency output (bats squeak louder).

Using folded capture bags as baffles that the bats might bite, we seldom were bitten ourselves. Everyone handling bats was protected with rabies vaccine.

Necklacing

Bats have proven difficult to mark without suffering serious injury. Wing and interfemoral membranes are sensitive and easily irritated. Toes are few and clipping them may interfere with the bat's ability to roost. Ear tags tear out and may limit maneuverability of the ear or obstruct reception of sonar signals. Body size is too small for effective use of heat or freeze brands. We needed a marking system that (1) would not interfere with normal behavior; (2) could be easily and quickly applied in the field; (3) could be easily and accurately interpreted upon recapture; and (4) was durable enough to

supply the desired information about the bat throughout its lifetime.

Previously, forearm banding was the method of choice for marking bats. Partly because the side effects of forearm banding proved even more disastrous to bats in the tropics than in temperate zones, few long-term studies based on marking tropical bats have been successful. Consequently, comparatively little is known about populations, longevity, movements, seasonal and annual variations, and other aspects of demography in tropical bats. This is a serious problem for biologists studying bats because the majority of the species of bats are exclusively tropical.

We tried and rejected a number of techniques for marking. Eventually we settled on stainless steel ball-chain necklaces carrying numbered aluminum bands as a safe and effective device for marking tropical bats. We rarely found signs of serious irritation from the necklace, and the loss rate of necklaces is so low (6.6%) that we didn't consider it a problem.

The chain consists of smooth hollow balls closely linked by dumbbell-shaped rods. An easily attached connector joins the ends of a length of chain into a necklace. Ball chain is strong, lightweight, and because it is flexible and swivels, it can not kink. Chain is available in continuous lengths up to thousands of feet. Although we used smooth, stainless steel ball chain, it is also manufactured from brass, monel, copper, aluminum, gilding metal, nickel-silver, and high carbon steel, all of which come in a variety of finishes. The smallest size available is no. 00 with balls measuring 1.575 mm in diameter and a tensile strength of 4.54 kg; the largest is no. 20, which has balls 9.525 mm in diameter and a tensile strength of 84 kg. We obtained our chain from Ball Chain Manufacturing Company, Inc., 741 South Fulton Ave., Mount Vernon, NY 10550; telephone (914) 664-7500.

Mainly we used no. 3 (2.388 mm, 9.08 kg) stainless steel, standard ball chain and no. 3 stainless steel connectors. No. 3 chain has 100 beads per foot. Five bats the size of *A. jamaicensis* (which take a 20-bead necklace) can be marked with one foot of chain. Five hundred feet of no. 3 chain will mark about 2,500 bats. For the smallest bats, whose necklaces would have less than 15 balls in no. 3 chain, we preferred no. 2 chain (2.083 mm, 7.264 kg).

Necklace lengths we used for each species are tabulated in Table 12-1. Geographic variation and sexual dimorphism in populations elsewhere may require adjustments in necklace lengths for these and other species. The advantage of necklacing is that it is harmless to bats. It should be emphasized that this advantage is lost if the necklace is improperly fitted. The necklace must not fit too snugly around a bat's neck. It should be possible to slip the points of a pair of hemostats (used to attach the necklace around the neck) under the closed necklace. On most bats, the necklace should be loose enough so that it can be slipped off over the bat's head with little difficulty. The bat itself, however, usually will not be able to remove the necklace because it will use only one foot or one thumb at a time, thus causing the necklace to bind diagonally. We have found that a bat will ignore the necklace within a few minutes after it is in place.

For obvious reasons we did not mark juvenile bats with necklaces smaller than those required for adults. Suckling young and even some volant young, because of their small size, cannot be necklaced. We also did not mark bats such as *Molossus* and *Centurio,* which have skin pads or gular glands on the throat. Chest glands are low enough on the torso not to interfere with necklacing in *Trachops, Tonatia,* and *Phylloderma.*

Anticipated numbers of necklaces of the lengths needed were cut with small sidecutter pliers and a connector was attached to one end of each before the capture operation. We found that slightly opening the unattached end of the connector with needle-nose pliers or hemostats made closure of the necklace when wrapped around a bat's neck much easier. We had two ways of storing necklaces. Our first method was to close (fasten) each necklace and string those of the same size on a large safety pin such as a diaper pin; each necklace has to be opened before use. Our preferred method was to store the unconnected necklaces by size in a flat, compartmentalized plastic box such as those designed to store buttons, fishing lures, or other small items.

At the time of marking, a numbered aluminum band was slipped onto each necklace so that individual bats could be recognized at recapture. The band should fit snugly on the necklace, e.g., a no. 1B band (inside diameter 2.769 mm) on a no. 3 necklace (outside diameter 2.388 mm), so as to cause a minimum of disturbance to the wearer. A large, loose-fitting band may move about on the necklace, pinching hair and otherwise irritating the bat's neck. Colored plastic bands may be useful for distant recognition of individual bats, but because

TABLE 12-1.—Necklace sizes for bats of Barro Colorado Island, Panamá. Numbers refer to the number of balls in length of chain. (* = a species that cannot be necklaced.)

Species	Size	
	#2 Chain	#3 Chain
Artibeus jamaicensis		20–21
Artibeus lituratus		23
Artibeus phaeotis	15	
Artibeus watsoni	15	
Carollia brevicauda		15
Carollia castanea	15	
Carollia perspicillata		15–16
*Centurio senex		
Chiroderma villosum		16–17
Chrotopterus auritus		23–24
Cormura brevirostris	14–15	
Desmodus rotundus		17
Glossophaga commissarisi	15	
Glossophaga soricina	15	
Lonchophylla robusta		15
Macrophyllum macrophyllum	15	
Mesophylla macconnelli	14	
Micronycteris brachyotis		15
Micronycteris hirsuta	15	
Micronycteris megalotis	13	
Micronycteris nicefori	15	
Micronycteris schmidtorum	13	
Mimon crenulatum		15–16
*Molossus coibensis		
*Molossus molossus		
Myotis albescens	13	
Myotis nigricans	13	
Noctilio albiventris		
female		15
male		17
Noctilio leporinus		
female		18
male		20
Phylloderma stenops		18
Phyllostomus discolor		18
Phyllostomus hastatus		23–25
Platyrrhinus helleri	15	
Pteronotus gymnonotus		15
Pteronotus parnellii		15–16
Rhogeessa tumida	13	
Rhynchonycteris naso	13	
Saccopteryx bilineata	15	
Saccopteryx leptura	14	
Tonatia bidens		18
Tonatia silvicola		18
Trachops cirrhosus		16
Uroderma bilobatum		15
Uroderma magnirostrum		15
Vampyressa nymphaea	15	
Vampyressa pusilla	13–14	
Vampyrodes caraccioli		17–18
Vampyrum spectrum		25

bands soon work around to a stable position under the bat's chin, they might not be readily visible. Some plastic bands become brittle and break in as few as three or four years, so they may not be as effective as the more durable aluminum bands for longer-term studies of populations. Custom-made aluminum bands can be obtained from Gey Band and Tag Co., Inc., Box 363, Norristown, PA 19404; telephone (215) 277-3280. Plastic bands suitable for necklacing are available from Messrs. I. Dennison and J. Warner, 116 Moor Crescent, High Grange Estate, Belmont, County Durham, DH1 1DL, England.

Bats were most easily necklaced by two persons working as a team. The processor removed the bat from the capture bag and called out its identification, age, sex, and reproductive condition. Usually grasped by its elbows between the fingers of one hand, the bat was then held out toward the necklacer. The person doing the necklacing recorded the data on a field form and then selected a chain of appropriate length, slipped it through a band, grasped the free end of the necklace between the second and third bead with the tips of a pair of small hemostats held in one hand, grasped the connector end of the necklace with the other hand and, placing the connector behind the bat's head, quickly passed the hemostats under the bat's chin, slipped the free bead into the connector, and snapped it shut. The processor then weighed the bat, released it, and checked its capture bag for feces. With a little experience, necklacing proceeded smoothly and quickly. When a third person was available to record data, the operation was even faster.

It is possible for one person to do the necklacing alone. The necklacer holds the bat on his knee with one hand while grasping the connector end of the necklace with fingers of that same hand, the free end of the necklace can be brought around the neck and fastened using hemostats held in the other hand. We also have used a bat restrainer "bat package" made by sewing 150 mm-long by 25 mm-wide, "hook" and "eye" strips of Velcro back to back to facilitate solo marking. When the Velcro is wrapped around a bat to bind its wings firmly to its sides, the animal usually ceases struggling and the necklacer can then fasten the necklace around the bat's neck.

When we recaptured a bat, we moved the band to the back of the neck to facilitate reading, and then rotated the band with fingers or hemostats to expose all of the numbers. To ensure accuracy, we always read the number at least twice, by different persons if possible, and the recorder called it back. Interpretation and recording of a band number can be the weakest link in the mark-recapture process. A misread number is a wasted capture.

Recording Data

Our recording system was designed for computer storage of the data. Through a process of trial and error, we developed data sheets that ensured standardized data, largely prevented accidental omissions, and eliminated transcription except at the computer terminal. All entries were made in waterproof black ink with a Rapidograph or equivalent pen. Field forms were on 216 × 280 mm (8$^{1}/_{2}$ × 11 inch) sheets held together on a clipboard, which we carried between the netting station and our home base in a zippered waterproof plastic pouch. We used three forms: (1) a nightly cover sheet for general information about the netting episode; (2) a field data sheet for individual band entries; and (3) a nightly summary sheet for marks and recaptures.

COVER SHEET.—We recorded general information about each netting episode on the Field Data Cover Sheet (Figures 12-2 and 12-3). This included date, location, release site (ordinarily the netting station), numbers and kinds of nets, opening and closing times of nets, net hours, light conditions, rain, fruit presence and evidence, presence of potential predators of bats, and remarks about peculiarities of the evening. This information was entered on the sheet whenever appropriate during a netting episode. Our standards for this form were as follows.

1. *Locality:* We identified capture localities to the nearest 0.1 km by their X and Y coordinates on a grid 100 km², with BCI at the center (Figure 12-4). This system allowed us to use the coordinates of the locality as the site's code name for record keeping. This system also made it easier to program for computer plotting and for calculation of areas and distances. By having BCI at the center of the grid (identified as point 50.0 × 50.0) we never had to deal with negative numbers. For example, the Laboratory Clearing is "50.7 × 50.9," Armour End on the opposite side of the island is "48.3 × 48.4," and Bohio Ridge on the mainland is "48.9 × 54.3." To simplify and speed the process we dropped the decimals from the six-digit code when we recorded the data, but we restored them when we entered the information in the computer. A constantly updated gazetteer (Table 12-2) provided the locality code for each site.

We caught bats at 106 localities on BCI (plus 15 others on the mainland and other islands)—far too many sites for convenient analysis. Consequently, we segregated the 106 localities on BCI into 20 regional groups (Figure 12-5).

2. *Time:* We recorded time of capture in 1-hour increments on a 24-hour clock. We defined an hour as from 30 minutes before zero to 30 minutes after (e.g., 2000 h = 1930 to 2030). Although we removed bats from nets as they were caught and tried to check all nets at the end of each hour, some errors undoubtedly occurred when we were catching many bats. At the marking station the bats from each hour were kept in separate lots.

Times of opening and closing the nets were recorded to the nearest quarter hour. If we required more than a quarter hour to open or close nets, we recorded the times in quarter-hour lots or recorded a median quarter-hour time.

For record keeping, the day began at daybreak and continued through the succeeding night. Thus, the date did not change (advance) at midnight.

3. *Net Hours:* Lapsed hours, minus rain time, equals catch time, and multiplied by the number of nets equals the number

TABLE 12-2.—BCI Bat Project gazetteer.

Locality	Code	Locality	Code
Allee Creek	506 × 510	Miller 8–10 and Nemesia 1	494 × 512
AMNH 1–2	503 × 493	Miller 11–13	493 × 513
AMNH 3–4	504 × 494	Miller 13–14	493 × 515
Armour 0 (AVA 0)	495 × 501	Miller 15–18	493 × 516
Armour–Zetek Junction (AVA 2.5)	495 × 500	Miller Cove (outer)	493 × 520
Armour 6	493 × 498	Mona Grita Point	512 × 474
Armour 9–11	490 × 497	Navigational Aid No. 5	599 × 465
Armour 14	487 × 494	Navigational Aid No. 7 (Darien Island)	575 × 445
Armour–Conrad Junction	486 × 494	Nemesia B (Upper Nemesia Creek)	490 × 511
Armour 20	484 × 490	Nemesia 2	492 × 512
Armour end	483 × 484	Nemesia 4	491 × 512
Balboa 2–3 (Balboa 1–4)	497 × 502	Orchid Island	497 × 528
Balboa 6 Creek (Balboa 4–6)	500 × 501	Orchid Island No. 2	496 × 527
Balboa 7 Ridge (Balboa 7–8)	501 × 500	Pearson 0.5–3.5	494 × 503
Balboa 12 Creek	504 × 497	Pearson Creek	489 × 511
Barbour–Lathrop–Miller Shortcut	499 × 510	Peña Blanca (No. 11 front marker)	461 × 522
Barbour–Lathrop 16	500 × 517	Power Line Hill	507 × 511
Barbour Stream Station A (B–L 3.5)	503 × 510	Punta Caño Quebrado	534 × 464
Barbour Stream Station B	502 × 514	Schneirla 1–2 (New Wheeler 9)	500 × 506
Barbour Stream Station C	503 × 515	Shannon 1–2 (creek)	503 × 506
Barro Colorado Island	000 × 000	Shannon 6	503 × 501
Bohio Trail	488 × 543	Shannon 10	503 × 499
Bohio Ridge (saddle)	489 × 543	Shannon 13.5	507 × 497
Buena Vista (No. 9 front marker)	501 × 535	Shannon 17	507 × 494
Cacao Cove	486 × 515	Snyder–Molino 0–2	506 × 508
Cat Creek Cove	489 × 515	Snyder–Molino–Lutz	505 × 507
Chapman 6–9	518 × 498	Snyder–Molino 4	504 × 507
Chapman 8–9 (FMC 8–10)	520 × 498	Snyder–Molino–Wheeler Junction	503 × 507
Chapman End (FMC 9–11)	522 × 498	Standley 5	483 × 500
Chilibrillo Caves	748 × 517	Standley 13 estero	479 × 511
Conrad 1–2	486 × 495	Standley 13 ridge (PCS 12.5–14)	479 X 510
Conrad 4	484 × 496	Standley 16 (PCS 15–16)	478 × 512
Conrad Creek	489 × 498	Standley 17 estero	479 × 513
Donato 1–3	508 × 508	Standley 19 estero	479 × 516
Donato 3–4 (Lutz Loop)	508 × 507	Standley 21 (Standley end)	476 × 516
Drayton 2–3	495 × 499	Barbour–Shannon cutoff	504 × 506
Drayton 16	498 × 488	Barbour 3 (TB 2–4)	505 × 505
Drayton 18.5	500 × 486	Barbour 5 (TB 4–6)	507 × 504
Drayton Estero	500 × 482	Barbour 7	508 × 504
Dump Cove	509 × 509	Barbour 9	510 × 504
Fairchild–Wheeler Junction	504 × 510	Barbour 11 (TB 10–12)	512 × 504
Fairchild 5–7	505 × 514	Barbour 14–16 (TB–Hood Junction)	515 × 505
Frijoles (Kidd's Place)	552 × 522	Barbour 20–21	520 × 505
Frijoles Road	543 × 538	Barbour end	526 × 506
Fuertes Cove	489 × 513	VanTyne 0–1 – Barbour 10	511 × 504
Fuertes House (Nemesia Creek)	489 × 512	VanTyne 2 – Big Tree	511 × 502
Fuertes Ridge	490 × 514	VanTyne–Hood–Chapman Junction	514 × 497
Gigante	514 × 463	VanTyne–Shannon end	508 × 493
Gigante Rancho	478 × 452	Wheeler 2–3	504 × 509
Giral Caves	667 × 603	Wheeler 4	503 × 508
Gross 9–10	500 × 526	Wheeler–Miller Junction	502 × 507
Gross Point	499 × 527	Wheeler 12–14	496 × 502
Harvard 10 (Harvard Cove)	517 × 492	Wheeler 15 (WMW–FD Junction)	496 × 499
Lab Clearing (Snyder–Molino 0)	507 × 509	Wheeler 20	497 × 498
Lake 4–6	497 × 507	Zetek 1–3	493 × 500
Lutz Creek (Lutz Creek–Lutz 3)	507 × 508	Zetek 3 (Zetek 2–4)	490 × 499
Lutz Creek mouth	508 × 510	Zetek–Armour shortcut	491 × 498
Miller Creek mouth	495 × 516	Zetek 7	487 × 499
Miller 1–2	501 × 507	Zetek 13	484 × 497
Miller 4–5	499 × 508	Zetek 21 (JZ 20–22)	477 × 501
Miller 7–8	495 × 511		

FIELD DATA COVER SHEET

DATE:	LOCATION:	LOCALITY CODE:

NUMBER OF NETS SET: 5(6) meter______, 10(12) meter______   RELEASE SITE:

NET HOURS:

______nets open_____, closed_____; lapsed hrs._____, minus rain time_____equals catch time, X_____nets = net hours_____

______nets open_____, closed_____; lapsed hrs._____, minus rain time_____equals catch time, X_____nets = net hours_____

______nets open_____, closed_____; lapsed hrs._____, minus rain time_____equals catch time, X_____nets = net hours_____

TOTAL NET HOURS_____

MOON PHASE
   New +_____, 1st Q +_____, Full +_____, 3rd Q +_____

CLOUD COVER:

LIGHT CONDITIONS:

RIPE FRUIT TREES at or near capture station:

FRUGIVORES PRESENT:

FRUIT CARRIED INTO NETS:

BAT FECES CONTENTS:

REMARKS & OTHER OBSERVATIONS:

RAIN (clock time):

   HEAVY ____________________

   MODERATE ____________________

   LIGHT ____________________

   DRIP____________________

RAIN TIME (hours)_____
(Lapsed time from onset of thrufall heavy enough to wet nets, until nets have been shaken dry enough to resume catching bats)

PREDATORS OBSERVED:

BAT DEATHS (number and cause):

FIGURE 12-2.—Field data cover sheet used to record general information about each netting episode.

FIELD DATA COVER SHEET

DATE: *19 Sep 80* | LOCATION: *STANDLEY 15* | LOCALITY CODE: *478 x 512*

NUMBER OF NETS SET: 5(6) meter ______, 10(12) meter *10* | RELEASE SITE:

NET HOURS:

10 nets open *1830*, closed *2315*; lapsed hrs. *4.75*, minus rain time *2* equals catch time, X *10* nets = net hours *27.5*

______ nets open ______, closed ______; lapsed hrs. ______, minus rain time ______ equals catch time, X ______ nets = net hours ______

______ nets open ______, closed ______; lapsed hrs. ______, minus rain time ______ equals catch time, X ______ nets = net hours ______

TOTAL NET HOURS *27.5*

MOON PHASE
New + ______, 1st Q + *2*, Full + ______, 3rd Q + ______

CLOUD COVER: *CLOUDY*

LIGHT CONDITIONS: *MODERATELY DARK*

RIPE FRUIT TREES at or near capture station:
5, MONGBW 200m AWAY, ON SHORE OF ESTERO

FRUGIVORES PRESENT: *ALOUATTA, POTOS*

FRUIT CARRIED INTO NETS: *QB*

BAT FECES CONTENTS: *QB and unidentified*

RAIN (clock time):

HEAVY ______________

MODERATE ______________

LIGHT *1930 — 2130 (Nets closed)*

DRIP *2130 — 2300*

RAIN TIME (hours) *2*
(Lapsed time from onset of thrufall heavy enough to wet nets, until nets have been shaken dry enough to resume catching bats)

PREDATORS OBSERVED: *—none*

BAT DEATHS (number and cause): *none*

REMARKS & OTHER OBSERVATIONS: *11 Ad repeats*
Bag = 28g

54 marks
24 recaps
78

FIGURE 12-3.—Completed field data cover sheet.

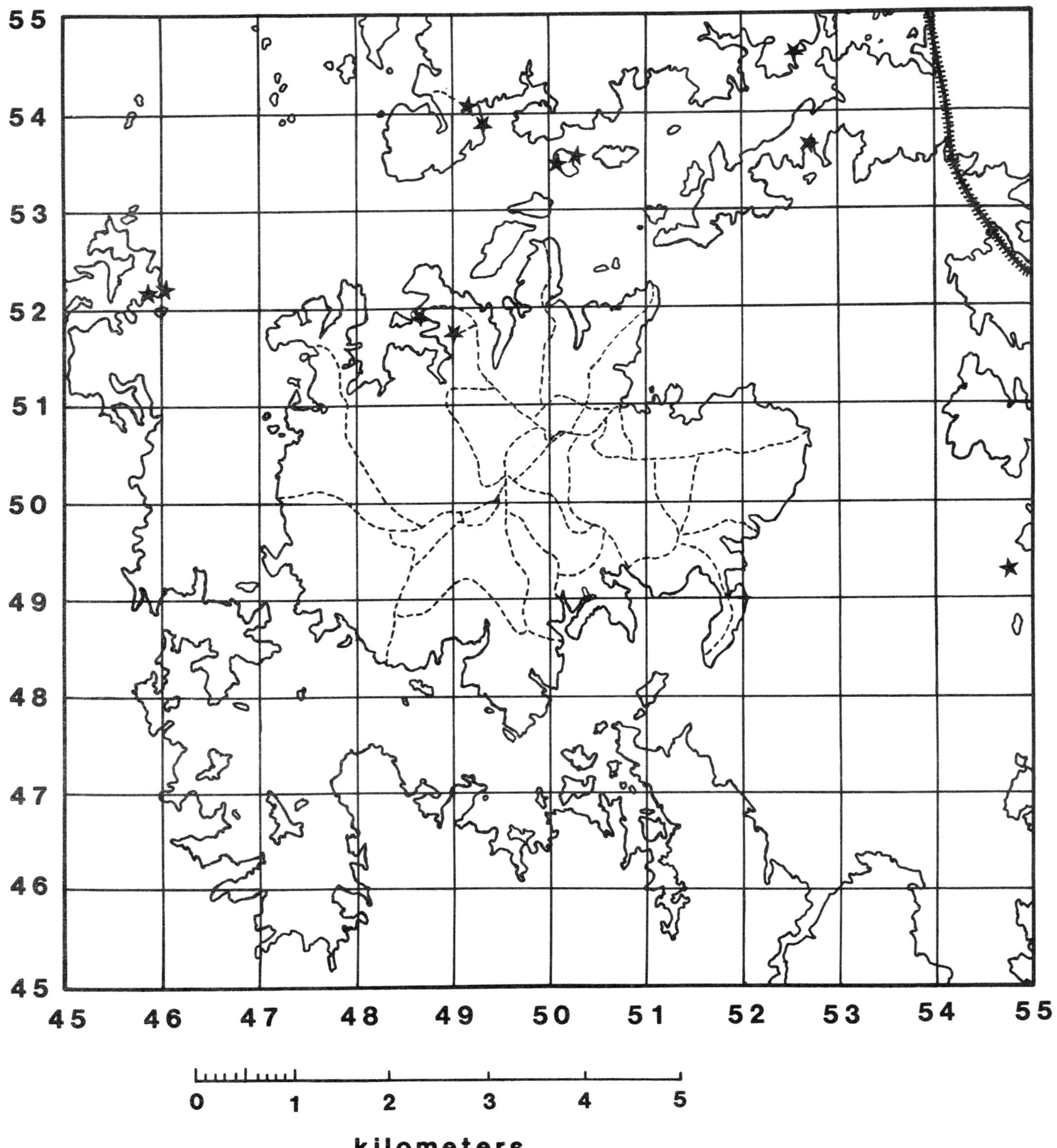

FIGURE 12-4.—Grid system of one kilometer squares used to identify localities on Barro Colorado Island and surrounding mainland. Stars indicate canal range light towers. Those mentioned in the text are Peña Blanca (near 46 × 52) and Buena Vista (50 × 53). Dashed lines on BCI trace approximate alignment of trails.

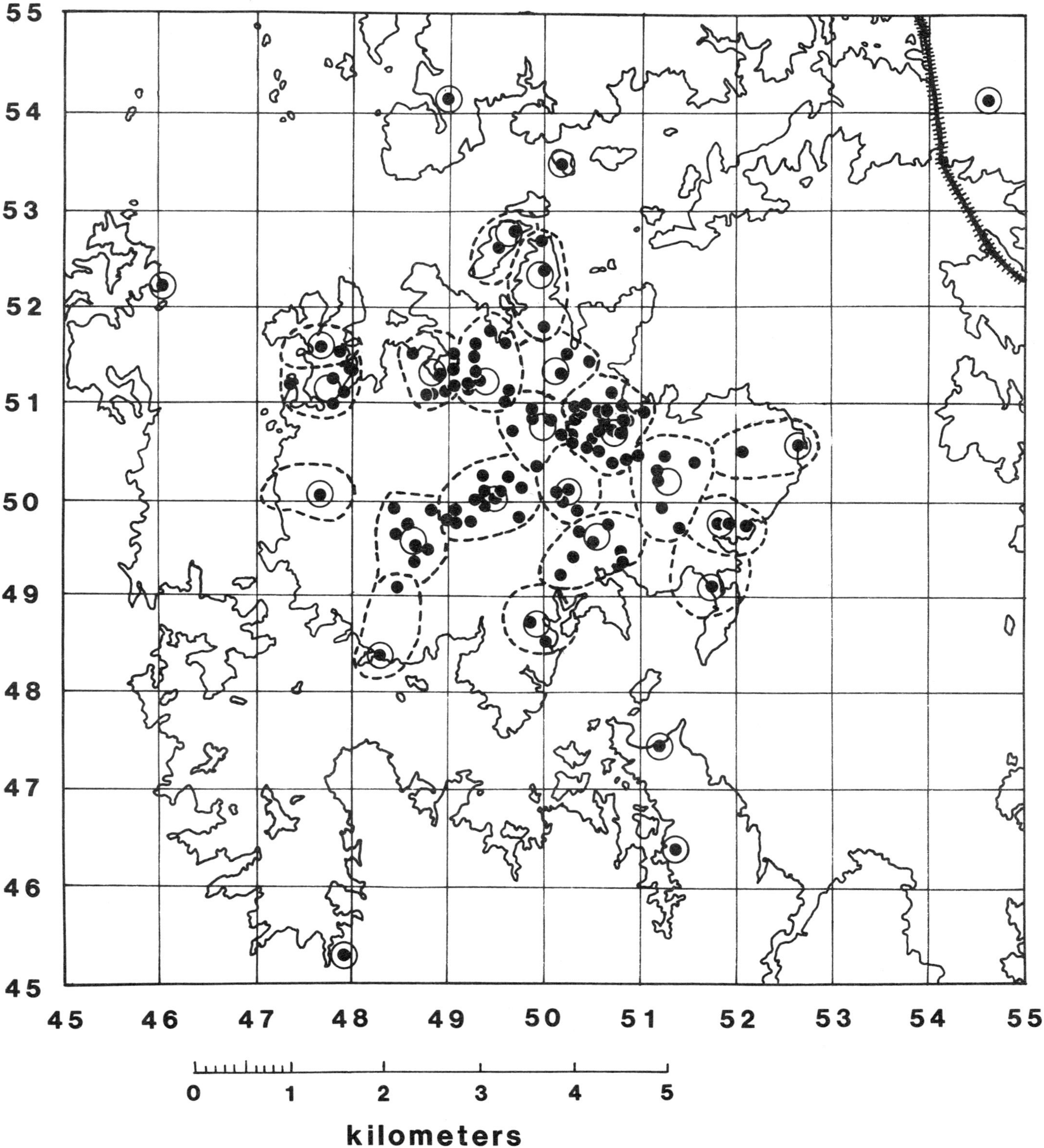

FIGURE 12-5.—Capture localities (black dots) and locality groups (large circles) on BCI and surrounding mainland used by the Bat Project, 1976–1980. Dashed lines show extent of locality groups.

of net hours. We treated a 5 or 6 m net as one half of a 10 or 12 m net.

4. *Moon Phase:* We recorded moon phase from the calendar, adding the number of nights since the beginning of the current phase.

5. *Cloud Cover:* CLEAR = sky cloudless; FEW CLOUDS = sky $1/4$ cloud covered; PARTLY CLOUDY = sky $1/2$ covered; MOSTLY CLOUDY = sky $3/4$ covered; CLOUDY = sky completely overcast.

6. *Light Conditions:* VERY DARK = too dark to see hands; MODERATELY DARK = light enough to distinguish objects; MODERATELY LIGHT = light enough to see a considerable distance; VERY LIGHT = light enough to read.

We estimated light conditions at the net station. Sometimes we recorded combinations such as "VERY DARK until moonrise, 2100, then VERY LIGHT;" or "MODERATELY DARK during rain, otherwise MODERATELY LIGHT."

7. *Rain:* We defined rain as throughfall heavy enough to wet the ground. We did not consider a fine mist, which contributes only leaf drip, as rain. The rain categories we used were: HEAVY = torrential, ground is covered with water, noise of the rain drowns out all other sounds; MODERATE = rain of average intensity, noisy, but not masking all other sounds; LIGHT = drizzle, quiet, individual leaf drips can be distinguished; DRIP = leaf drip during mist or after a rain (noted in our records but not qualifying as rain).

FIELD DATA SHEETS.—Data for individual band numbers were entered on Field Data Sheets (Figures 12-6 and 12-7)—new marks on one set of sheets; recaptures on another. Our form has 15 double-line spaces and we used each couplet of lines for the record of one bat; the upper line for field data and the lower line for a computer code. Data that allowed mnemonic codes (e.g., age, sex, date, hour, and band number) were entered as they were noted directly on the code line. The remaining data (such as locality, species identification, and female reproductive stage) were coded upon return to our operations base, preferably during the same evening.

The columns on this form are organized into 17 data fields separated by spaces rather than zeros. Computer entry for this form was programmed to accept blanks in nonapplicable spaces such as in items that applied only to one sex or the other. Information standards for this form are as follows.

1. *Identification of Species:* From the outset we required that inexperienced field personnel spend a day in the museum reviewing a synoptic set of bats before going to Panamá. The banding kit carried each night included a key to the bats of the lowlands of Panamá and a synopsis of the species known to occur on BCI, with detailed descriptions and key characters of each.

Personnel were instructed to save the first of any species not known to occur on BCI as a voucher specimen to substantiate its occurrence. With the exception of constant (but understandable) confusion of *A. phaeotis* and *A. watsoni*, data from

recaptures revealed surprisingly few inconsistencies, which strengthens our confidence in the identifications.

For coding purposes, we alphabetized the list of species known to occur on BCI and applied a simple numeric code, 1 to 56 to each kind. Fortuitously, *A. jamaicensis* was number 1 and several of the most common species had easily recallable single digit numbers. Hindsight shows, however, that we should have used more easily remembered alpha codes, which quickly evolved in field parlance, such as AJ (= *A. jamaicensis*), CC (= *Carollia castanea*), CP (= *C. perspicillata*), CV (= *Chiroderma villosum*), UB (= *Uroderma bilobatum*), VC (= *Vampyrodes caraccioli*), etc.

2. *Sex:* Sex was coded F for female and M for male.

3. *Age:* We based age standards on observation of maturation in several generations of *A. jamaicensis* in our NZP colony. These proved applicable to other species as well. Although we looked at a number of characters, the most important were ossification of the digital epiphyses to distinguish subadults from juveniles; evidence of reproduction to distinguish adult females from subadults; and enlarged testes to set apart adult males. Our full protocol for determining age in *A. jamaicensis* was as follows.

Juvenile: Individuals that are 1 to 3 months of age are considered juveniles. The prime factor that distinguishes juveniles from the other age groups is the epiphyses of the fingers are open (unossified), with the joints swollen and tapering. Epiphyses appear translucent when viewed before a light. When the epiphyses of the last joint (usually on finger V) ossify, the bat is classed as a subadult. Other important characteristics of juveniles include a body size that is small at first but then large; a gray pelage that is downy initially but becomes smooth; tiny (5×3 mm or less), pale testes, with the skin of the testicular region hairy (males); and tiny, pale nipples that are hidden in hair (females).

Subadult: Individuals that are 3 to 12 months of age are termed subadults. In subadults the epiphyses of the fingers are closed (ossified) and knobby. In males, the testes are small (3×2 mm to 7×5 mm) and pale, with the testicular region hairy. The nipples of females are tiny, pale, and hidden in hair. Females are nulliparous. Other distinguishing characteristics include a large body size and a smooth, usually gray pelage.

Adult: Individuals are considered to be adults when they are 8 to 12 months of age or older. The epiphyses of the fingers are closed and the joints knobby. In males, the testes are medium-size to large (7×4 mm or more) with the overlying skin sometimes pigmented and naked. In females, the nipples are small (never tiny), medium-size, or large and may be dark and naked. Females are parous. By these criteria the infrequent nulliparous female beyond a year in age would be recorded as a subadult. Other important characteristics include large body size, dark gray or brownish pelage, and tooth wear may be obvious.

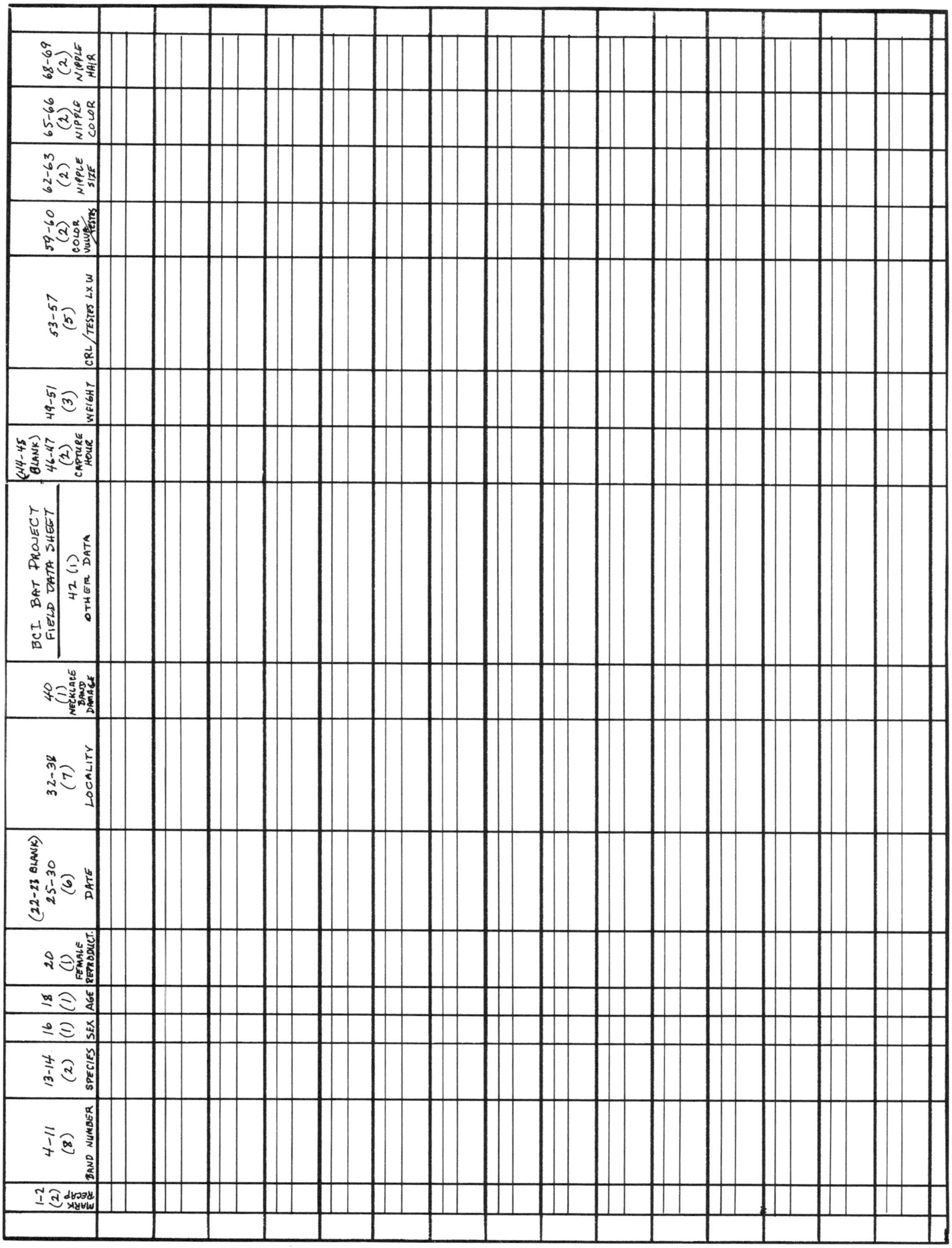

BCI BAT PROJECT
FIELD DATA SHEET

1-2 (2) MARK RECAP	4-11 (8) BAND NUMBER	13-14 (2) SPECIES	16 (1) SEX	18 (1) AGE	20 (1) FEMALE REPRODUCT.	(22-23 BLANK) 25-30 (6) DATE	32-38 (7) LOCALITY	40 (1) NECKLACE BAND DAMAGE	42 (1) OTHER DATA	(44-45 BLANK) 46-47 (2) CAPTURE HOUR	49-51 (3) WEIGHT	53-57 (5) CRL/TESTES LXW	59-60 (2) COLOR VULVA/TESTES	62-63 (2) NIPPLE SIZE	65-66 (2) NIPPLE COLOR	68-69 (2) NIPPLE HAIR

FIGURE 12-6.—Field data sheet used to record data on each capture.

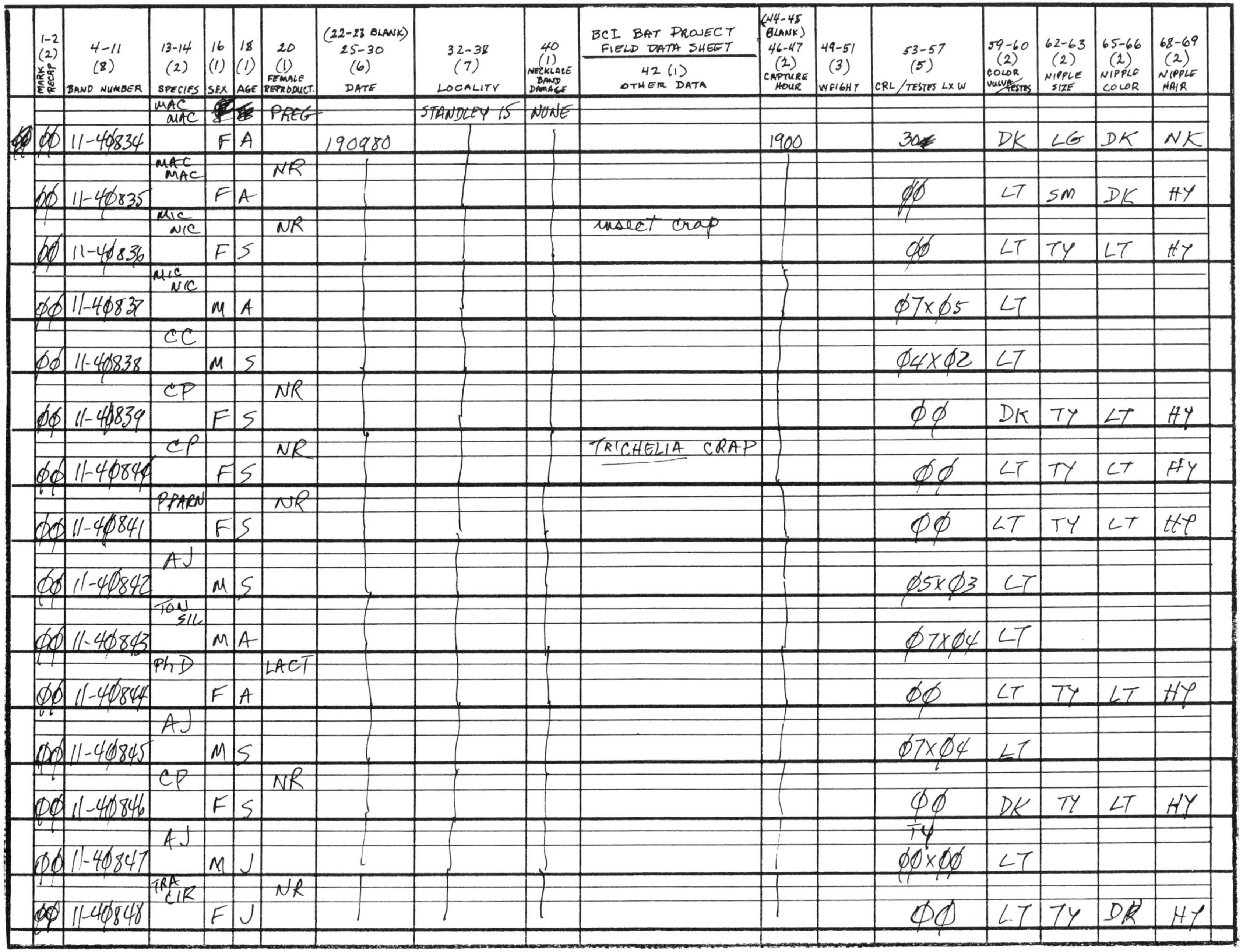

1-2 (2) MARK RECAP	4-11 (8) BAND NUMBER	13-14 (2) SPECIES	16 (1) SEX	18 (1) AGE	20 (1) FEMALE REPRODUCT.	(22-23 BLANK) 25-30 (6) DATE	32-38 (7) LOCALITY	40 (1) NECKLACE BAND DAMAGE	BCI BAT PROJECT FIELD DATA SHEET 42 (1) OTHER DATA	(44-45 BLANK) 46-47 (2) CAPTURE HOUR	49-51 (3) WEIGHT	53-57 (5) CRL/TESTES L×W	59-60 (2) COLOR VULVA/TESTES	62-63 (2) NIPPLE SIZE	65-66 (2) NIPPLE COLOR	68-69 (2) NIPPLE HAIR
ØØ	11-40834	MAC MAC	F	A	PREG	190980	STANDLEY IS	NONE		1900	30#		DK	LG	DK	NK
ØØ	11-40835	MAC MAC	F	A	NR							ØØ	LT	SM	DK	HY
ØØ	11-40836	NIC NIC	F	S	NR				insect crap			ØØ	LT	TY	LT	HY
ØØ	11-40837	NIC NIC	M	A								Ø7×Ø5	LT			
ØØ	11-40838	CC	M	S								Ø4×Ø2	LT			
ØØ	11-40839	CP	F	S	NR							ØØ	DK	TY	LT	HY
ØØ	11-40840	CP	F	S	NR				TRICHELIA CRAP			ØØ	LT	TY	LT	HY
ØØ	11-40841	PPARN	F	S	NR							ØØ	LT	TY	LT	HY
ØØ	11-40842	AJ	M	S								Ø5×Ø3	LT			
ØØ	11-40843	TON SIL	M	A								Ø7×Ø4	LT			
ØØ	11-40844	PhD	F	A	LACT							ØØ	LT	TY	LT	HY
ØØ	11-40845	AJ	M	S								Ø7×Ø4	LT			
ØØ	11-40846	CP	F	S	NR							ØØ	DK	TY	LT	HY
ØØ	11-40847	AJ	M	J								ØØ×ØØ	LT			
ØØ	11-40848	TRA CIR	F	J	NR							ØØ	LT	TY	DR	HY

FIGURE 12-7.—Completed field data sheet.

4. *Reproductive Categories:* Females: We used these reproductive categories for females: no reproduction, pregnant, lactating, postlactating, pregnant and lactating, and young on nipple.

Pregnancy: The abdomen of every female, regardless of age and reproductive condition, was palpated to detect pregnancy in the following manner. The bat was grasped by its elbows with the right hand, with its abdomen facing left. The left thumb and index finger were used to gently palpate the abdomen. A fetus could be felt by gentle pressure (down, back, and inward) on the abdomen, just posterior to the ribs. If the head of a fetus (a hard lump) could be felt, the bat was palpably pregnant. Abdominal distention also may be caused by food or gas, so it was necessary to feel the fetal skull to be sure of a pregnancy.

In *A. jamaicensis,* a fetus with a crown-rump (CR) length of 30 to 35 mm or more will distend the abdomen, and the CR length can be measured with surprising accuracy (tested by later dissection and direct measurement of the fetus in bats that died in handling) with a millimeter rule. When a fetus is too small to distend the abdomen its CR length can be estimated. It is possible to be misled by the left kidney, which can be felt easily and mistaken for a small fetus. The left kidney is about 10×5 mm in its greater dimensions and can be felt and moved by gentle pressure on the abdomen close to the backbone. Familiarity with the position and shape of the kidney can be gained by palpating male bats. A bat was not recorded as pregnant unless *both* kidney and fetus could be felt.

Lactation: Lactation was determined by expression of milk from a nipple. A bat was presumed to be postlactating if nipples would not express milk, but were large, flabby, and surrounded by naked skin. This was the only criterion used early in the project. Later on, we made a more accurate determination. Postlactation was recorded if no milk was present, areolas were naked or new hair was growing in, and nipples were any size.

Nipple Condition: The parameters of nipple condition we recorded were size, color, and presence of areolar hair. Because these vary with age and season, as we learned from our NZP bats, they provide accurate clues to the reproductive state. For example, TINY-LIGHT-HAIRY can only be a nulliparous bat; MEDIUM-LIGHT-HAIRY is a preparous adult bat; LARGE-DARK-NAKED goes with lactation; MEDIUM-DARK-NEW HAIR is postlactating.

Nipple size was measured with a template (Figure 12-8) based on measurements taken on the *A. jamaicensis* in the NZP colony. TINY nipples are unmeasurable, and they cannot be pinched. They are characteristic of juvenile and subadult (nulliparous) bats. SMALL nipples vary from 0.5 to 3 mm in length and from 1 to 6 mm in diameter at the base. MEDIUM nipples vary from 4 to 6 mm in length and from 5 to 7 mm in diameter. LARGE nipples vary from 7 to 10 mm in length and from 5 to 10 mm in diameter.

Nipple color was recorded as LIGHT or DARK. All

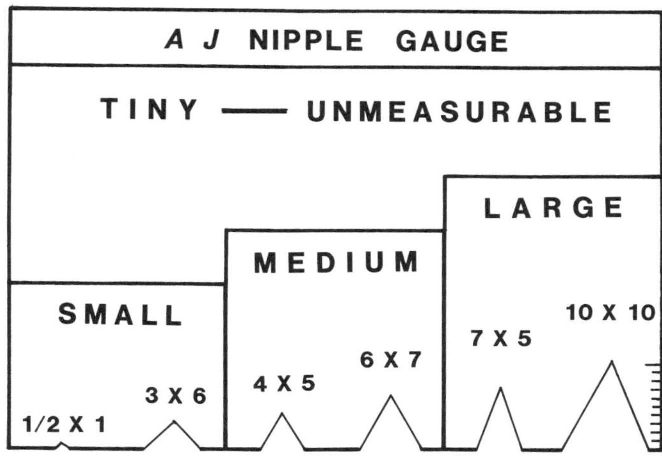

FIGURE 12-8.—Gauge used to determine size of nipples of *Artibeus jamaicensis* on BCI.

nulliparous and some adult *A. jamaicensis* have pale, unpigmented, i.e., LIGHT, nipples. Some parous individuals have nipples colored with tan to fuscous pigmentation, which may vary from a few flecks of color to the entire nipple. Any pigmentation, regardless of amount, was recorded as DARK. Although the distribution and intensity of pigment might vary seasonally, any bat with pigmented nipples was assumed to be adult.

Nipple hair refers to the areolar region, which we described as HAIRY, NAKED, or having NEW HAIR. The areola is HAIRY in all nulliparous bats, and NAKED just before, during, and following lactation. NEW HAIR, usually conspicuous because its gray color contrasts with the drab brown older hair of the chest, grows in as the nipples recede in size following lactation.

Vulvar Coloration: We recorded the color of the vulva as LIGHT, DARK, or BLACK. Nulliparous bats usually have a pale (LIGHT) vulva. At parturition and during estrus the vulva sometimes becomes heavily pigmented and BLACK. Following estrus the coloration fades to fuscous (DARK) and eventually only the vulval margin is pigmented (LIGHT).

5. *Reproductive Categories:* Males: Size and appearance of the testes are clues to age and reproductive condition (Figure 12-9). We measured the approximate length and width of testes with a millimeter ruler. Size of testes in *A. jamaicensis* varies in juveniles from too small to locate, up to about 5×3 mm. Our most frequently recorded dimensions on juveniles were 3×2 and 4×2 mm. Size of testes in subadults varies from 3×2 mm to 7×5 mm (most frequently noted measurements are 5×3 mm and 6×3 mm), overlapping juvenile size at one extreme and adult size at the other. Size of testes in adults varies from 7×4 mm to 13×7 mm and occasionally larger. In adults, size varies seasonally, being largest when females are in estrus.

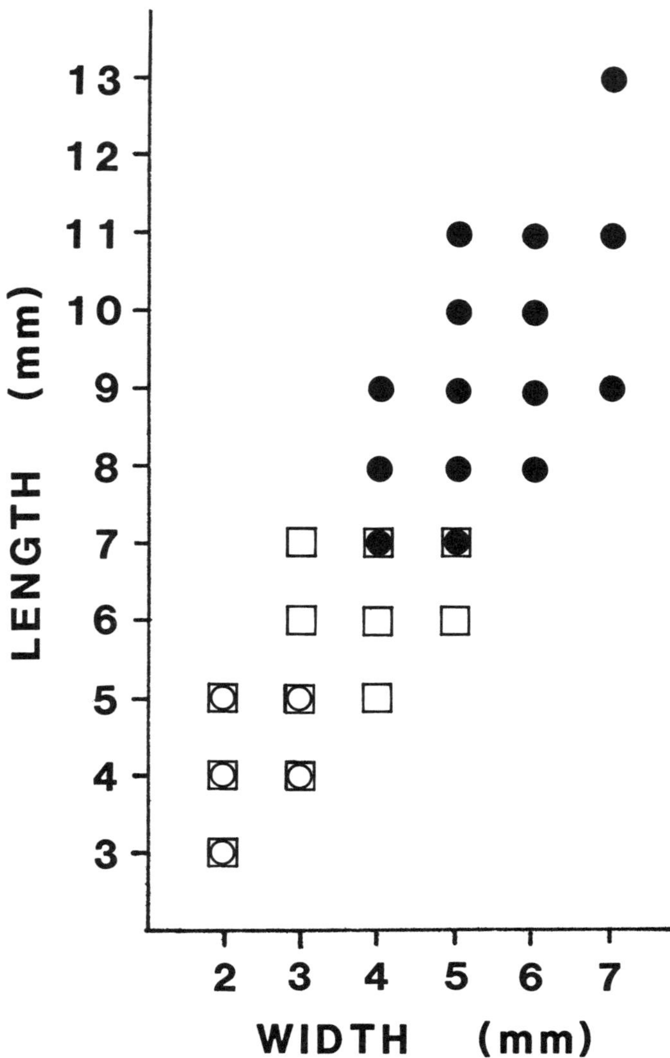

FIGURE 12-9.—Variation in testis size with age in a captive colony of Panamanian *Artibeus jamaicensis*. Circles represent juveniles; squares, subadults; and dots, adults.

Testes in the shrinking (regressed) stage sometimes resemble a deflated football. Loss of hair on the skin overlying the testes (not a true scrotum) and increased pigmentation of the testicular skin might reflect sexual activity. We recorded the skin color as LIGHT, DARK, or BLACK.

6. *Mass:* We weighed bats with a Pesola spring balance having a 300 g capacity. The bats were restrained in a clean bag, which we weighed empty several times during the evening to make certain the mass had not changed. We recorded all masses only to the nearest gram, assuming that greater accuracy was unrealistic.

7. *Necklace Band Damage:* Because a bat can reach its own necklace only with its feet and thumbs, we have attributed damage to necklace bands to be the result of grooming by a roost partner. Because grooming behavior varies from species to species, it is worthwhile to record the extent of damage to bands. Damage was described as NONE, LITTLE (few scratches), MODERATE (definitely chewed), EXTENSIVE (almost illegible), or ILLEGIBLE.

8. *Other Data:* We recorded notes on wounds, hair loss, feces contents, apparent stress, and other information on an individual in column 42 (other data).

NIGHTLY SUMMARY SHEET.—This summary sheet (Figures 12-10 and 12-11) lists the nightly catch by new marks and recaptures by hour and species. New marks were recorded in the upper triangle of a pair and recaptures in the lower triangle. Recaptures were always circled to emphasize and highlight them. Totals were computed by species and by hour, and bats per net hour was calculated by dividing total number of bats captured by total number of net hours. For *A. jamaicensis,* the catch was further subdivided by age and sex, and adult recapture rate was calculated by dividing the number of adults recaptured by the total number of adults caught.

NIGHTLY SUMMARY OF BAT CATCH

DATE: LOCATION: NETS SET:

Hours begin on half hour
Capture/Recapture

NIGHTLY CATCH SUMMARY	1800	1900	2000	2100	2200	2300	2400	0100	0200	0300	0400	0500	0600	Total
ARTIBEUS JAMAICENSIS														
ARTIBEUS LITURATUS														
ARTIBEUS PHAEOTIS														
ARTIBEUS WATSONI														
CAROLLIA CASTANEA														
CAROLLIA PERSPICILLATA														
MICRONYCTERIS HIRSUTA														
PHYLLOSTOMUS DISCOLOR														
PTERONOTUS PARNELLII														
TONATIA BIDENS														
TONATIA SILVICOLA														
URODERMA BILOBATUM														
VAMPYRESSA NYMPHAEA														
VAMPYRODES CARACCIOLI														
HOURLY TOTAL														

Catch summary, A. jamaicensis (Capture/Recapture)

	AD	SAD	JUV	Mark/Rec total	Catch total
FEMALE					
MALE					
Mark/Rec Total					
Species Total					

ADULT RECAPTURE RATE:

TOTAL BATS CAPTURED:

BATS PER NET HOUR:

RECAPTURES:

REPEATS:

FIGURE 12-10.—Nightly summary sheet for a netting episode.

NIGHTLY SUMMARY OF BAT CATCH

DATE: 19 Sep 80 LOCATION: Stundley 15 NETS SET: 10

Hours begin on half hour
Capture/Recapture

	1800	1900	2000	2100	2200	2300	2400	0100	0200	0300	0400	0500	0600	Total
ARTIBEUS JAMAICENSIS		26/(16)		4/(4)	2/(1)									32/(21)
~~ARTIBEUS LITURATUS~~														
ARTIBEUS PHAEOTIS		(1)		1										1/(1)
~~ARTIBEUS WATSONI~~														
CAROLLIA CASTANEA		1			1									2
CAROLLIA PERSPICILLATA		4/(1)		2										6/(1)
MICRONYCTERIS ~~HIRSUTA~~ NICEFORI		2												2
PHYLLOSTOMUS DISCOLOR		1												1
PTERONOTUS PARNELLII		1		(1)										1/(1)
TONATIA BIDENS		1												1
TONATIA SILVICOLA		2												2
~~URODERMA BILOBATUM~~														
~~VAMPYRESSA NYMPHAEA~~														
~~VAMPYRODES CARACCIOLI~~														
MACROPHYLLUM MACROPH.		2												2
TRACHOPS CIRRHOSUS		1			1									2
MIMON CRENULATUM				2										2
HOURLY TOTAL		41/(18)	-/-	9/(5)	4/(1)									54/(24)

Catch summary, A. jamaicensis (Capture/Recapture)

	AD	SAD	JUV	Mark/Rec total	Catch total
FEMALE	-/(2)	1/(2)	6/(2)	7/(6)	13
MALE	2/(12)	8/(3)	15/-	25/(15)	40
Mark/Rec Total	2/(14)	9/(5)	21/(2)	32/(21)	53
Species Total	16	14	23	53	

ADULT RECAPTURE RATE: 87.5%

TOTAL BATS CAPTURED: 78

BATS PER NET HOUR: 2.1

RECAPTURES: (24)

REPEATS: 2

NIGHTLY CATCH SUMMARY

FIGURE 12-11.—Completed nightly summary sheet.

Literature Cited

Altman, P.L., and D.S. Dittmer
1961. *Blood and Other Body Fluids.* xvii + 540 pages. Washington, D.C.: Federation of American Societies for Experimental Biology.

Armstrong, R.B., C.D. Ianuzzo, and T.H. Kunz
1977. Histochemical and Biochemical Properties of Flight Muscle Fibers in the Little Brown Bat, *Myotis lucifugus. Journal of Comparative Physiology,* 119:141–154.

August, P.V.
1979. Distress Calls in *Artibeus jamaicensis:* Ecology and Evolutionary Implications. *In* J.F. Eisenberg, editor, *Vertebrate Ecology in the Northern Neotropics,* pages 151–159. Washington, D.C.: Smithsonian Institution Press.
1981. Fig Fruit Consumption and Seed Dispersal by *Artibeus jamaicensis* in the Llanos of Venezuela. *Reproductive Botany,* supplement to *Biotropica,* 13(2):70–76.

Baker, R.J., J.K. Jones, Jr., and D.C. Carter, editors
1976. Biology of Bats of the New World Family Phyllostomatidae, Part I. *Special Publications of The Museum, Texas Tech University,* 10:1–218.
1977. Biology of Bats of the New World Family Phyllostomatidae, Part II. *Special Publications of The Museum, Texas Tech University,* 13:1–364.
1979. Biology of Bats of the New World Family Phyllostomatidae, Part III. *Special Publications of The Museum, Texas Tech University,* 16:1–441.

Baker, W.W., S.G. Marshall, and V.B. Baker
1968. Autumn Fat Deposition in the Evening Bat (*Nycticeius humeralis*). *Journal of Mammalogy,* 49(2):314–317.

Barlow, J.C., and J.R. Tamsitt
1968. Twinning in American Leaf-nosed Bats (Chiroptera: Phyllostomatidae). *Canadian Journal of Zoology,* 46(2):290–292.

Bartholomew, G.A., C.M. Vleck, and T.L. Bucher
1983. Energy Metabolism and Nocturnal Hypothermia in Two Tropical Passerine Frugivores, *Manacus vitellinus* and *Pipra mentalis. Physiological Zoology,* 56(3):370–379.

Bassett, J.E., and J.E. Wiebers
1979. Urine Concentration Dynamics in the Post Prandial and the Fasting *Myotis lucifugus lucifugus. Comparative Biochemistry and Physiology,* 64A(3):373–379.

Beenakkers, A.M.T.
1969. Carbohydrate and Fat as a Fuel for Insect Flight; A Comparative Study. *Journal of Insect Physiology,* 15(3):353–361.

Beenakkers, A.M.T., A.T.M. Van den Broek, and T.J.A. de Ronde
1975. Development of Catabolic Pathways in Insect Flight Muscles; A Comparative Study. *Journal of Insect Physiology,* 21(4):849–859.

Ben Shaul, D.M.
1962. The Composition of the Milk of Wild Animals. *International Zoo Yearbook,* 4:333–342.

Bernard, R.T.F., and J.A.J. Meester
1982. Female Reproduction and the Female Reproductive Cycle of *Hipposideros caffer caffer* (Sundevall, 1846) in Natal, South Africa. *Annals of the Transvaal Museum,* 33:131–144.

Bhatnagar, K.P.
1978. Head Presentation in *Artibeus jamaicensis* with Some Notes on Parturition. *Mammalia,* 42(3):359–363.

Bhide, S.A.
1980. Observations on the Stomach of the Indian Fruit Bat, *Rousettus leschenaulti* (Desmarest). *Mammalia,* 44(4):571–579.

Black, H.L., G. Howard, and R. Stjernstedt
1979. Observations on the Feeding Behavior of the Bat Hawk (*Macheiramphus alcinus*). *Biotropica,* 11(1):18–21.

Blair-West, J.R., J.P. Coghlan, D.A. Denton, J.F. Nelson, E. Orchard, B.A. Scoggins, R.D. Wright, K. Myers, and C.L. Junqueira
1968. Physiological, Morphological and Behavioral Adaptations to a Sodium Deficient Environment by Wild Native Australian and Introduced Species of Mammals. *Nature,* 217(5132):922–928.

Bonaccorso, F.J.
1975. Foraging and Reproduction Ecology of a Community of Bats in Panama. 122 pages. Unpublished doctoral dissertation, Department of Zoology, University of Florida, Gainesville.
1979. Foraging and Reproductive Ecology in a Panamanian Bat Community. *Bulletin of the Florida State Museum, Biological Sciences,* 24(4):359–408.

Bonaccorso, F.J., and N. Smythe
1972. Punch-marking Bats: An Alternative to Banding. *Journal of Mammalogy,* 53(2):389–390.

Bradbury, J.W.
1977. Social Organization and Communication. *In* W.A. Wimsatt, editor, *Biology of Bats,* volume 2, pages 1–72. New York: Academic Press.

Bradbury, J.W., and S.L. Vehrencamp
1977a. Social Organization and Foraging in Emballonurid Bats, III: Mating Systems. *Behavioral Ecology and Sociobiology,* 2(1):1–17.
1977b. Social Organization and Foraging in Emballonurid Bats, IV: Parental Investment Patterns. *Behavioral Ecology and Sociobiology,* 2(1):19–29.

Bradley, S.R., and D.R. Deavers
1980. A Reexamination of the Relationship Between Thermal Conductance and Body Weight in Mammals. *Comparative Biochemistry and Physiology,* 65A(4):465–476.

Bradshaw, G.V.R.
1962. Reproductive Cycle of the California Leaf-nosed Bat, *Macrotis californicus. Science,* 136:645.

Brenner, F.J.
1968. A Three-year Study of Two Breeding Colonies of the Big Brown Bat, *Eptesicus fuscus. Journal of Mammalogy,* 49(4):775–778.

Brody, S.
1945. *Bioenergetics and Growth, with Special Reference to the Efficiency Complex in Domestic Animals.* xii + 1023 pages. New York: Reinhold Publishing Corporation.

Brown, P.
1976. Vocal Communication in the Pallid Bat, *Antrozous pallidus. Zeitschrift für Tierpsychologie,* 41(1):34–54.

Burgen, A.S.V., and N.G. Emmelin
1961. *Physiology of the Salivary Glands.* vii + 279 pages. London: Edward Arnold.

Burleigh, I.G., and R.T. Schimke
1969. The Activities of Some Enzymes Concerned with Energy Metabolism in Mammalian Muscles of Differing Pigmentation. *Biochemical Journal,* 113(1):157–166.

Burns, R.J.
1970. Twin Vampire Bats Born in Captivity. *Journal of Mammalogy,* 51(2):391–392.

Carpenter, R.E.
1968. Salt and Water Metabolism in the Marine Fish-eating Bat, *Pizonyx vivesi. Comparative Biochemistry and Physiology,* 24(3):951–964.
1969. Structure and Function of the Kidney and the Water Balance of Desert Bats. *Physiological Zoology,* 42(3):288–302.

Case, T.J.
1978. On the Evolution and Adaptive Significance of Postnatal Growth Rates in Terrestrial Vertebrates. *The Quarterly Review of Biology,* 53(3):243–282.

Chapman, F.M.
1938. *Life in an Air Castle.* ix + 250 pages. New York: D. Appleton-Century Company.

Choi, S.C.
1978. *Introductory Applied Statistics in Science.* x + 278 pages. Englewood Cliffs, New Jersey: Prentice-Hall.

Cowan, I.M., and V.C. Brink
1949. Natural Game Licks in the Rocky Mountain National Parks of Canada. *Journal of Mammalogy,* 30(4):379–387.

Crespo, R.F., S.B. Linhart, R.J. Burns, and G.C. Mitchell
1972. Foraging Behavior of the Common Vampire Bat Related to Moonlight. *Journal of Mammalogy,* 53(2):366–368.

Croat, T.B.
1978. *Flora of Barro Colorado Island.* vx + 943 pages. Stanford, California: Stanford University Press.

Dalke, P.D., R.D. Beeman, F.J. Kindel, R.J. Robel, and T.R. Williams
1965. Use of Salt by Elk in Idaho. *Journal of Wildlife Management,* 29(2):319–332.

Dalquest, W.W.
1953. Mammals of the Mexican State of San Luis Potosi. *Biological Sciences Series, Louisiana State University,* 1:1–229.

Dalquest, W.W., H.J. Werner, and J.H. Roberts
1952. The Facial Glands of a Fruit-eating Bat, *Artibeus jamaicensis* Leach. *Journal of Mammalogy,* 33(1):102–103.

Davis, R.
1969a. Wing Loading in Pallid Bats. *Journal of Mammalogy,* 50(1):140–144.
1969b. Growth and Development of Young Pallid Bats, *Antrozous pallidus. Journal of Mammalogy,* 50(4):729–736.
1970. Carrying of Young by Flying Female North American Bats. *The American Midland Naturalist,* 83(1):186–196.

Davis, R.B., C.F. Herreid, II, and H.L. Short
1962. Mexican Free-tailed Bats in Texas. *Ecological Monographs,* 32:311–346.

Davis, W.H.
1967. Survival of *Eptesicus fuscus. Bat Research News,* 8(1):18–19.

Davis, W.H., R.W. Barbour, and M.D. Hassell
1968. Colonial Behavior of *Eptesicus fuscus. Journal of Mammalogy,* 49(1):44–50.

Diem, K., editor
1962. *Scientific Tables.* Sixth edition, 502 pages. Ardsley, New York: Geigy Pharmaceuticals.

Dinerstein, E.
1983. Reproductive Ecology of Fruit Bats and Seasonality of Fruit Production in a Costa Rican Cloud Forest. 136 pages. Unpublished doctoral dissertation, University of Washington, Seattle.

Ditmars, R.L.
1936. A *Vampyrum spectrum* is Born. *Bulletin of the New York Zoological Society,* 39:162–163.

Dowdeswell, W.H., R.A. Fisher, and E.B. Ford
1940. The Quantitative Study of Populations in the Lepidoptera, I: *Polyommatus icarus. Annals of Eugenics,* 10:123–136.

Dwyer, P.D.
1963. The Breeding Biology of *Miniopterus schreibersi blepotis* (Temminck) (Chiroptera) in North-eastern New South Wales. *Australian Journal of Zoology,* 11(2):219–140.
1966. The Population Pattern of *Miniopterus schreibersii* (Chiroptera) in North-eastern New South Wales. *Australian Journal of Zoology,* 14:1073–1137.
1970. Latitude and Breeding Season in a Polyestrus Species of *Myotis. Journal of Mammalogy,* 51(2):405–410.

Eckert, H.G.
1974. Der Einfluss des Mondlichtes auf die Aktivitätsperiodik nachtaktiver Säugetiere. *Oecologia,* 14:269–287.
1982. Activity Rhythms. *In* T.H. Kunz, editor, *Ecology of Bats,* pages 201–242. New York: Plenum Publishing Corporation.

Eisenberg, J.F.
1981. *The Mammalian Radiations.* xx + 610 pages. Chicago: The University of Chicago Press.

Eisenberg, J.F., and D.E. Wilson
1978. Relative Brain Size and Feeding Strategies in the Chiroptera. *Evolution,* 32(4):740–751.

Emlen, S.T., and L.W. Oring
1977. Ecology, Sexual Selection and the Evolution of Mating Systems. *Science,* 197(4300):215–223.

Ewing, W.G., E.H. Studier, and M.J. O'Farrell
1970. Autumn Fat Deposition and Gross Body Composition in Three Species of *Myotis. Comparative Biochemistry and Physiology,* 36(1):119–129.

Fenton, M.B.
1969. The Carrying of Young by Females of Three Species of Bats. *Canadian Journal of Zoology,* 47(1):158–159.

Fenton, M.B., N.G.H. Boyle, T.M. Harrison, and D.J. Oxley
1977. Activity Patterns, Habitat Use, and Prey Selection by Some African Insectivorous Bats. *Biotropica,* 9:73–85.

Fenton, M.B., D.H.M. Cumming, and D.J. Oxley
1977. Prey of Bat Hawks and Availability of Bats. *Condor,* 79(4):495–497.

Findley, J.S., E.H. Studier, and D.E. Wilson
1972. Morphologic Properties of Bat Wings. *Journal of Mammalogy,* 53(3):429–444.

Fisher, R.A., and E.B. Ford
1947. The Spread of a Gene in Natural Conditions in a Colony of the Moth *Panaxia dominula* L. *Heredity* 1:143–174.

Fleming, T.H.
1971. *Artibeus jamaicensis:* Delayed Embryonic Development in a Neotropical Bat. *Science,* 171(3969):402–404.
1988. *The Short-tailed Fruit Bat.* 365 pages. Chicago: University of Chicago Press.

Fleming, T.H., and E.R. Heithaus
1986. Seasonal Foraging Behavior of the Frugivorous Bat *Carollia perspicillata. Journal of Mammalogy,* 67:660–671.

Fleming, T.H., E.T. Hooper, and D.E. Wilson
1972. Three Central American Bat Communities: Structure, Reproductive Cycles, and Movement Patterns. *Ecology,* 53(4):555–569.

Forman, G.L.
1972. Comparative Morphological and Histochemical Studies of Stomachs of Selected American Bats. *The University of Kansas Science Bulletin,* 49(10):591–729.

Foster, G.W., S.R. Humphrey, and P.P. Humphrey
1978. Survival Rate of Young Southeastern Brown Bats, *Myotis austroriparius,* in Florida. *Journal of Mammalogy,* 59(2):299–304.

Foster, R.B.
1982. The Seasonal Rhythm of Fruitfall on Barro Colorado Island. *In* E.G. Leigh, Jr., A.S. Rand, and D.M. Windsor, editors, *The Ecology of a Tropical Forest,* pages 151–172. Washington, D.C.: Smithsonian Institution Press.

Foster, R.B, and N.V.L. Brokaw
1982. Structure and History of the Vegetation of Barro Colorado Island. *In* E.G. Leigh, Jr., A.S. Rand, and D.M. Windsor, editors, *The Ecology of a Tropical Forest,* pages 21–46. Washington, D.C.: Smithsonian Institution Press.

Gaertner, R.A., J.S. Hart, and O.Z. Roy
1973. Seasonal Spontaneous Torpor in the White-footed Mouse, *Peromyscus leucopus. Comparative Biochemistry and Physiology,* 45(1A):169–181.

Gardner, A.L.
1977. Feeding Habits. *In* R.J. Baker, J.K. Jones, Jr., and D.C. Carter, editors, Biology of Bats of the New World Family Phyllostomatidae, Part II. *Special Publications of The Museum, Texas Tech University,* 13:293–350.

Geluso, K.N.
1975. Urine Concentration Cycles of Insectivorous Bats in the Laboratory. *Journal of Comparative Physiology,* 99(4):309–319.
1978. Urine Concentrating Ability and Renal Structure of Insectivorous Bats. *Journal of Mammalogy,* 59(2):312–323.
1980. Renal Form and Function in Bats: An Ecophysiological Appraisal. *In* D.E. Wilson and A.L. Gardner, editors, *Proceedings Fifth International Bat Research Conference,* pages 403–414. Lubbock, Texas: Texas Tech Press.

Goodwin, G.C., and A.M. Greenhall
1961. A Review of the Bats of Trinidad and Tobago. *Bulletin of the American Museum of Natural History,* 122(3):187–302.

Gould, E.
1971. Studies of Maternal-infant Communication and Development of Vocalizations in the Bats *Myotis* and *Eptesicus. Communications in Behavioral Biology,* Part A, 5(5):263–313.
1975. Neonatal Vocalizations in Bats of Eight Genera. *Journal of Mammalogy,* 56(1):15–29.

Greenhall, A.M.
1976. Care in Captivity. *In* R.J. Baker, J.K. Jones, Jr., and D.C. Carter, editors, Biology of Bats of the New World Family Phyllostomatidae, Part I. *Special Publications of The Museum, Texas Tech University,* 10:89–131.

Griffin, D.R.
1958. *Listening in the Dark.* xviii + 413 pages. New Haven: Yale University Press.

Hainsworth, F.R., and L.L. Wolf
1972. Energetics of Nectar Extraction in a Small, High Altitude, Tropical Hummingbird, *Selasphorus flammula. Journal of Comparative Physiology,* 80:377–387.

Häussler, U., and H.G. Eckert
1978. Different Direct Effects of Light Intensity on the Entrained Activity Rhythm in Neotropical Bats (Chiroptera, Phyllostomidae). *Behavior Proceedings,* 3:223–239.

Heideman, P.D.
1988. The Timing of Reproduction in *Haplonycteris fischeri* Lawrence (Pteropodidae): Geographic Variation and Delayed Development. *Journal of Zoology* [London], 215(4):577–595.

Heinrich, B.
1979. *Bumblebee Economics.* viii + 245 pages. Cambridge, Massachusetts: Harvard University Press.

Heinz International Research Center
1964. *Heinz Nutritional Data.* Fifth edition, vi + 143 pages. Pittsburgh, Pennsylvania: H.J. Heinz Company.

Heithaus, E.R., and T.H. Fleming
1978. Foraging Movements of a Frugivorous Bat, *Carollia perspicillata* (Phyllostomatidae). *Ecological Monographs,* 48:127–143.

Heithaus, E.R., T.H. Fleming, and P.A. Opler
1975. Foraging Patterns and Resource Utilization in Seven Species of Bats in a Seasonal Tropical Forest. *Ecology,* 56(4):841–854.

Heithaus, E.R., P.A. Opler, and H.G. Baker
1974. Bat Activity and Pollination of *Bauhinia pauletia:* Plant–pollinator Coevolution. *Ecology,* 55:412–419.

Helwig, J.T., and K.A. Council
1979. *SAS User's Guide.* 494 pages. Cary, North Carolina: SAS Institute, Inc.

Herbert, F., and I.M. Cowan
1970. Natural Salt Licks as a Part of the Ecology of the Mountain Goat. *Canadian Journal of Zoology,* 49(5):605–610.

Herbst, L.H.
1988. Methods of Nutritional Ecology. *In* T.H. Kunz, editor, *Ecological and Behavioral Methods for the Study of Bats,* pages 233–246. Washington, D.C.: Smithsonian Institution Press.

Hill, R.W.
1977. Body Temperature in Wisconsin *Peromyscus leucopus:* A Reexamination. *Physiological Zoology,* 50(2):130–141.

Hladik, A., and C.M. Hladik
1969. Rapports Trophiques entre végétation et primates dans la Foret de Barro Colorado (Panama). *La Terre et la Vie,* 23:25–117.

Hladik, C.M., A. Hladik, J. Bousset, P. Valdebouze, G. Viroben, and J. Delort-LaVal
1971. Le Regime alimentaire de primates de l'ile de Barro-Colorado (Panama). *Folia Primatologia,* 16(1–2):85–122.

Howe, H.F.
1979. Fear and Frugivory. *The American Naturalist,* 114(6):925–931.

Howell, D.J.
1974. Bats and Pollen: Physiological Aspects of the Syndrome of Chiropterophily. *Comparative Biochemistry and Physiology,* 48(2A):263–276.
1979. Flock Foraging in Nectar-feeding Bats: Advantages to the Bats and to the Host Plants. *American Naturalist,* 114:23–49.

Hulme, A.C., editor
1970. *The Biochemistry of Fruits and Their Products.* ix + 620 pages. New York: Academic Press.

Humphrey, S.R., and F.J. Bonaccorso
1979. Population and Community Ecology. *In* R.J. Baker, J.K. Jones, Jr., and D.C. Carter, editors, Biology of Bats of the New World Family Phyllostomatidae, Part III. *Special Publications of The Museum, Texas Tech University,* 16:409–437.

Humphrey, S.R., and J.B. Cope
1970. Population Samples of the Evening Bat, *Nycticeius humeralis. Journal of Mammalogy,* 51(2):399–401.
1976. Population Ecology of the Little Brown Bat, *Myotis lucifugus,* in Indiana and North-central Kentucky. *Special Publication, The American Society of Mammalogists,* 4:vii + 81 pages.
1977. Survival Rates of the Endangered Indiana Bat, *Myotis sodalis. Journal of Mammalogy,* 58(1):32–36.

Izutsu, K.T.
1981. Theory and Measurement of the Buffer Value of Bicarbonate in Saliva. *Journal of Theoretical Biology,* 90:397–403.

Jenness, R., and E.H. Studier
1976. Lactation and Milk. *In* R.J. Baker, J.K. Jones, Jr., and D.C. Carter, editors, Biology of Bats of the New World Family Phyllostomatidae, Part I. *Special Publications of The Museum, Texas Tech University,* 10:201–218.

Jerison, H.J.
1973. *Evolution of the Brain and Intelligence.* xiv + 482 pages. New York: Academic Press.

Jimbo, S., and H.O. Schwassmann
1967. Feeding Behavior and Daily Emergence Pattern of "Artibeus jamaicensis" Leach (Chiroptera, Phyllostomidae). *In* H. Lent, editor, *Atas do Simpósio Sôbre a Biota Amazonica,* volume 5 (Zoologia), pages 239–254. Rio de Janeiro: Conselho Nacional de Pesquisas.

Jones, C.
1967. Growth, Development, and Wing Loading in the Evening Bat, *Nycticeius humeralis* (Rafinesque). *Journal of Mammalogy,* 48(1):1–19.

Jones, T.S.
1945. Unusual State at Birth of a Bat. *Nature,* 156(3960):365.
1946. Parturition in a West Indian Fruit Bat (Phyllostomidae). *Journal of Mammalogy,* 27(4):327–330.

Jordan, C.F., and R. Herrera
1981. Tropical Rain Forests: Are Nutrients Really Critical? *The American Naturalist,* 117(2):167–180.

Jordan, P.A., D.B. Botkin, A.S. Dominski, H.S. Jowendorf, and R.E. Bolovsky
1973. Sodium as a Critical Nutrient for Moose of Isle Royal. *Proceedings of the North American Moose Conference Workshop*, 9:13–42.

Keen, R., and H.B. Hitchcock
1980. Survival and Longevity of the Little Brown Bat (*Myotis lucifugus*) in Southeastern Ontario. *Journal of Mammalogy*, 61(1):1–7.

Kirkham, W.R., and V.G. Allfrey
1972. Animal and Plant Cells and Cell Parts: Chemical Composition. *In* P.L. Altman and D.S. Dittmer, editors, *Biology Data Book*, second edition, volume 1, pages 387–392. Bethesda, Maryland: Federation of American Societies for Experimental Biology.

Kleiber, M.
1932. Body Size and Metabolism. *Hilgardia*, 6(11):315–353.
1975. *The Fire of Life: An Introduction to Animal Energetics*. xxii + 453 pages. New York: R.E. Krieger Publishing Company.

Kleiman, D.G.
1969. Maternal Care, Growth Rate, and Development in the Noctule (*Nyctalus noctula*), Pipistrelle (*Pipistrellus pipistrellus*), and Serotine (*Eptesicus serotinus*) Bats. *Journal of Zoology* (London), 157(2):187–211.
1980. The Sociobiology of Captive Propagation. *In* M.E. Soule and B.A. Wilcox, editors, *Conservation Biology*, pages 243–261. Sunderland, Massachusetts: Sinauer Associates, Incorporated.

Kleiman, D.G., and T.M. Davis
1974. Punch-mark Renewal in Bats of the Genus *Carollia*. *Bat Research News*, 15(4):29–30.
1979. Ontogeny and Maternal Care. *In* R.J. Baker, J.K. Jones, Jr., and D.C. Carter, editors, Biology of Bats of the New World Family Phyllostomatidae, Part III. *Special Publications of The Museum, Texas Tech University*, 16:387–402.

Knight, D.H.
1975. A Phytosociological Analysis of Species-rich Tropical Forest on Barro Colorado Island, Panama. *Ecological Monographs*, 45(3):259–284.

Krátky, J.
1970. Postnatale Entwicklung des Grossmausohrs, *Myotis myotis* (Borkhausen, 1797). *Vestnik Ceskoslovenska Spolecnost Zoologická*, 34:202–218.

Kunz, T.H.
1973. Population Studies of the Cave Bat (*Myotis velifer*): Reproduction, Growth, and Development. *Occasional Papers of the Museum of Natural History, The University of Kansas*, 15:1–43.
1974. Reproduction, Growth, and Mortality of the Vespertilionid Bat, *Eptesicus fuscus*, in Kansas. *Journal of Mammalogy*, 55(1):1–13.
1980. Daily Energy Budgets of Free-living Bats. *In* D.E. Wilson and A.L. Gardner, editors, *Proceedings of the Fifth International Bat Research Conference*, pages 369–392. Lubbock, Texas: Texas Tech Press.
1982. Roosting Ecology of Bats. *In* T.H. Kunz, editor, *Ecology of Bats*, pages 1–55. New York: Plenum Publishing Corporation.

Kunz, T.H., editor
1988. *Ecological and Behavioral Methods for the Study of Bats*. 533 pages. Washington, D.C.: Smithsonian Institution Press.

Kunz, T.H., and K.A. Nagy
1988. Methods of Energy Budget Analysis. *In* T.H. Kunz, editor, *Ecological and Behavioral Methods for the Study of Bats*, pages 277–302. Washington, D.C.: Smithsonian Institution Press.

Lamb, D.R.
1984. *Physiology of Exercise: Responses and Adaptations*. Second edition, xx + 489 pages. New York: Macmillian Publishing Company.

Likens, G.E., and F.H. Bormann
1970. Chemical Analysis of Plant Tissues from the Hubbard Brook Ecosystem in New Hampshire. *Yale School of Forestry Bulletin*, 79:1–25.

Linzel, J.L.
1972. Milk Yield, Energy Loss in Milk, and Mammary Gland Weight in Different Species. *Dairy Science Abstracts*, 34(5):351–360.

Maeda, K.
1972. Growth and Development of Large Noctule, *Nyctalus lasiopterus* Schreber. *Mammalia*, 36(2):269–278.

Martin, M.M., and T.J. Lieb
1979. Patterns of Fuel Utilization by the Thoracic Muscles of Adult Worker Ants; The Use of Lipid by a Hymenopteran. *Comparative Biochemistry and Physiology*, 64B(4):387–390.

May, R.M.
1974. On the Theory of Niche Overlap. *Theoretical Population Biology*, 5:297–332.

May, R.M., and R.H. MacArthur
1972. Niche Overlap as a Function of Environmental Variability. *Proceedings of the National Academy of Sciences, USA*, 69:1109–1113.

Maynard, L.A., and O.K. Loosli
1969. *Animal Nutrition*. Sixth edition, x + 613 pages. New York: McGraw-Hill.

McCracken, G.F., and J.W. Bradbury
1977. Paternity and Genetic Heterogeneity in the Polygynous Bat, *Phyllostomus hastatus*. *Science*, 198(4314):303–306.
1981. Social Organization and Kinship in the Polygynous Bat, *Phyllostomus hastatus*. *Behavioral Ecology and Sociobiology*, 8(1):11–34.

McManus, J.J., and D.W. Nellis
1972. Ontogeny of Wing-loading in the Jamaican Fruit-eating Bat, *Artibeus jamaicensis*. *Journal of Mammalogy*, 53(4):866–868.

McNab, B.K.
1969. The Economics of Temperature Regulation in Neotropical Bats. *Comparative Biochemistry and Physiology*, 31(2):227–268.
1973. Energetics and the Distribution of Vampires. *Journal of Mammalogy*, 54(1):131–144.
1976. Seasonal Fat Reserves of Bats in Two Tropical Environments. *Ecology*, 57(2):332–338.
1982. Evolutionary Alternatives in the Physiological Ecology of Bats. *In* T.H. Kunz, editor, *Ecology of Bats*, pages 151–200. New York: Plenum Publishing Corporation.
1983. Energetics, Body Size, and the Limits to Endothermy. *Journal of Zoology* (London), 199:1–29.

Medway, L.
1971. Observations of Social and Reproductive Biology of the Bent-winged Bat *Miniopterus australis* in Northern Borneo. *Journal of Zoology* (London), 165:261–273.
1972. Reproductive Cycles of the Flat-headed Bats *Tylonycteris pachypus* and *T. robustula* (Chiroptera: Vespertilionidae) in a Humid Equatorial Environment. *Zoological Journal of the Linnaean Society*, 51(1):33–61.

Meyer, J.H., R.R. Grunert, M.T. Zepplin, R.H. Grummer, G. Bohstedt, and P.H. Phillips
1950. Effect of Dietary Levels of Sodium and Potassium on Growth and on Concentrations in Blood Plasma and Tissues of the White Rat. *American Journal of Physiology*, 162(1):182–188.

Mills, R.S., G.W. Barrett, and M.P. Farrell
1975. Population Dynamics of the Big Brown Bat (*Eptesicus fuscus*) in Southwestern Ohio. *Journal of Mammalogy*, 56(3):591–604.

Milton, K.
1978. Behavioral Adaptations to Leaf-eating by the Mantled Howler Monkey (*Alouatta palliata*). *In* G.G. Montgomery, editor, *The Ecology of Arboreal Folivores*, pages 525–547. Washington, D.C.: Smithsonian Institution Press.

Morrison, D.W.
1975. The Foraging Behavior and Feeding Ecology of a Neotropical Fruit Bat, *Artibeus jamaicensis*. 94 pages. Unpublished doctoral disserta-

tion, Division of Biological Sciences, Cornell University, Ithaca, New York.

1978a. Lunar Phobia in a Neotropical Fruit Bat, *Artibeus jamaicensis* (Chiroptera: Phyllostomidae). *Animal Behaviour*, 26(3):852–855.

1978b. Infuence of Habitat on the Foraging Distances of the Fruit Bat, *Artibeus jamaicensis*. *Journal of Mammalogy*, 59(3):622–624.

1978c. On the Optimal Searching Strategy for Refuging Predators. *The American Naturalist*, 112(987):925–934.

1978d. Foraging Ecology and Energetics of the Frugivorous Bat *Artibeus jamaicensis*. *Ecology*, 59(4):716–723.

1979. Apparent Male Defense of Tree Hollows in the Fruit Bat, *Artibeus jamaicensis*. *Journal of Mammalogy*, 60(1):11–15.

1980a. Foraging and Day-roosting Dynamics of Canopy Fruit Bats in Panama. *Journal of Mammalogy*, 61(1):20–29.

1980b. Efficiency of Food Utilization by Fruit Bats. *Oecologia*, 45(1):270–273.

1980c. Flight Speeds of Some Tropical Forest Bats. *The American Midland Naturalist*, 104(1):189–192.

Morrison, D.W., and S.H. Morrison

1981. Economics of Harem Maintenance by a Neotropical Bat. *Ecology*, 62(3):864–866.

Morrison, P., and B.K. McNab

1967. Temperature Regulation in Some Brazilian Phyllostomid Bats. *Comparative Biochemistry and Physiology*, 21(1):207–221.

Muller, B.D., and J. Baldwin

1978. Biochemical Correlates of Flying Behaviour in Bats. *Australian Journal of Zoology*, 26(1):29–37.

Myers, P.

1978. Sexual Dimorphism in Size of Vespertilionid Bats. *The American Naturalist*, 112(986):701–711.

Nagy, K.A., and K. Milton

1979. Aspects of Dietary Quality, Nutrient Assimilation and Water Balance in Wild Howler Monkeys (*Alouatta palliata*). *Oecologia*, 39(3):249–258.

National Research Council

1978. *Nutrient Requirements of Laboratory Animals*. Number 10, third revised edition, vii + 96 pages. Washington, D.C.: National Academy of Sciences.

Novick, A.

1960. Successful Breeding in Captive *Artibeus*. *Journal of Mammalogy*, 41(4):508–509.

Oates, J.F.

1978. Water-plant and Soil Consumption by Guereza Monkeys (*Colobus guereza*): A Relationship with Minerals and Toxins in the Diet? *Biotropica*, 10(4):141–153.

O'Farrell, M.J., and E.H. Studier

1973. Reproduction, Growth, and Development in *Myotis thysanodes* and *M. lucifugus* (Chiroptera: Vespertilionidae). *Ecology*, 54(1):18–30.

O'Farrell, M.J., E.H. Studier, and W.G. Ewing

1971. Energy Utilization and Water Requirements of Captive *Myotis thysanodes* and *Myotis lucifugus* (Chiroptera). *Comparative Biochemistry and Physiology*, 39(3A):549–552.

Okon, E.E., R.M. Umukoro, and A. Ajuda

1978. Diurnal Variations of the Glycogen and Fat Stores in the Liver and Breast Muscle of the Fruit Bat, *Eidolon helvum* (Kerr). *Physiology and Behavior*, 20(2):121–123.

Orr, R.T.

1970. Development: Prenatal and Postnatal. *In* W.A. Wimsatt, editor, *Biology of Bats*, volume 1, pages 217–231. New York: Academic Press.

Pagels, J.F.

1975. Temperature Regulation, Body Weight and Changes in Total Body Fat of the Free-tailed Bat, *Tadarida brasiliensis cynocephala* (Le Conte). *Comparative Biochemistry and Physiology*, 50(2A):237–246.

Pagels, J.F., and C. Jones

1974. Growth and Development of the Free-tailed Bat, *Tadarida brasiliensis cynocephala* (Le Conte). *The Southwestern Naturalist*, 19(3):267–276.

Pearson, O.P., M.R. Koford, and A.K. Pearson

1952. Reproduction of the Lump-nosed Bat (*Corynorhinus rafinesquei*) in California. *Journal of Mammalogy*, 33(3):273–320.

Phillips, C.J., G.W. Grimes, and G.L. Forman

1977. Oral Biology. *In* R.J. Baker, J.K. Jones, Jr., and D.C. Carter, editors, Biology of Bats of the New World Family Phyllostomatidae, Part II. *Special Publications of The Museum, Texas Tech University*, 13:121–146.

Phillips, C.J., K.M. Studholme, and G.L. Forman

1984. Results of the Alcoa Foundation Suriname Expeditions; VIII: Comparative Ultrastructure of Gastric Mucosae in Four Genera of Bats (Mammalia: Chiroptera), with Comments on Gastric Evolution. *Annals of Carnegie Museum*, 53(4):71–117.

Racey, P.A.

1982. Ecology of Bat Reproduction. *In* T.H. Kunz, editor, *Ecology of Bats*, pages 57–104. New York: Plenum Publishing Corporation.

1988. Reproductive Assessment in Bats. *In* T.H. Kunz, editor, *Ecological and Behavioral Methods for the Study of Bats*, pages 31–46. Washington, D.C.: Smithsonian Institution Press.

Rakhmatulina, I.K.

1972. The Breeding, Growth, and Development of Pipistrelles in Azerbaidzhan. *Soviet Journal of Ecology*, 2(2):131–136.

Ralls, K.

1976. Mammals in which Females are Larger than Males. *The Quarterly Review of Biology*, 51(3):245–276.

Rand, A.S., and W.M. Rand

1982. Variation in Rainfall on Barro Colorado Island. *In* E.G. Leigh, Jr., A.S. Rand, and D.M. Windsor, editors, *The Ecology of a Tropical Forest*, pages 47–60. Washington, D.C.: Smithsonian Institution Press.

Rasweiler, J.J., IV

1975. Maintaining and Breeding Neotropical Frugivorous, Nectarivorous and Pollenivorous Bats. *International Zoo Yearbook*, 15:18–30.

1977. The Care and Management of Bats as Laboratory Animals. *In* W.A. Wimsatt, editor, *Biology of Bats*, volume 3, pages 519–617. New York: Academic Press.

Rasweiler, J.J., IV, and H. de Bonilla

1972. Laboratory Maintenance Methods for Some Nectarivorous and Frugivorous Phyllostomatid Bats. *Laboratory Animal Science*, 22(5):658–663.

Rasweiler, J.J., IV, and V. Ishiyama

1973. Maintaining Frugivorous Phyllostomatid Bats in the Laboratory: *Phyllostomus*, *Artibeus*, and *Sturnira*. *Laboratory Animal Science*, 23(1):56–61.

Rice, D.W.

1957. Life History and Ecology of *Myotis austroriparius* in Florida. *Journal of Mammalogy*, 38(1):15–32.

Ridgely, R.S.

1976. *A Guide to the Birds of Panama*. xv + 394 pages. Princeton, New Jersey: Princeton University Press.

Rouk, C.S., and B.P. Glass

1970. Comparative Gastric Histology of Five North and Central American Bats. *Journal of Mammalogy*, 51(3):455–472.

Sauchelli, V.

1969. *Trace Elements in Agriculture*. viii + 248 pages. New York: Van Nostrand Reinhold Company.

Sazima, I., and M. Sazima

1977. Solitary and Group Foraging: Two Flower-visiting Patterns of the

Lesser Spear-nosed Bat *Phyllostomus discolor*. *Biotropica*, 9:213–215.

Schmidt, U., A.M. Greenhall, and W. Lopez-Forment

1971. Ökologische Untersuchungen der Vampirfledermäuse (*Desmodus rotundus*) im Staate Puebla, Mexiko. *Zeitschrift für Säugetierkunde*, 36:360–370.

Schmidt, U., and U. Manske

1973. Die Jugendentwicklung der Vampirfledermäuse (*Desmodus rotundus*). *Zeitschrift für Säugetierkunde*, 38:14–33.

Short, H.L.

1961. Growth and Development of Mexican Free-tailed Bats. *The Southwestern Naturalist*, 6(3–4):156–163.

Silva Taboada, G.

1979. *Los Murciélagos de Cuba*. xiv + 423 pages. La Habana: Editorial Academia.

Smythe, N.

1974. Rainfall. *In* R.W. Rubinoff, editor, *Environmental Monitoring and Baseline Data*, pages 26–27. Washington, D.C.: Smithsonian Institution Press.

Sokal, R.R., and F.J. Rohlf

1969. *Biometry*. xxii + 776 pages. San Francisco: W.H. Freeman and Company.

Staaland, H., R.G. White, J.R. Luick, and D.F. Holleman

1980. Dietary Influences on Sodium and Potassium Metabolism of Reindeer. *Canadian Journal of Zoology*, 58(10):1728–1734.

Stebbings, R.E.

1966. A Population Study of Bats of the Genus *Plecotus*. *Journal of Zoology* (London), 150:53–75.

Stevenson, D.E., and M.D. Tuttle

1981. Survivorship in the Endangered Gray Bat (*Myotis grisescens*). *Journal of Mammalogy*, 62(2):244–257.

Stockstad, D.S., M.S. Morris, and E.C. Lory

1953. Chemical Characteristics of Natural Licks Used by Big Game Animals in Western Montana. *Transactions of the North American Wildlife Conference*, 18:247–257.

Stones, R.C., and J.E. Wiebers

1967. Temperature Regulation in the Little Brown Bat, *Myotis lucifugus*. *In* K.C. Fisher, A.R. Dawe, C.P. Lyman, E. Schonbaum, and F.E. South, Jr., editors, *Mammalian Hibernation III*, pages 97–109. New York: American Elsevier Publishing Company.

Studier, E.H.

1970. Evaporative Water Loss in Bats. *Comparative Biochemistry and Physiology*, 35(4):935–943.

1981. Energetic Advantages of Slight Drops in Body Temperature in Little Brown Bats, *Myotis lucifugus*. *Comparative Biochemistry and Physiology*, 70A(4):537–540.

Studier, E.H., B.C. Boyd, A.T. Feldman, R.W. Dapson, and D.E. Wilson

1983. Renal Function in the Neotropical Bat, *Artibeus jamaicensis*. *Comparative Biochemistry and Physiology*, 74A(2):199–209.

Studier, E.H., V.L. Lysengen, and M.J. O'Farrell

1972. Biology of *Myotis thysanodes* and *M. lucifugus* (Chiroptera: Vespertilionidae)–II: Bioenergetics of Pregnancy and Lactation. *Comparative Biochemistry and Physiology*, 44:467–471.

Studier, E.H., and M.J. O'Farrell

1972. Biology of *Myotis thysanodes* and *M. lucifugus* (Chiroptera: Vespertilionidae)–I: Thermoregulation. *Comparative Biochemistry and Physiology*, 41(3A):567–595.

Studier, E.H., and D.E. Wilson

1970. Thermoregulation in Some Neotropical Bats. *Comparative Biochemistry and Physiology*, 34(2):251–262.

1979. Effects of Captivity on Thermoregulation and Metabolism in *Artibeus jamaicensis* (Chiroptera: Phyllostomatidae). *Comparative Biochemistry and Physiology*, 62A(2):347–350.

1983. Natural Urine Concentrations and Composition in Neotropical Bats. *Comparative Biochemistry and Physiology*, 75A(4):509–515.

Studier, E.H., S.J. Wisniewski, A.T. Feldman, R.W. Dapson, B.C. Boyd, and D.E. Wilson

1983. Kidney Structure in Neotropical Bats. *Journal of Mammalogy*, 64(3):445–452.

Talesara, C.L., and P. Kumar

1974. Histo-enzymological Evidence for the Aneuronal Ontogenesis of the Pectoralis Major Muscle of Bat and Sparrow into Atypical Red Muscle. *Annales d'Histochimie*, 18(4):277–287.

Tamsitt, J.R., and D. Valdivieso

1961. Notas sobre actividades nocturnas y estados de reproducción de algunos Quiróteros de Costa Rica. *Revista Biologia Tropical*, 9:219–225.

Thomas, M.E.

1974. Bats as a Food Source for *Boa constrictor*. *Journal of Herpetology*, 8(2):188.

Thomas, S.P.

1975. Metabolism during Flight in Two Species of Bats, *Phyllostomus hastatus* and *Pteropus gouldii*. *Journal of Experimental Biology*, 63:273–293.

Trune, D.R., and C.N. Slobodchikoff

1976. Social Effects of Roosting on the Metabolism of the Pallid Bat (*Antrozous pallidus*). *Journal of Mammalogy*, 57(4):656–663.

Turner, D.C.

1975. *The Vampire Bat*. xii + 145 pages. Baltimore: The Johns Hopkins University Press.

Tuttle, M.D.

1968. Feeding Habits of *Artibeus jamaicensis*. *Journal of Mammalogy*, 49(4):787.

1975. Population Ecology of the Gray Bat (*Myotis grisescens*): Factors Influencing Early Growth and Development. *Occasional Papers of the Museum of Natural History, The University of Kansas*, 36:1–24.

1976. Collecting Techniques. *In* R.J. Baker, J.K. Jones, Jr., and D.C. Carter, editors, Biology of Bats of the New World Family Phyllostomatidae, Part I. *Special Publications of The Museum, Texas Tech University*, 10:71–88.

Tuttle, M.D., and D.E. Stevenson

1982. Growth and Survival of Bats. *In* T.H. Kunz, editor, *Ecology of Bats*, pages 105–150. New York: Plenum Publishing Corporation.

Van der Westhuyzen, J.

1978. The Diurnal Cycle of Some Energy Substrates in the Fruit Bat *Rousettus aegyptiacus*. *South African Journal of Science*, 74(3):99–101.

Vaughn, T.A.

1959. Functional Morphology of Three Bats: *Eumops*, *Myotis*, *Macrotus*. *University of Kansas Publications, Museum of Natural History*, 12(1):1–153.

Vehrencamp, S.L., F.G. Stiles, and J.W. Bradbury

1977. Observations on the Foraging Behavior and Avian Prey of the Neotropical Carnivorous Bat, *Vampyrum spectrum*. *Journal of Mammalogy*, 58(4):469–478.

Villa-R, B.

1966. *Los Murciélagos de México*. 491 pages. Mexico City: Universidad Nacional Autónoma México.

Ward, P., and A. Zahavi

1973. The Importance of Certain Assemblages of Birds as "Information-centres" for Food-finding. *Ibis*, 115(4):517–534.

Watt, B.K., and A.L. Merrill

1963. *Composition of Foods*. Agricultural Handbook number 8, ii + 189 pages. Washington, D.C.: U.S. Government Printing Office.

Weber, N.S., and J.S. Findley

1970. Warm Season Changes in Fat Content of *Eptesicus fuscus*. *Journal of Mammalogy*, 51(1):160–162.

Weeks, H.P., Jr.
1978. Variation in the Sodium and Potassium Content of Food Plants of Wild Indiana Herbivores. *Agricultural Experimental Station Research Bulletin, Purdue University,* 957:1–13.
Weeks, H.P., Jr., and C.M. Kirkpatrick
1976. Adaptations of White-tailed Deer to Naturally Occurring Sodium Deficiencies. *Journal of Wildlife Management,* 40(4):610–625.
1978. Salt Preferences and Sodium Drive Phenology in Fox Squirrels and Woodchucks. *Journal of Mammalogy,* 59(3):531–542.
Williams, D.F., and J.S. Findley
1979. Sexual Size Dimorphism in Vespertilionid Bats. *The American Midland Naturalist,* 102(1):113–126.
Wilson, D.E.
1979. Reproductive Patterns. *In* R.J. Baker, J.K. Jones, Jr., and D.C. Carter, editors, Biology of Bats of the New World Family Phyllostomatidae, Part III. *Special Publications of The Museum, Texas Tech University,* 16:317–378.

1988. Maintaining Bats for Captive Studies. *In* T.H. Kunz, editor, *Ecological and Behavioral Methods for the Study of Bats,* pages 247–266. Washington, D.C.: Smithsonian Institution Press.
Wilson, D.E., and E.L. Tyson
1970. Longevity Records for *Artibeus jamaicensis* and *Myotis nigricans. Journal of Mammalogy,* 51(1):203.
Wimsatt, W.A.
1956. Histological and Histochemical Observations on the Parotid, Submaxillary, and Sublingual Glands of the Tropical American Fruit Bat *Artibeus jamaicensis* Leach. *Journal of Morphology,* 99:169–210.
1969. Transient Behavior, Nocturnal Activity Patterns, and Feeding Efficiency of Vampire Bats (*Desmodus rotundus*) under Natural Conditions. *Journal of Mammalogy,* 50(1):233–244.
Yacoe, M.E., J.W. Cummings, P. Myers, and G.K. Creighton
1982. Muscle Enzyme Profile, Diet, and Flight in South American Bats. *American Journal of Physiology,* 242(3):R189–R194.